Lecture Notes in Mathematics

2064

Editors:
J.-M. Morel, Cachan
B. Teissier, Paris

Editors *Mathematical Biosciences Subseries:*
P.K. Maini, Oxford

Jerry J. Batzel • Mostafa Bachar • Franz Kappel
Editors

Mathematical Modeling and Validation in Physiology

Applications to the Cardiovascular
and Respiratory Systems

 Springer

Editors
Jerry J. Batzel
University of Graz
Institute for Mathematics
and Scientific Computing
and Medical University of Graz
Institute of Physiology
Graz, Austria

Franz Kappel
University of Graz
Institute for Mathematics and Scientific
Computing
Graz, Austria

Mostafa Bachar
King Saud University
College of Sciences
Department of Mathematics
Riyadh, Saudi Arabia

ISBN 978-3-642-32881-7   ISBN 978-3-642-32882-4 (eBook)
DOI 10.1007/978-3-642-32882-4
Springer Heidelberg New York Dordrecht London

Lecture Notes in Mathematics ISSN print edition: 0075-8434
                               ISSN electronic edition: 1617-9692

Library of Congress Control Number: 2012951624

Mathematics Subject Classification (2010): 34-XX, 92-XX, 92C30, 92C42, 92C50, 93A30

Printed on acid-free paper

Springer is part of Springer Science+Business Media (www.springer.com)

Preface

The focus of this volume is on mathematical modeling techniques essential for studying human physiology and biomedical and clinical problems at primarily the organ and system level. The areas of focus for physiological modeling are the cardiovascular and respiratory systems, central systems which impact all aspects of human physiological function. Particular emphasis is placed on control mechanisms and clinical problems arising from deficiencies in these control mechanisms.

Mathematical modeling of physiological systems is a complicated enterprise. Decisions must be made on model complexity in the context of the problem of acquiring sufficient data for model validation. As a consequence, close collaboration is essential between scientists from the mathematical/engineering side and the physiological/medical side. This requires developing a new perspective based on a common language and common understanding of diverse scientific principles. New research insight and improved research protocols can evolve in response to this merging of several scientific disciplines and the skills, training, and experience of each.

Important state-of-the-art questions are included in this volume. For example, modeling of physiological systems often leads to the design problem that these models have to be detailed enough in order to represent the complex physiological mechanisms acting at various functional levels. Consequently such models usually contain a large number of parameters. Based on these factors, valid models are often difficult to apply in the clinical setting where noninvasive tests and measurements restrict the availability of data from which parameters can be estimated. In addition, interindividual variation in physiological function and in particular the individualized combination of a physiological system's inventory of control responses add further complication to model application for diagnosis and treatment design.

Goal

The goal of this volume is to provide the reader with an understanding of the theory and practice of modeling of physiological control systems with a primary application of studying clinical problems related to the cardiovascular and respiratory control systems.

Scientific Content

The content of this volume is focused on principles of modeling and applications of mathematical methods for model development, model validation, and model application related to important physiological questions and clinical conditions with the focus on application to the cardiovascular and respiratory systems. Cardiovascular and respiratory control mechanisms receive particular attention and represent the vehicle for exemplifying the theoretical ideas presented. The role of time delay in feedback control is also examined theoretically and in terms of clinical manifestations of instabilities induced by delay.

Interdisciplinary Collaboration

A key theme in this volume is the development of interdisciplinary collaboration between mathematicians, engineers, clinicians, and life scientists (both theoretical and experimental). Authors from all these disciplines have contributed to this volume. As a consequence, important mathematical and physiological concepts and perspectives are included and merged into the volume content. In particular significant attention is given to the practical side of modeling, especially in regard to assessing model parameter identifiability as well as insight into problems and methods of experimental design and data collection.

This approach illustrates interdisciplinary research in a concrete way. Model development and experimental design go hand in hand: models must contain the necessary degree of detail to reflect the problem under study but must allow for model validation in the context of the data that can be collected from available measurement tools and experiments. The merging of both the modeling and experimental perspectives and coordinating their contributions can lead to improved adaptation of models to individual patients leading to potential diagnostic and treatment tools.

In the spirit of these observations the text is divided into the following two parts:

- Part I entitled Theory focuses on key concepts and methods needed for the development of models with a special emphasis on the problems of parameter estimation and model validation.
- Part II entitled Practice provides a number of modeling studies which employ the concepts discussed in the Theory part of the volume. These studies also exemplify the parameter identification tools discussed in the theory section such as sensitivity analysis and subset selection. In addition this part seeks to provide insight into data collection and experimental design.

Advancement of knowledge in the focus area of this volume impacts important areas of clinical research including orthostatic intolerance (a key problem in the elderly), hypotension during hemodialysis, the interaction of respiratory and cardiovascular mechanisms in sleep apnea, congestive heart failure, and many other areas.

The book is intended for applied mathematicians, engineers, biophysicists, physicians, physiologist and others interested in learning about mathematical modeling of physiological systems in general and the cardiovascular and respiratory systems, in particular. The reader who has at least a background in differential and integral calculus, ordinary differential equations, can profit from this book. While the application focus is on cardiovascular and respiratory modeling, the more general reader will find these applications accessible and illuminating since important topics such as physiological control mechanisms, feedback delays in such systems, experimental design, model design and development and model parameter estimation are illustrated in the concrete cases examined.

Chapters

Chapter 1 provides an overview of modeling issues and modeling strategies. An overall framework for the topics discussed in the book is presented, including issues of interdisciplinary collaboration.

Chapter 2 provides a detailed examination of the modeling process using the application of the principles developed to the cardiovascular control systems as template.

Chapter 3 provides an in-depth analysis of subset selection, a sensitivity identifiability-based approach introduced in Chaps. 1 and 2. This procedure is applied to assess the parameter estimation problem and improve the parameter estimation results.

Chapter 4 takes a closer look at the process of parameter estimation and provides an application of the Kalman filter to refine the estimation process.

Chapter 5 exemplifies how delays in model equations can arise and be treated in parameter estimation.

Chapter 6 looks at various levels of modeling and perspectives of the modeling process. This chapter acts as a bridge to the second part of the book.

Chapter 7 considers the modeling of the respiratory system illustrating and using a minimal model approach. The chapter provides a concrete example of the stability issues of respiratory control discussed in Chap. 8, namely the important clinical problem of sleep apnea.

Chapter 8 provides a careful explication of the steps involved in deriving a model of the respiratory system from physiological information and principles. This chapter exemplifies many of the concepts described in Part I and illustrates the role of experimental design and data collection in model development.

Chapter 9 is devoted entirely to the experimentation process, examining how physiological information about the baroreflex, a key control of the cardiovascular system, has been derived. This chapter also provides the physiological background for the modeling studies of Chap. 10.

Chapter 10 provides an extensive study of the many issues involved in cardiovascular blood pressure control and how model development proceeds from physiology to the problem of patient-specific model adaptation.

Chapter 11 examines another key component of the cardiovascular control system, namely the quantity unstressed volume, and illustrates how sensitivity analysis and subset selection proceed in studying this control using invasive and noninvasively collected data.

This volume has its origin in a series of combined summer school/workshop events supported by funds from the Marie Curie Training Series. This program involved four events which examined key methods for modeling, data utilization, and clinical application of complex models of important physiological systems. Taken as a whole, the sequence of events spanned four years and represented a multisided examination of mathematical modeling for biomedical application.

The general web page linking and reflecting all four events can be found at:

www.uni-graz.at/biomedmath/info.html

The addresses of the individual web pages for the four events are:

Graz: www.uni-graz.at/mc_training_schools/graz/summerschool/
Copenhagen: www.math.ku.dk/~susanne/SummerSchool2008/
Lipari: www.biomatematica.it/lipari2009/
Dundee: www.maths.dundee.ac.uk/summerschool2010/

The first event in the series from which the current volume draws its inspiration was held in Graz and was entitled *Biomedical Modeling and Cardiovascular–Respiratory Control : Theory and Practice*. It was held in Schloss Seggau in Leibnitz near Graz. This event consisted of 11 days of summer school followed by a 3-day scientific workshop on the same theme as the school focus. The summer school introduced new researchers (pre to early post doc) to critical mathematical and physiological ideas and methods by experts in both fields. The workshop which followed directly after the summer school was designed to be equivalent to a stand-alone workshop involving current researchers in the field. As part of their training, students from the summer school attended the workshop. In this way, students introduced to new methods and concepts in the school could apply this knowledge in learning state-of-the-art issues in cardiovascular–respiratory system modeling research.

Acknowledgments

This book would never have existed without the commitment and work of the contributors of this volume and we would like to thank them all for their efforts and for sharing their insights with us. In particular we acknowledge support by the European Commission under Contract MSCF-CT-2006-045961 Biomathtech 07–10. For each chapter at least one contributor also participated in the first event (Graz) of our Marie Curie training event series. We are grateful to all those who helped with organizing the four Marie Curie training events. We wish to also thank all the people who attended the summer school and workshop and especially the teachers and workshop presenters of the summer school/workshop events. We also thank Springer Verlag and especially Ute McCrory for their support during the production of the book.

Finally, this volume would not have been possible without funds from different sources. The European Union under the program Marie Curie Conferences and Training Course program provided the support for the BiomathTech project which encompassed four combined summer school and workshop events. We were also supported by the European Society for Mathematical and Theoretical Biology, the research by Jerry Batzel was partially funded by FWF (Austria) project P18778-N13, and Mostafa Bachar was partially supported by Deanship of Scientific Research, College of Science Research Center, King Saud University.

Graz and Riyadh   Jerry J. Batzel
July 2011   Mostafa Bachar
                                                                                              Franz Kappel

Contents

List of Contributors

Ariel Cintrón-Arias Department of Mathematics and Statistics, East Tennessee State University, Johnson City, TN, USA

Adam Attarian Department of Mathematics, North Carolina State University, USA

Mostafa Bachar Department of Mathematics, College of Sciences, King Saud University, Riyadh, Saudi Arabia

H. T. Banks Department of Mathematics, North Carolina State University, USA

Jerry J. Batzel Institute for Mathematics and Scientific Computing, University of Graz, and Institute of Physiology, Medical University of Graz, Austria

Clive M. Brown Department of Medicine, Division of Physiology, University of Fribourg, Switzerland

Eugene Bruce Center for Biomedical Engineering, University of Kentucky Lexington, KY, USA

James Duffin Departments of Anaesthesia and Physiology, University of Toronto, Canada

Raffaello Furlan Internal Medicine IV, Humanitas Clinical and Research Center, University of Milan, Rozzano (Milan), Italy

Ferenc Hartung Department of Mathematics, University of Pannonia, Veszprém, Hungary

Thomas Heldt Computational Physiology and Clinical Inference Group, Research Laboratory of Electronics, Massachusetts Institute of Technology, Cambridge, MA, USA

Franz Kappel Institute for Mathematics and Scientific Computing, University of Graz, Graz, Austria

John M. Karemaker Department of Systems Physiology, Academic Medical Center at the University of Amsterdam, Amsterdam, The Netherlands

Roger G. Mark Harvard-MIT Division of Health Sciences and Technology, Massachusetts Institute of Technology, Cambridge, MA, USA

Brett Matzuka Department of Mathematics, North Carolina State University, USA

Vera Novak SAFE Laboratory, Beth Israel Deaconess Medical Center and Harvard Medical School, Boston, USA

Mette Olufsen Department of Mathematics, North Carolina State University, USA

Chi-Sang Poon Institute for Medical Engineering and Science, Massachusetts Institute of Technology, Cambridge, MA, USA

Chung Tin Department of Mechanical and Biomedical Engineering, City University of Hong Kong, Hong Kong, China

Karl Thomaseth ISIB-CNR, Corso Stati Uniti 4, 35127 Padova, Italy

Hien T. Tran Department of Mathematics, North Carolina State University, USA

Janos Turi Department of Mathematical Sciences, University of Texas at Dallas, Dallas, USA

George C. Verghese Computational Physiology and Clinical Inference Group, Research Laboratory of Electronics, Massachusetts Institute of Technology, Cambridge, MA, USA

Acronyms

AIC	Akaike's Information Criterion
ANOVA	Analysis of variance
CBF	Cerebral blood flow
COPD	Chronic obstructive pulmonary disease
CSF	Cerebral spinal fluid
CVP	Central venous pressure
CVS	Cardiovascular system
ECG	Electrocardiogram
EKF	Extended Kalman filter
EPCA	Equations with piecewise constant argument
G + W	Whole brain
GLS	Generalized least squares
GM	Grey matter
GRV	Gaussian random variables
GSF	Generalized sensitivity function
HIV	Human immunodeficiency virus
HR	Heart rate
HUT	Head up tilt
LBNP	Lower body negative pressure
LQP	Least squares problem
LTP	Long term potentiation
M	Muscle
MIMO	Multiple-input-multiple-output
MLE	Maximum likelihood estimator
MSNA	Muscle sympathetic nerves activation
NIRS	Near infrared spectroscopy
NLWLS	Nonlinear weighted least squares
NM	Non-muscle
NMDA	N-methyl D-aspartate
NREM	Non-REM sleep
ODE	Ordinary differential equations

OLS	Ordinary least squares
PDE	Partial differential equation
PI	Protease inhibitor
RTI	Reverse transcriptase inhibitor
SID	Strong ion difference
SISO	Single-input-single-output
SMC	Sequential Monte Carlo
STS	Sitting to standing
SVD	Singular value decomposition
TSF	Traditional sensitivity function
UKF	Unscented Kalman filter
UT	Unscented transformation
WM	White matter

Part I
Theory

Chapter 1
Merging Mathematical and Physiological Knowledge: Dimensions and Challenges

Jerry J. Batzel, Mostafa Bachar, John M. Karemaker, and Franz Kappel

Abstract This chapter introduces the main theoretical and practical topics which arise in the mathematical modeling of the human cardiovascular–respiratory system. These topics and ideas, developed in detail in the text, also represent a template for considering interdisciplinary research involving mathematical and life science disciplines in general. The chapter presents a multi-sided view of the modeling process which synthesizes the mathematical and life science viewpoints needed for developing and validating models of physiological systems. Particular emphasis is placed on the problem of coordinating model design and experimental design and methods for analyzing the model identification problem in the light of restricted data. In particular a variety of approaches based on information derived from parameter sensitivity are examined. The themes presented seek to provide a coordinated view of modeling that can aid in considering the current problem of patient-specific model adaptation in the clinical setting where data is in particular typically limited.

J.J. Batzel (✉)
Institute for Mathematics and Scientific Computing, University of Graz, and Institute of Physiology, Medical University of Graz, A 8010 Graz, Austria
e-mail: jerry.batzel@uni-graz.at

M. Bachar
Department of Mathematics, King Saud University, Riyadh, Saudi Arabia
e-mail: mbachar@ksu.edu.sa

J.M. Karemaker
Department of Systems Physiology, Academic Medical Center at the University of Amsterdam, 1105 AZ Amsterdam, The Netherlands
e-mail: j.m.karemaker@amc.uva.nl

F. Kappel
Institute for Mathematics and Scientific Computing, University of Graz, A 8010 Graz, Austria
e-mail: franz.kappel@uni-graz.at

J.J. Batzel et al. (eds.), *Mathematical Modeling and Validation in Physiology*, Lecture Notes in Mathematics 2064, DOI 10.1007/978-3-642-32882-4_1, © Springer-Verlag Berlin Heidelberg 2013

1.1   Introduction

Given the complexity of physiological systems, and the intricacies of the interactions between systems and subsystems, mathematical modeling is needed to
quantify interactions and predict responses resulting from various stresses, perturbations, as well as diseases which alter system function. In addition to theoretical
studies of system function, the adaptation to the clinical setting of mathematical
models to assess patient-specific physiological function is an important problem
because models must be complex enough to reflect real conditions but data for
patient-specific model parameter estimation are typically limited to minimally
invasive measurements. Consequently, improved methods and algorithms to address
both model design and model validation are needed. A broad overview of these
issues are presented in this chapter within the context of modeling studies of the
cardiovascular and respiratory control systems.

1.2   The Problem Setting

One of the very first examples of modeling in biology can be found in the book
that shaped our understanding of how blood moves through the vasculature, i.e.,
William Harvey's famous treatise "De motu cordis . . ." (About the movement of the
heart and blood in animals) published in 1628 [12]. It challenged the set of intricate
explanations of what was happening to the blood that was supposed to be formed
continually in the liver out of the chyle that arrived there from the food; views that
were held from the time of Galen.[1]

Why had Galen missed the point of how the circulation circulates? After all,
among one of his accomplishments he had been physician to the Emperor of Rome,
and must have observed many wounds and accidents in Roman gladiators. Spurting
of blood out of an arterial wound should not have been strange to him. The main
reason, probably, why he did go wrong and after him many of his followers, was
the fact that anatomy of the human body was, of course, learned from corpses. In
a dead body nearly all the blood is collected in the veins, very little remaining in
the—collapsed—arteries.

1.2.1   Conceptual Models

Harvey showed by experiments that the blood stream in the veins was directed
to the (right half of the) heart, never away from it, putting an end to the myth

---

[1] Aelius Galenus, born 129 or 131 AD in Pergamon, died 199, 201 or 216 in Rome(?).

that blood in the veins would flow back and forth. From the right ventricle it could with low resistance pass through the lungs to the left heart, rather than by hypothetical pores through the ventricular septum. That ventricle would, in turn, distribute the blood via the arteries over all body parts, from where it would, via—at the time not yet observable—porosities, flow back to the veins. Harvey's "Anatomical disputation" as he called it, gave the world a new model of how to think of the blood in the body, actually in a continuous circulation. His theory put all earlier observations into a new framework, refuted some myths along the way that had been around but had never been tested experimentally and he postulated a new part of this circulation, the capillaries, which were only discovered when Antoni van Leeuwenhoek started to look through his high-power microscope lenses. Moreover, Harvey used a mathematical reasoning to show the absurdity of the idea that blood would be formed continually by the liver and delivered to the right heart. He calculated how much blood would be pumped out of the ventricles in half an hour, assuming some realistic numbers for the end-diastolic volume as he, himself had observed. Even at the lowest estimates of the propelled amount of blood per beat, multiplying by the number of beats (more than one thousand in half an hour, two to three, even up to four thousand in some) one comes to a larger quantity of blood than is contained in the whole body.

We are used to look at the world that surrounds us by way of conceptual models. No longer do we believe that the Sun is drawn across the skies on a carriage pulled by fierce horses driven by Apollo himself. Still it is one of the hardest things to change from one model to a new one that explains the same old phenomena, but in a way that is more in line with current understanding. The discovery of the circulation by Harvey shows a number of typical steps in the development of a new conceptual model in biology: (1) a new, logically consistent theory, (2) new experiments that falsify old theories, (3) historical observations that require an explanation, and (4) if possible, formalization of the new theory in a mathematical model. Furthermore (5), the theory should withstand "stress-tests": new experiments designed to falsify it, either in the real world or in the mathematical model. Finally (6), if the new theory requires assumptions about not-yet known entities, it should predict properties and values for these unknowns.

1.2.2 Mathematical Models

In the biomedical literature mathematical models have been developed mainly to formalize existing knowledge, to test its completeness and internal consistency. The advantage of a model is, of course, the ease by which "new" experiments can be done: no need for the nitty-gritty complications of a biological experiment; the subject of study may die many lives in the computer without any consequence. This is also why computer models of physiological functions are used more and more in education. Alternative reasons to construct mathematical models in biology are the need to obtain parameters that cannot (or not as easily) be gotten to in other ways.

An example of this is the "Modelflow" computation of stroke volume from an arterial pulse wave [23]: by using knowledge from in vitro experiments on the elastic properties from the human aorta, and adapting the internal computations to the gender, age, height and weight of the subject, a model-supported guess of the volume that the ventricle has output to generate this particular pulse wave can be made. Such a model can be used in everyday practice in the hospital, where keeping track of cardiac performance is of paramount importance.

The conceptual model of the circulation introduced by Harvey does not require much mathematics. Not so for the control systems which fill the books on medical physiology. To model the workings of hormones like insulin on blood glucose levels in conditions from rest to heavy exercise, much a priori knowledge of the various involved system components is required. Even if such knowledge exists to the degree of precision necessary to construct a model, the engineer who is called in for help, will find in many cases that the available knowledge is incomplete since not all interrelationships with other regulatory systems are known—let alone quantified. This leaves the engineer no alternative but to work with "educated guesses" using first principles like conservation of mass and adding "black boxes" with input–output relationships that approximate what is observed in reality. Here views of the physicist and physician often tend to diverge: the physician preferring "grey boxes" or even "white boxes" (also known as "glass boxes") where the internal workings are known and modeled as such, the physicist being satisfied when input–output relationships match the real world, no matter what is inside the box. In this process, just the finding of such lack of basic knowledge may help in designing appropriate experiments to fill the gap and turn black boxes into white.

In view of the general incompleteness of biological/medical knowledge, one may well ask how much detail is required to make a "good model." The answer is, in the first place, determined by the purpose of the model. The computational speed of the system running the model should help in understanding reality. If one heart beat takes a whole night to be computed, it may be a feat in itself to show that the mechanisms underlying it are now well-understood from the level of genomics to the generation of pressure inside the ventricle. However, it is of little help in projects that require fast turn-around to give answers to "what if" questions about parameter changes. As James Bassingthwaighte et al. put it [3]: "One does not build a body out of atoms or a truck out of quarks." Alternatively, if the model is intended to be used in a feedback loop to steer an insulin pump, replacing the patients own pancreatic function, demands of patient safety may put restrictions on what can and what should be modeled.

Notwithstanding these considerations, the future of systems biology and systems physiology is, of course, in the area of more and faster computing. Once we are able to tie the various levels and scales of biology together, from the genome to the whole organism and back, then we can start to learn how matter may lead to what we call mind and how mind influences matter. This level of complexity is far beyond the scope of this or any introductory text. Still it should not keep us from striving to understand how minimal differences in the germ cells, be it in the DNA or in other essential parts of the cell machinery, may lead to the diversity of biology and mankind as we see it around us. Multi-scale modeling requires that building blocks can be tied together; as we once learned to move away from "spaghetti-programs" to modular programming, so must we now learn to make modular models, that can be easily incorporated into larger systems. Computer modeling of biological function may turn into a discipline that combines mathematics, physics and biology in search for the essence of life. However, first the basics of the trade should be learned, taught and disseminated.

1.2.3 Interdisciplinary Research

The construction of a model of some intricate biological system, in particular one that extends beyond basic science into the medical arena, requires the combined knowledge from all relevant disciplines. This may stretch from the expert in information and communication technology (ICT) to build the required hard- and software to the medical professional who has "his brain picked" to generate input for an expert system. Such interdisciplinary research should be an "alliance of the willing," where proficiency in one's own discipline is not the only or most important requirement. Rather all team members should have an interest beyond their own discipline and the ability to communicate in a way that can be understood by professionals in other disciplines. In this respect it is awkward to observe the tendency in some medical faculties to shy away from the teaching and practice of basic science and to narrow the view to applied statistics and "evidence based medicine."

1.2.4 Overcoming Barriers to Collaboration

When one author of this chapter (J.M. Karemaker) started to study medicine in 1963 he had to buy a Dutch medical dictionary which turned out to contain some 45,000 entries. Today's 31st edition of Dorland's illustrated medical dictionary has over 120,000. Learning the medical trade is, for the better part, learning a new language. Consequently, when one has become more or less fluent in that jargon, communication about medical matters with non-speakers may become increasingly difficult, as many patients discover when they leave the doctor's office, having had their diagnosis delivered.

In the collaboration between engineers and physicians the jargon is the most difficult barrier to overcome. Both the medical and the engineering jargon act as barriers to those who wish to enter the respective fields. Of course it is not the jargon, but the world of implicit knowledge behind it that makes the language of a particular trade for what it is: an efficient way to communicate between colleagues. Therefore those who wish to close the gap will have to take a few extra steps and invest time and effort to learn the basics of the other trade. The addition of an "interpreter" to the team will, of course, be an enormous boost. As such upcoming new professions as that of "medical engineer" may fill the ever widening gap with the evolution of biomedical science.

1.3 Aspects of Parameter Estimation and Model Validation

The necessity to supplement qualitative knowledge about the working of real systems by quantitative statements leads inevitably to the development of mathematical models of various types. Mathematical models may just describe numerical relations between quantities and variables of a system which have been obtained from experiments. That can range from simple empirical formulas to statistical models of considerable complexity. Such models which one would call descriptive models can be extremely useful for classification purposes including diagnostic procedures. Another class of models tries to describe the working of a real system on the basis of fundamental physical, chemical, or biological mechanisms. Such models could be called explanatory. Besides the quantitative representation of the real system the development of an "explanatory" model which is a rather structured approach to the problem at hand usually leads also to a gain of knowledge about the real system which is considered. Of course, the distinction between descriptive and explanatory models is not a strict classification for mathematical models. Almost every model intended to be explanatory contains also descriptive components as can be seen for the models shown is this volume. Advantages of mathematical modeling were already mentioned in Sect. 1.2.2. There the importance of the goals respectively of the purpose of a mathematical modeling process is already mentioned. For one and the same real system different goals for the modeling process can lead to quite different models. Modeling goals are strongly related to the concept of validity of a model. The *domain of validity* of a model can be roughly characterized as that functional, spatial or temporal part of the real world which is described by the model with sufficient accuracy. On the other hand the modeling goals define that part of the real world which should be the domain of validity of the model right from the beginning. Since development of the model typically involves simplifying assumptions of various types, at the end of the modeling process it is not clear what the domain of validity of the model will be. Therefore it is necessary to conduct a validation procedure in order to make sure that the domain of validity of the model conforms with that domain characterized by the goals of the modeling process. If the validation procedure gives a negative result then one has to restart the modeling process and to modify the model.

It is useful to make the following remark. If one tries to describe principles or guidelines for mathematical modeling one usually does it in a sequential scheme which provides a step by step approach, e.g., first defining the goals, then collecting what is known in this area from experiments and previous investigations, then formalizing this knowledge in mathematical structures etc. Of course, such a scheme is useful as a guideline but in a real modeling process we conduct more a network of these steps rather than a sequence. On the basis of new knowledge the modeling goals may be changed at some point, investigations of the structural properties of the model may make it necessary to go back and redo a previous step, validation tests will be done already during the development of the model etc.

The enormous progress of mathematical modeling in the life sciences which we can observe nowadays is characterized by some important aspects:

- The use of mathematical models as diagnostic tools or as the basis for treatment strategies requires rather comprehensive models in order to describe the specific situation of individual patients with the necessary accuracy.
- A comprehensive model will have a large number of a prioriunknown parameters which have to be determined (as part of the validation process) by parameter estimation techniques on the basis of available measurements. The problem is that the data collection in a clinical environment is usually restricted to noninvasive measurements and thus provides in general only a very limited set of data for parameter estimation techniques.
- The discrepancy between a large number of wanted parameters versus the limited amount of available data for parameter estimation and the requirement to adapt the model to individual patients poses serious challenges for the mathematical methodologies for parameter estimation and model validation.

The chapters of this volume illustrate these aspects by presenting concrete modeling processes.

The challenges mentioned above led to a number of developments of mathematical methodologies in recent years which constitute an important theme in the contributions to this volume:

- Structural sensitivity and sensitivity identifiability.
- Generalized sensitivities.
- Subset selection.

1.3.1 Model Outputs and Measurements

In order to explain some of the basic ideas we start with a simple output model

$$\eta(t) = f(t, \theta), \quad 0 \le t \le T, \ \theta \in \mathscr{A}, \tag{1.1}$$

where $\eta(\cdot) \in \mathbb{R}^m$ is the vector of outputs of the model, which have to correspond to *measurable outputs* of the real system. The vector $\theta \in \mathbb{R}^p$ is the parameter vector and $\mathscr{A} \subset \mathbb{R}^p$ is the set of admissible parameters. For the following considerations we assume that the function f is sufficiently smooth. This output model may arise from an ODE-model for the dynamics of the real system,

$$\dot{x}(t) = g(t, x(t), \theta), \quad x(0) = x_0(\theta),$$
$$\eta(t) = h(t, x(t), \theta), \quad 0 \le t \le T, \ \theta \in \mathscr{A}. \tag{1.2}$$

Here, $x(\cdot) \in \mathbb{R}^n$ is the vector of the system states and g, h are sufficiently smooth functions of their arguments. Note that parameters are not only present in the differential equations of the model but they can also be initial conditions and they can also show up in the output function. The output model (1.1) may also arise from different dynamical models as for instance from a model consisting of delay-differential equations.

Given *sampling times* $0 \le t_1 < \cdots < t_N \le T$ on some fixed time interval $[0, T]$,[2] we denote by $y(t_j)$, $j = 1, \ldots, N$, the vector of measurements for the outputs of the real system at t_j. As stated above, the measurements $y(t_j)$ correspond to the model outputs $\eta(t_j)$ (for some unknown parameter vector θ). In order to formulate this correspondence mathematically on frequently assumes that

$$y(t_j) = f(t_j, \theta_0) + \epsilon_j, \quad j = 1, \ldots, N, \tag{1.3}$$

where θ_0 is the so-called "true" or nominal parameter vector and ϵ_j represents the measurement errors at time t_j. Assuming (1.3) for the measurements is basically unrealistic, because it assumes that in case of very precise measurements the model output for the nominal parameter vector θ_0 coincides (at least at the sampling times) with the measurable quantities of the real system, which because of inevitable modeling errors is extremely unrealistic. However, if our model captures the essential features of the real system then we may have that $y(t_j) \approx f(t_j, \theta_0) + \epsilon_j$, $j = 1, \ldots, N$, and the results based on (1.3) may be sufficiently accurate. If things go wrong we may have to improve our model.

The measurement errors ϵ_j are assumed to be representations of random variables $\mathscr{E}_j$, the measurement noise at t_j. The standard assumptions usually assumed for the measurement noise $\mathscr{E}_j$, $j = 1, \ldots, N$, are unbiasedness (i.e., the expected values $\mathrm{E}(\mathscr{E}_j)$ are zero), independence and identical distribution (i.i.d.). However, one should be aware that in case of very frequent measurements they may be autocorrelated. Since the ϵ_j are representations of random variables, the same is true for the $y(t_j)$, which are representations of the measurement process $\mathscr{Y}(t)$ at $t = t_j$. From (1.3) we see that

$$\mathscr{Y}(t_j) = f(t_j, \theta_0) + \mathscr{E}_j, \quad j = 1, \ldots, N.$$

This relation is usually called the *statistical model* corresponding to the output model (1.1).

---

[2]For simplicity of presentation we assume that the sampling times are the same for all measurable system outputs which in general is not case.

1.3.2 Parameter Estimation, Least Squares Formulation

Given the output model (1.1) and the data $y(t_j)$ our main task is to determine a parameter vector $\hat{\theta} \in \mathscr{A}$ such that the model outputs $f(t_j, \hat{\theta})$ are as close as possible to the measurements $y(t_j)$. This is the *parameter estimation* or *parameter identification problem*. This amounts to optimizing a measure for the difference of the model outputs for a parameter vector θ and the given measurements. We shall restrict ourselves to the simple case of least squares estimates. In order to simplify notation we assume that we have just one scalar output, i.e., $m = 1$, and that $\mathrm{Var}\epsilon_j = \sigma_0^2$, $j = 1, \ldots, N$. Then given a parameter vector $\theta \in \mathscr{A}$ and measurements (1.3) we define the weighted quadratic cost functional

$$J(Y(\theta_0), \theta) = \frac{1}{2\sigma_0^2} \sum_{j=1}^{N} \big(y(t_j) - f(t_j, \theta)\big)^{\mathsf{T}} \big(y(t_j) - f(t_j, \theta)\big), \qquad (1.4)$$

where $Y(\theta_0) = (y(t_1), \ldots, y(t_N))^{\mathsf{T}}$. Of course, J is quadratic in the residuals $y(t_j) - f(t_j, \theta)$ and not in the parameters. The parameter estimate $\hat{\theta}$ is obtained by minimizing J,

$$\hat{\theta} = \underset{\theta \in \mathscr{A}}{\mathrm{argmin}}\, J(Y(\theta_0), \theta). \qquad (1.5)$$

Identifiability of the parameter vector $\theta \in \mathscr{A}$ with respect to the available measurements Y means that (1.5) has a unique solution $\hat{\theta}$ in $\mathscr{A}$.

Here it is important to observe that the estimate $\hat{\theta}$ is also a representation of a random variable $\hat{\Theta}$, because for different representations ϵ_j we get different values of $\hat{\theta}$. The random variable $\hat{\Theta}$ is called the *least squares estimator* for our parameter estimation problem. Of course, we are interested in the statistical properties of the estimator. But before we discuss these we have to look closer to the optimization problem (1.5). It is clear that the geometrical properties of the surface defined by $\theta \to J(Y(\theta_0), \theta)$ for θ in a neighborhood of $\hat{\theta}$ is of interest. If $\hat{\theta}$ exists then the necessary conditions are satisfied, i.e.,

$$\nabla_\theta J(Y(\theta_0), \theta)\big|_{\theta = \hat{\theta}} = 0 \text{ and } H(\hat{\theta}) \geq 0,$$

where $H(\theta) = \big(\partial^2 J(Y(\theta_0), \theta)/\partial\theta_i \partial\theta_j\big)_{i,j=1,\ldots,p}$ denotes the Hessian of J. Usually it is assumed that $H(\hat{\theta})$ is positive definite. In order to establish identifiability we need to prove existence of a minimizer $\hat{\theta}$ and uniqueness of this minimizer. Existence is not difficult. Since J is continuous it is enough to assume that $\mathscr{A}$ is compact. Even if we are satisfied with local identifiability, i.e., uniqueness of $\hat{\theta}$ in a neighborhood of $\hat{\theta}$, uniqueness of $\hat{\theta}$ is a far more difficult problem, because we would have to make assumptions on $\theta \to J(Y(\theta_0), \theta)$ in a neighborhood of $\hat{\theta}$ which in general we don't know. On the other hand, we have introduced in (1.3) the "true" or nominal parameter vector θ_0 which we don't know either, but which in practice is our best guess for the true parameter vector. If $\hat{\theta}$ would be sufficiently close to θ_0, we could infer properties of $J(Y(\theta_0), \theta)$ in a neighborhood of $\hat{\theta}$ from properties of $J(Y(\theta_0), \theta)$ in at $\theta = \theta_0$. This leads to the concept of *consistency*, which means that as the number of measurements tends to infinity that the corresponding estimates $\hat{\theta}$ tend to θ_0. For precise definitions and sufficient conditions for consistency we refer to [20, Chap. 12]. Assuming consistency (the assumptions guaranteeing this may not easily be verified in a concrete case) and also assuming that the number of available measurements is sufficiently large, one investigates the cost functional J in a neighborhood of θ_0 hoping that $\hat{\theta}$ is sufficiently close to θ_0. In doing this one should always have in mind that this is based on some asymptotic theory and linearization of the output model around θ_0. In particular, we can prove the following results on the least squares estimator $\hat{\Theta}$:

$$E(\hat{\Theta}) \approx \theta_0 \text{ and } \text{Cov}(\hat{\Theta}) \approx \mathscr{F}(\theta_0)^{-1}. \tag{1.6}$$

Here $E(\cdot)$ and $\text{Cov}(\cdot)$ denote the expected value and the covariance matrix of a vector valued random variable, whereas "$\approx$" means asymptotically and neglecting higher order terms. The matrix (we set $F(\theta) = \left(f(t_1, \theta), \ldots, f(t_N, \theta) \right)^{\mathsf{T}}$)

$$\mathscr{F}(\theta_0) = \frac{1}{\sigma_0^2} \nabla_\theta F(\theta_0)^{\mathsf{T}} \nabla_\theta F(\theta_0) \tag{1.7}$$

is the *Fisher information matrix* for our parameter identification problem, which also can be written as

$$\mathscr{F}(\theta_0) = \frac{1}{\sigma_0^2} \sum_{j=1}^{N} \nabla_\theta f(t_j, \theta_0)^{\mathsf{T}} \nabla_\theta f(t_j, \theta_0).$$

Note that the gradient $\nabla_\theta f(t_j, \theta)$ is a p-dimensional row vector. It should be noted that $\mathscr{F}(\theta_0)$ depends not only on the nominal parameter vector θ_0 but also on the sampling mesh $(t_j)_{j=1,\ldots,N}$. The Fisher information matrix is of central importance in the context of parameter identification. It will show up frequently in our presentation (see the subsequent sections).

1.3.3 Sensitivity Analysis

Parameter estimation is just the inverse problem of determining the parameter values in our model from the data, i.e., inverting the mapping $\theta \rightarrow Y(\theta)$, θ in a neighborhood of θ_0, with the additional difficulty that $Y(\theta)$ is the representation of a random variable. We can expect that the matrix

$$S(\theta) = \frac{\partial Y(\theta)}{\partial \theta} = \frac{\partial F(\theta)}{\partial \theta} \in \mathbb{R}^{N \times p}, \tag{1.8}$$

which is called the *sensitivity matrix* of the output model, plays an important role. Note that $S(\theta)$ also depends on the sampling mesh $(t_j)_{j=1,\ldots,N}$. The linearization of $Y(\theta)$ around θ_0 is given by $Y(\theta) = Y(\theta_0) + (\partial F/\partial \theta)(\theta_0)(\theta - \theta_0)$. Injectivity of the linearized mapping holds if and only if

$$\text{rank} S(\theta_0) = p. \tag{1.9}$$

Consequently, under condition (1.9) we have injectivity of $\theta \to Y(\theta)$ in a neighborhood of θ_0, i.e., local identifiability (compare [1, Appendix]). Condition (1.9) is easily seen to be equivalent to

$$\det S(\theta_0)^{\mathsf{T}} S(\theta_0) \neq 0. \tag{1.10}$$

We refer the reader to Chap. 2, Sect. 2.4.1, for a discussion of the relation between the Hessian H of the least squares functional J as defined in (1.4). From the definition of the Fisher information matrix (see (1.7)) we obtain

$$\mathscr{F}(\theta_0) = \frac{1}{\sigma_0^2} S(\theta_0)^{\mathsf{T}} S(\theta_0).$$

Of course, we should expect difficulties when solving the parameter estimation problem numerically if the condition number of the Fisher information matrix $\mathscr{F}(\theta_0)$ is large. Since the singular values of $S(\theta_0)$ are the eigenvalues of $S(\theta_0)^{\mathsf{T}} S(\theta_0)$, a singular value decomposition of $S(\theta_0)$ would reveal if the parameter identification problem is ill-conditioned.

The function $t \to \partial f(t, \theta)/\partial \theta_i$, $i = 1, \ldots, p$, $0 \leq t \leq T$, is called the *sensitivity function* of the output with respect to the parameter θ_i. Large values of $s_i(t, \theta_0) = \partial f(t, \theta)/\partial \theta_i|_{\theta = \theta_0}$ for t in some interval $I \subset [0, T]$ indicate that, in a neighborhood of θ_0 the parameter θ_i has a large influence influence on the output $\eta(t, \theta)$, $t \in I$, i.e., the model output is sensitive with respect to the parameter θ_i on the interval I.[3] In this case one would expect that the parameter θ_i identifiable using data sampled in the time interval I. This is indeed true if the parameter θ_i is the only parameter to be identified. Already for two parameters θ_i and θ_j, $i, j \in \{1, \ldots, p\}$, $i \neq j$, we could have large values of the sensitivities $s_i(t, \theta_0)$ and $s_j(t, \theta_0)$ for $t \in I$, but θ_i, θ_j are not identifiable simultaneously with data sampled in I. A trivial case would be if the dependence of f on θ_i, θ_j is only via $\theta_i + \theta_j$, for instance. Such a situation is related to the concept of *parameter redundancy* (see [5, 6, 14])

---

[3]In abuse of language we frequently find the statement that the parameter θ_i is sensitive in a neighborhood of θ_0.

The use of the functions $s_i(t, \theta)$, $i = 1, \ldots, p$, as a measure for the influence of the parameters θ_i onto the model output has the disadvantage that these functions are not dimensionless quantities and thus depend on the units for the quantities involved. Moreover, it is difficult to compare sensitivities for different parameters in a meaningful way. The derivatives $s_i(t, \theta_0)$ are limits of the quotients of the absolute errors $\Delta\theta_i = \theta_i - \theta_{0,i}$ and $f(t, \theta_0 + \Delta\theta_i e_i) - f(t, \theta)$, where e_i is the ith vector of the canonical basis of $\mathbb{R}^p$. Instead of this we take the limit of the relative errors and get the sensitivities $\sigma_i(t, \theta_0)$,

$$\sigma_i(t, \theta_0) = \lim_{\Delta\theta_i \to 0} \frac{\left(f(t, \theta_0 + \Delta\theta_i e_i) - f(t, \theta_0)\right)/f(t, \theta_0)}{\Delta\theta_i/\theta_{0,i}} = \frac{\theta_{0,i}}{f(t, \theta_0)} s_i(t, \theta_0).$$

Consequently one also uses the "normalized" sensitivity matrix

$$\tilde{S}(\theta_0) = \left(\sigma_i(t_j, \theta_0)\right)_{i=1,\ldots,p,\ j=1,\ldots,N}$$

instead of $S(\theta_0)$.

Concerning parameter identification we want to draw attention to the use of the Kalman filter from linear control theory and its extensions to nonlinear systems in order to estimate parameters of a system together with the system states. We refer here to Chap. 4.

1.3.4   Generalized Sensitivities

As we have seen in the previous section, sensitivity functions—also called classical sensitivity functions in order to distinguish them from generalized sensitivity functions discussed in this section—characterize the influence of a model parameter onto the model output. The question behind generalized sensitivity functions is different: How to characterize the influence of the measurements onto the estimated parameters? According to assumption (1.3) on the structure of the measurements these are apart from measurements noise determined by the nominal parameter vector θ_0. Therefore we consider, instead of the influence of the measurements, the influence of the nominal parameters $\theta_{0,i}$, $i = 1, \ldots, p$, onto the estimated parameters $\hat{\theta}_j$, $j = 1, \ldots, p$. But we have be aware that a parameter estimate $\hat{\theta}$ given by (1.5) is a realization of the random variable $\hat{\Theta}(\theta_0)$, the so-called list squares estimator (see Sect. 1.3.2).

As generalization of the parameter estimation problem considered in Sect. 1.3.2 we consider, for $k = 1, \ldots, N$, the problems with the least squares functionals

$$J^{(k)}(\theta, \tau) = \frac{1}{\sigma_0^2} \sum_{j=1}^{k} \left(y(t_j, \theta) - f(t_j, \tau)\right)^2 + \frac{1}{\sigma_0^2} \sum_{j=k+1}^{N} \left(y(t_j, \theta_0) - f(t_j, \tau)\right)^2$$

for $\theta \in \mathcal{U}$, $\tau \in \mathcal{A}$, where $\mathcal{U}$ is a neighborhood of θ_0. We see that the measurements at sampling times $j = 1, \ldots, k$ may vary with the nominal parameter vector θ, whereas the measurements at sampling times $j = k+1, \ldots, N$ remain fixed to their values corresponding to the parameter vector θ_0. For a nominal parameter $\theta \in \mathcal{U}$ and $k \in \{1, \ldots, N\}$ we denote by $\hat{\theta}^{(k)}(\theta)$ the parameter estimate obtained by

$$\hat{\theta}^{(k)}(\theta) = \operatorname*{argmin}_{\tau \in \mathcal{A}} J^{(k)}(\theta, \tau).$$

Assuming unique local identifiability for nominal parameter vectors $\theta \in \mathcal{U}$ and sufficient smoothness of f we see that

$$\nabla_\tau J^{(k)}(\theta, \tau)\big|_{\tau = \hat{\theta}^{(k)}} = 0, \quad \theta \in \mathcal{U}.$$

Taking the derivative of this expression at $\theta = \theta_0$ provides a linear equation for the Jacobian $\partial \hat{\theta}^{(k)}(\theta)/\partial \theta\big|_{\theta=\theta_0}$. From this equation we see that this Jacobian is the realization of a matrix valued random variable which we denote by

$$\frac{\partial \hat{\Theta}^{(k)}}{\partial \theta}(\theta_0).$$

Under assumptions which we do not formulate here we can take expected values and get

$$\mathrm{E}\left(\frac{\partial \hat{\Theta}^{(k)}}{\partial \theta}(\theta_0)\right) \approx \mathscr{F}(\theta_0)^{-1} \mathscr{F}_k(\theta_0) =: \mathscr{G}(t_k, \theta_0), \quad k = 1, \ldots, N, \qquad (1.11)$$

where $\mathscr{F}_k(\theta_0)$ is the Fisher information matrix corresponding to the cost functional $J^{(k)}$,

$$\mathscr{F}_k(\theta_0) = \frac{1}{\sigma_0^2} \sum_{j=1}^{k} \nabla_\theta f(t_j, \theta_0)^\mathsf{T} \nabla_\theta f(t_j, \theta), \quad k = 1, \ldots, N.$$

We call $t_k \to \mathscr{G}(t_k, \theta_0)$, $k = 1, \ldots, N$, the *generalized sensitivity matrix* for our parameter identification problem at θ_0. The diagonal elements $t_k \to g_i(t_k, \theta_0) = \mathscr{G}(t_k, \theta_0)_{i,i}$, $i = 1, \ldots, p$, are called the *generalized sensitivity functions* for the parameters θ_i at θ_0. These functions were introduced by K. Thomaseth and C. Cobelli in [21] (see also [4, Appendix A.4] and [1, Sect. 3.2]). The interpretation (1.11) was elaborated in [1] and for multiple output systems in [16]. The general sensitivity matrix can give useful information concerning experimental design. For instance, measurements taken at sampling times at which the generalized sensitivity function $g_i(t_k, \theta_0)$ is steeply increasing to 1 contain more information on the parameter θ_i than measurements taken at sampling times where $g_i(t_k, \theta_0)$ is rather flat (see [21] or [4]). Oscillatory behavior of the generalized sensitivity functions is interpreted to indicate correlations between the corresponding parameters (see, for instance, [4,21]). However, more careful investigations are needed in order to achieve a better understanding of the meaning of these oscillations. Definitely the off-diagonal elements of the generalized sensitivity matrix play an important role in this context.

1.3.5 Experimental Design

Mathematical modeling in the Life Sciences is characterized by the discrepancy between complex models with many parameters and data collection limited several constraints. In this context methods of *experimental design* become very important in order to organize data collection such that the measurements provide as much information on the system (i.e., on the parameters) as possible. Questions of key importance are:

- How long should an experiment be extended, i.e., how large should $T > 0$ be chosen for the sampling interval $[0, T]$?
- How many measurements should one take in in the sampling interval?
- What is the optimal location of the sampling times in $[0, T]$?

Of course, these questions have to be answered with respect to some *design criterion*. Most design criteria are in terms of the Fisher information matrix at the nominal parameter vector θ_0 (see, for instance, [10, 17]:

- Maximize $\det \mathscr{F}(\theta_0)$ (D-optimal design).
- Maximize the smallest eigenvalue of $\mathscr{F}(\theta_0)$ (E-optimal design).
- Minimize the asymptotic standard errors of the parameter estimates (SE-optimal design), i.e., minimize

$$\left\| (\mathrm{SE}(\hat{\theta}_1), \ldots, \mathrm{SE}(\hat{\theta}_p)) \right\|,$$

where $\| \cdot \|$ is some norm on $\mathbb{R}^p$ and the asymptotic standard error $\mathrm{SE}(\hat{\theta}_k)$ for the estimate $\hat{\theta}_k$ is given by

$$\mathrm{SE}(\hat{\theta}_k) = \left(\left(\mathscr{F}\big((t_j)_{j=1,\ldots,N}, \theta_0\big)_{k,k} \right)^{-1} \right)^{1/2}.$$

Here we indicated that the Fisher information matrix also depends on the sampling mesh.

SE-optimal design was introduced in [1] (see also [2]).

When applying optimal design methods we should have in mind that additional restrictions have to be observed. Depending on the setting and application, data can be collected non- or minimally invasively or invasively. In the experimental setting for studies both forms are common. However, in the clinical setting data, at least in early clinical screening tests, are collected non-invasively, due to cost and practicality constraints. In addition data can sometimes only be collected at fixed times while in some cases essentially continuous data are available. Blood glucose levels would exemplify the former type of measurement frequency while heart rate, blood pressure via a finger cuff, and transcutaneous blood gases are examples of data that are essentially continuously measured. If the distance between consecutive sampling times becomes very short one can no longer assume that the corresponding measurements are independent, which means that we may have autocorrelated measurements. Concerning experimental design in case of autocorrelated data, see [22].

1.4   Model Validation

As stated at the beginning of Sect. 1.3, model validation has to guarantee that the developed model has a domain of validity such that the goals of the modeling process are achieved. That is, a model cannot be a valid model as such, it can only be valid in relation to the goals which should be achieved with the model. Of course, it is always tempting to use a model in situations it was originally not designed for. In such a case we may leave the domain of validity of the model. It is also obvious that a model will be more insensitive against variations of the original goals if the components of the model are—as much as possible—based on the underlying physiological mechanisms. However, we should be aware that a complex model usually also has components which represent black or grey box situations, because either the underlying physiological mechanisms are not known or are too complex in order to be modeled adequately.

Model validation can be a rather complicated process [8,9]. The following list of tasks we have to accomplish in a validation process is by far not complete:

- It is necessary to assess how well the model exhibits observed steady state and dynamic behavior of observable states and whether *all* modeled states reflect reasonable values and behavior using reasonable parameter values. This can be referred to as internal consistency of the model.
- It is necessary to test the model response to a variety of perturbations and experimental conditions within the context of the model assumptions.
- Model identifiability should be assessed. Some comprehensive models are so complex that it is impossible to identify all parameters from typically measured data (e.g. [11, 13, 18]). These models may have the potential to be perfectly accurate but cannot be adapted to individual subjects to assess individual system status. These models are not valid models when the modeling goal was to design a model which can be individualized.
- To test if models can represent observed and measured behavior in individual subjects requires that model parameters be identified from the data. Model identifiability refers to the issue of ascertaining unambiguous and accurate parameter estimation.

For the process of model validation it may be necessary to use additional information provided by basic research which may include results from animal studies and by additional experimental studies. For instance, mechanisms for control responses can be examined by measuring data from healthy subjects during protocols that stress the system such as using head-up-tilt to induce orthostatic stress (see Chap. 10). These data can be compared with data from patients with varying degrees/levels of system dysfunction.

As we have seen in the previous sections there exists a number of tools which can be used for model validation:

- Classical sensitivity analysis (see Sect. 1.3.3, Chaps. 10 and 11).
- Subset selection (see Chaps. 2, 3, 10 and 11).
- Generalized sensitivity analysis (Sect. 1.3.4).

The concepts explained up to now in Sect. 1.3 are local in the sense that they require knowledge of the nominal parameter vector θ_0. The results obtained by various methods can drastically change when θ_0 is varied due to the non-linearities in the model. Methods are being developed to test the local characteristics over a wide (global) range in the parameter space. This issue is currently receiving a great deal of attention (see, for instance, [7, 15, 19]).

1.5   A General Validation Protocol

In general the process of model validation requires the interaction of several dimensions of design and recursive application of several tools to refine that design. These methods are discussed in the modeling applications of this volume and in general consist of the following:

- Classical sensitivity analysis.
- Generalized sensitivity analysis.
- Subset Selection.
- Investigation of the local characteristics of parameter estimation characteristics.
- Experimental design.
- Data acquisition and assessment of the quality of the data.
- Inter-disciplinary collaboration to maximize experimental design, data acquisition, and model reduction to relevant mechanisms.
- Global analysis.

A flow chart of the decision making process is presented and illustrated in Chap. 11.

References

1. Banks, H.T., Dediu, S., Ernstberger, S.L., Kappel, F.: A new approach to optimal design problems. J. Inverse Ill Posed Probl. **18**, 25–83 (2010)
2. Banks, H.T., Holm, K., Kappel, F.: Comparison of optimal design methods in inverse problems. Inverse Probl. **27**(7), 075002(31pp.) (2011). doi: 10.1088/0266-5611/27/7/75002
3. Bassingthwaighte, J.B., Chizeck, H.J., Atlas, L.E., Qian, H.: Multiscale modeling of cardiac cellular energetics. Ann. N Y Acad. Sci. **1047**, 395–424 (2005)
4. Batzel, J.J., Kappel, F., Schneditz, D., Tran, H.T.: Cardiovascular and Respiratory Systems: Modeling, Analysis and Control. Frontiers in Applied Mathematics, vol. FR34. SIAM, Philadelphia (2006)
5. Catchpole, E.A., Morgan, B.J.T.: Detecting parameter redundancy. Biometrika **84**, 187–196 (1997)
6. Catchpole, E.A., Morgan, B.J.T., Freeman, S.N.: Estimation in parameter redundant models. Biometrika **85**, 462–468 (1998)
7. Chu, Y., Hahn, J.: Parameter set selection for estimation of nonlinear dynamic systems. J. AIChE **53**(11), 2858–2870 (2007)
8. Cobelli, C., Carson, E.R., Finkelstein, L., Leaning, M.S.: Validation of simple and complex models in physiology and medicine. Am. J. Physiol. Regul. Integr. Comp. Physiol. **246**(2), R259–R266 (1984)
9. Cobelli, C., DiStefano 3rd, J.J.: Parameter and structural identifiability concepts and ambiguities: A critical review and analysis. Am. J. Physiol. **239**(1), R7–24 (1980)
10. Fedorov, V.V., Hackel, P.: Model-Oriented Design of Experiments. Springer, New York (1997)
11. Guyton, A.C., Coleman, T.G., Granger, H.J.: Circulation: Overall regulation. Annu. Rev. Physiol. **34**, 13–46 (1972)
12. Harveius, G.: Exercitatio anatomica de motu cordis et sanguinis in animalibus. G. Fitzerus, ed. (1628). Facsimile in: Begründer der experimentellen Physiologie, Blasius, W., Boylan, J., Kramer, K. (eds.) J.F. Lehmanns Verlag, München (1971)
13. Hester, R.L., Iliescu, R., Summers, R.L., Coleman, T.G.: Systems biology and integrative physiological modeling. J. Physiol. **589**, 1053–1060 (2011)
14. Morgan, B.J.T.: Applied Stochastic Modelling. Arnold Texts in Statistics. Arnold, London (2000)
15. Morris, M.: Factorial samplig plans for preliminary computational experiments. Technometrics **33**(2), 161–174 (1991)
16. Munir, M.: Generalized Sensitivity Functions in Physiological Modelling. PhD-Thesis, University of Graz (April 2010)
17. Pukelsheim, F.: Optimal Design of Experiments. Wiley, New York (1993)
18. Rothe, C.F., Gersting, J.M.: Cardiovascular interactions: An interactive tutorial and mathematical model. Adv. Physiol. Educ. **26**(2), 98–109 (2002)
19. Saltelli, A., Ratto, M., Andres, T., Campolongo, F., Cariboni, J., Gatelli, D., Saisana, M., Tarantola, S.: Global Sensitivity Analysis: The Primer. Wiley-Interscience, New York (2008)
20. Seber, G.A.F., Wild, C.J.: Nonlinear Regression. Wiley, Chichester (2003)
21. Thomaseth, K., Cobelli, C.: Generalized sensitivity functions in physiological system identification. Ann. Biomed. Eng. **27**(5), 607–616 (1999)
22. Ucinski, D., Atkinson, A.C.: Experimental design for time-dependent models with correlated observations. Studies in Nonlinear Dynamics and Econometrics **8**(2), Article 13: The Berkeley Electronic Press, Berkeley, CA (2004)
23. Wesseling, K.H., Jansen, J.R.C., Settels, J.J., Schreuder, J.J: Computation of aortic flow from pressure in humans using a nonlinear, three-element model. J. Appl. Physiol. **74**, 2566–2573 (1993)

Chapter 2
Mathematical Modeling of Physiological Systems

Thomas Heldt, George C. Verghese, and Roger G. Mark

Abstract Although mathematical modeling has a long and very rich tradition in physiology, the recent explosion of biological, biomedical, and clinical data from the cellular level all the way to the organismic level promises to require a renewed emphasis on *computational physiology*, to enable integration and analysis of vast amounts of life-science data. In this introductory chapter, we touch upon four modeling-related themes that are central to a computational approach to physiology, namely simulation, exploration of hypotheses, parameter estimation, and model-order reduction. In illustrating these themes, we will make reference to the work of others contained in this volume, but will also give examples from our own work on cardiovascular modeling at the systems-physiology level.

2.1 Introduction

Mathematical modeling has a long and very rich history in physiology. Otto Frank's mathematical analysis of the arterial pulse, for example, dates back to the late nineteenth century [12]. Similar mathematical approaches to understanding the mechanical properties of the circulation have continued over the ensuing decades, as recently reviewed by Bunberg and colleagues [5]. By the middle of the last century,

T. Heldt (✉) · G.C. Verghese
Computational Physiology and Clinical Inference Group, Research Laboratory of Electronics, Massachusetts Institute of Technology, 10-140L, 77 Massachusetts Avenue, Cambridge, MA 02139, USA
e-mail: thomas@mit.edu; verghese@mit.edu

R.G. Mark
Laboratory for Computational Physiology, Harvard-MIT Division of Health Sciences and Technology, Massachusetts Institute of Technology, E25-505, 77 Massachusetts Avenue, Cambridge, MA 02139, USA
e-mail: rgmark@mit.edu

J.J. Batzel et al. (eds.), *Mathematical Modeling and Validation in Physiology*,
Lecture Notes in Mathematics 2064, DOI 10.1007/978-3-642-32882-4_2,
© Springer-Verlag Berlin Heidelberg 2013

Hodgkin and Huxley had published their seminal work on neuronal action-potential initiation and propagation [25], from which models of cardiac electrophysiology readily emerged and proliferated [33]. To harness the emergent power of first analog and later digital computers, mathematical modeling in physiology soon shifted from analytical approaches to computational implementations of governing equations and their simulation. This development allowed for an increase in the scale of the problems addressed and analyzed. In the late 1960s, Arthur Guyton and his associates, for example, developed an elaborate model of fluid-electrolyte balance that still impresses today for the breadth of the physiology it represents [16].

Since the days of Guyton's initial work, the widespread availability of relatively low-cost, high-performance computer power and storage capacity has enabled physiological modeling to move from dedicated—and oftentimes single-purpose—computers to the researcher's desktop, as even small-scale computer clusters can be assembled at comparatively little expense. The technological advancements in computer power and digital storage media have also permitted increasingly copious amounts of biological, biomedical, and even clinical data to be collected and archived as part of specific research projects or during routine clinical management of patients. Our ability to collect, store, and archive large volumes of data from all biological time and length scales is therefore no longer a rate-limiting step in scientific or clinical endeavors. Ever more pressing, however, is the concomitant need to link characteristics of the observed data streams mechanistically to the properties of the system under investigation and thereby turn—possibly in real-time as required by some clinical applications [23]—otherwise overwhelming amounts of biomedical data into an improved understanding of the biological systems themselves. This link is the mechanistic, mathematical and computational modeling of biological systems at all physiological length and time scales, as envisioned by the Physiome project [3, 8, 26].

Mechanistic mathematical models reflect our present-level understanding of the functional interactions that determine the overall behavior of the system under investigation. By casting our knowledge of physiology in the framework of dynamical systems (deterministic or stochastic), we enable precise quantitative predictions to be made and to be compared against results from suitably chosen experiments. Mechanistic mathematical models often allow us to probe a system in much greater detail than is possible in experimental studies and can therefore help establish the cause of a particular observation [22]. When fully integrated into a scientific program, mathematical models and experiments are highly synergistic, in that the existence of one greatly enhances the value of the other: models depend on experiments for specification and refinement of parameter values, but they also illuminate experimental observations, allow for differentiation between competing scientific hypotheses, and help aid in experimental design [22]. Analyzing models rigorously, through sensitivity analyses, formal model-order reduction, or simple simulations of what-if scenarios also allow for identification of crucial gaps in our knowledge and therefore help motivate the design of novel experiments. Finally, mathematical models serve as important test beds against which estimation and identification algorithms can be evaluated, as the true target values are precisely known and controllable [20]. It seems therefore that a renewed emphasis on *computational physiology* is not merely a positive development, but an essential step toward increasing our knowledge of living systems in the twenty-first century.

In this chapter, we will touch upon four main themes of mathematical modeling, namely simulation, exploration of hypotheses, parameter estimation, and model-order reduction. In addition to drawing upon our own work to illustrate these application areas, we will point the reader to the work of others, some of which is represented in this volume.

2.2 Simulation

Given a chosen model structure and a nominal set of parameter values, a central application of mathematical modeling is the simulation of the modeled system. Closely related to the simulation exercise is the comparison of the simulated model response to experimental data. In the area of respiratory physiology, the contributions by Bruce (Chap. 7) and Duffin (Chap. 8) in this volume are examples of such applications of mathematical modeling. The contributions by Tin and Poon (Chap. 5) and Ottesen and co-workers (Chap. 10) focus on modeling the respiratory control system and the cardiovascular response to orthostatic stress, respectively.

Our own interest in the cardiovascular response to changes in posture led us to develop a detailed lumped-parameter model of the cardiovascular system [17]. The model consists of a 21-compartment representation of the hemodynamic system, shown in Fig. 2.1, coupled to set-point controllers of the arterial baroreflex and the cardiopulmonary reflex, as depicted in Fig. 2.2, that mimic the short-term action of the autonomic nervous system in maintaining arterial and right-atrial pressures constant (blood pressure homeostasis) [17, 22].

In the context of cardiovascular adaptation to orthostatic stress, numerous computational models have been developed over the past 40 years [4, 9–11, 15, 24, 27–29, 31, 32, 34, 35, 38, 39, 41–43, 48, 49, 51]. Their applications range from simulating the physiological response to experiments such as head-up tilt or lower body negative pressure [4, 9, 10, 15, 27–29, 31, 32, 38, 43, 50, 51], to explaining observations seen during or following spaceflight [29, 35, 42, 44, 48, 51]. The spatial and temporal resolutions with which the cardiovascular system has been represented are correspondingly broad. Several studies have been concerned with changes in steady-state values of certain cardiovascular variables [35, 41, 43, 48], others have investigated the system's dynamic behavior over seconds [15, 24, 27, 28, 34], minutes [4, 9, 10], hours [29, 42, 51], days [29, 39, 42], weeks [29], or even months [38]. The spatial representations of cardiovascular physiology range from simple two- to four-compartment representations of the hemodynamic system [4, 15, 31, 32, 48] to quasi-distributed or fully-distributed models of the arterial or venous system [28, 35, 41, 43].

In choosing the appropriate time scale of our model, we were guided by the clinical practice of diagnosing orthostatic hypotension, which is usually based on average values of hemodynamic variables measured a few minutes after the onset

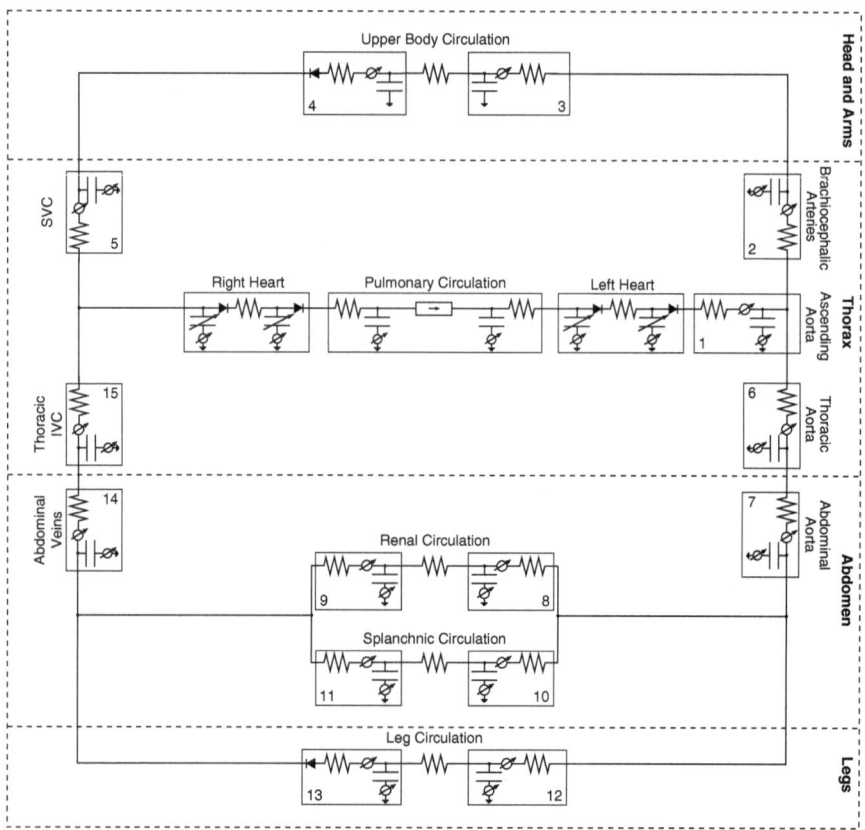

Fig. 2.1 Circuit representation of the hemodynamic system. *IVC*: inferior vena cava; *SVC*: superior vena cava

of gravitational stress [7]. The spatial resolution of our model was dictated by our desire to represent the prevailing hypotheses of post-spaceflight orthostatic intolerance (see Sect. 2.3). To determine a set of nominal parameter values, we searched the medical literature for appropriate studies on healthy subjects. In cases in which direct measurements could not be found, we estimated nominal parameter values on the basis of physiologically reasonable assumptions [17, 22]. We tested our simulations against a series of experimental observations by implementing a variety of stress tests, such as head-up tilt, supine to standing, lower-body negative pressure, and short-radius centrifugation, all of which are commonly used in clinical or research settings to assess orthostatic tolerance [17, 52].

Figure 2.3 shows simulations (solid lines) of the steady-state changes in mean arterial blood pressure and heart rate in response to head-up tilts to varying angles of elevation [17, 19], along with experimental data taken from Smith and co-workers [40]. (The dashed lines in this and later figures from simulations indicate the 95 % confidence limits of the nominal simulation on the basis of representative population simulations [18].) In Fig. 2.4, we show the dynamic

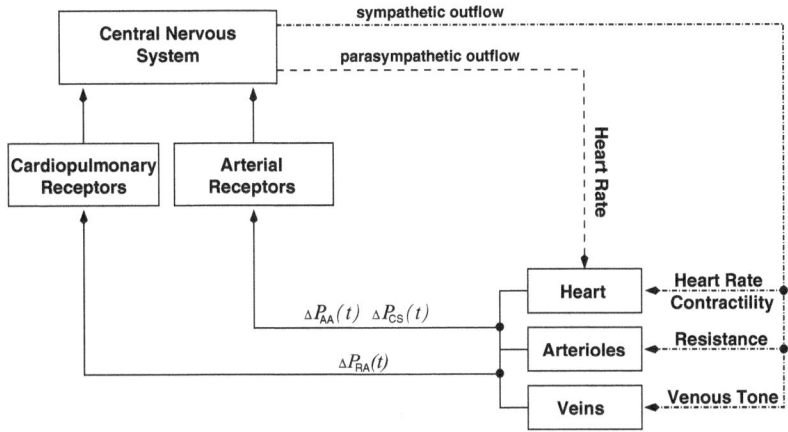

Fig. 2.2 Schematic representation of the cardiovascular control model. $\Delta P_{AA}(t)$, $\Delta P_{CS}(t)$, $\Delta P_{RA}(t)$: aortic arch, carotid sinus, and right atrial transmural pressures, respectively

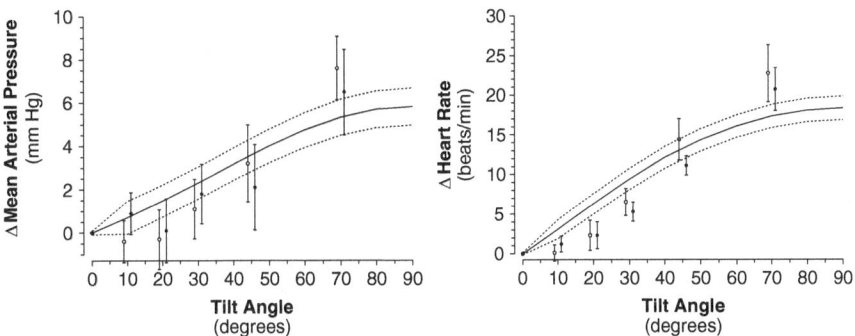

Fig. 2.3 Simulated steady-state changes (*solid lines*) and 95 % confidence intervals (*dashed lines*) in mean arterial pressure (*left*) and heart rate (*right*), in response to head-up tilt maneuvers to different angles of elevation. Data for young subjects (*open circles*) and older subjects (*filled circles*) from Smith et al. [40]

responses of measured mean arterial blood pressure and heart rate (lower panels) and the respective simulated responses (upper panels) to a rapid head-up tilt experiment [17,21]. Figure 2.5 shows the dynamic behavior of the same variables in response to standing up from the supine position. The simulations of Figs. 2.3–2.5 were all performed with the same set of nominal parameter values, and the same population distribution of parameter values.

Similar dynamic responses in arterial blood pressure and heart rate to orthostatic challenges have been reported by van Heusden [24] and Olufsen et al. [34], and are reported by Ottesen et al. in this volume (Chap. 11) for the transition from sitting to standing.

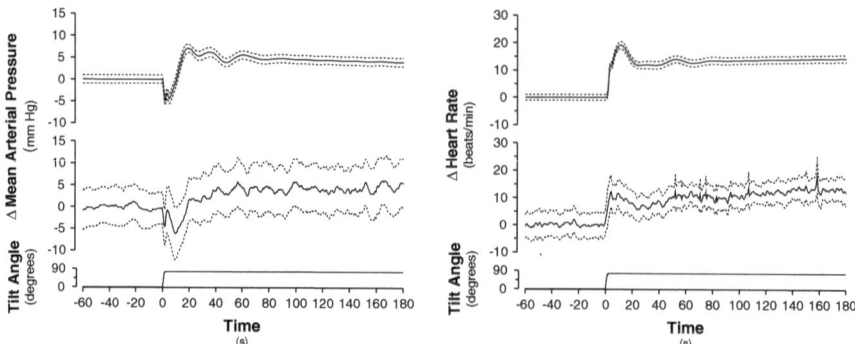

Fig. 2.4 Dynamic responses in mean arterial pressure (*left*) and heart rate (*right*) to a sudden head-up tilt maneuver. *Bottom panels* show experimental recordings [21]; *upper panels* show simulated responses [17]

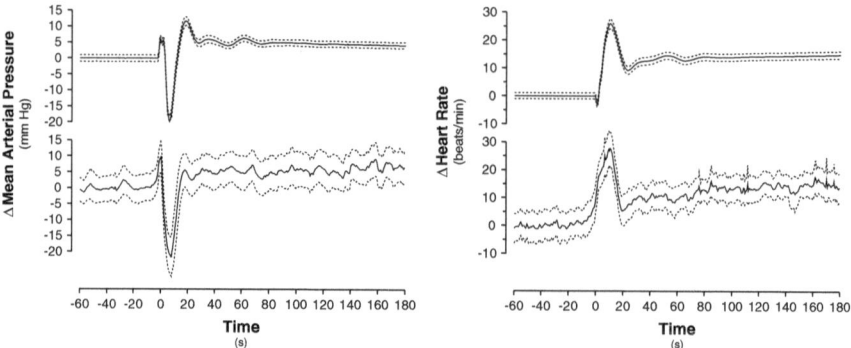

Fig. 2.5 Dynamic responses (*solid lines*) and 95 % confidence intervals (*dashed lines*) in mean arterial pressure (*left*) and heart rate (*right*) to standing up. *Bottom panels* show recordings [21]; *upper panels* show simulated responses [17]

Once a particular model structure has been chosen and simulations have been calibrated and validated against suitable sets of experimental data, the ensuing scientific step usually involves exploration of particular physiological hypotheses, or detailed sensitivity analyses as pursued by Banks (Chap. 3) or Ottesen (Chap. 10) in this volume.

2.3 Exploration of Hypotheses

Using the model of the previous section, we were interested in gaining insight into the cardiovascular system's failure to adapt to the upright posture following spaceflight. By simulating the system-level hemodynamic response to a tilt or a

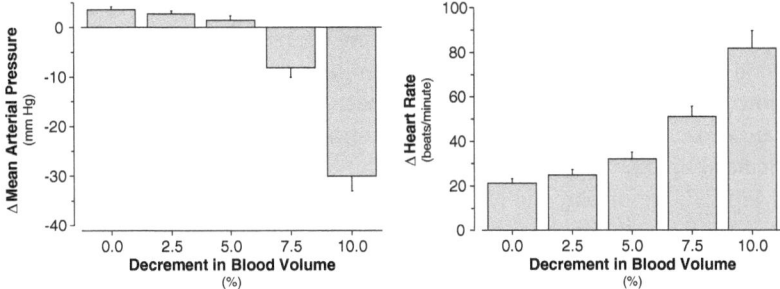

Fig. 2.6 Mean arterial pressure and heart rate changes induced by head-up tilt to 75°. Dependence on volume status. Mean response ± SE based on 20 simulations

stand test under varying parameter profiles, we sought to identify which of the prevailing physiological hypotheses lead to the system-level hypotension seen in affected astronauts upon assumption of the upright posture. This approach can be viewed as a targeted sensitivity analysis that differs from the more general explorations presented by Banks (Chap. 3), in that the parameters to be varied are selected based on *a priori* physiological considerations. Furthermore, the parameter values will be subjected to larger perturbations than in the more local analysis of Chap. 3.

In our analysis, we choose to include those parameters that have been implicated in contributing to the post-flight orthostatic intolerance phenomenon [17]. Our analysis therefore includes total blood volume, the venous compliance of the legs, the end-diastolic compliance of the right ventricle, and the static gain values (both arterial and cardiopulmonary) of arteriolar resistance and venous tone. We assess the impact of parameter perturbations by analyzing the changes they induce in the mean arterial pressure and heart rate responses to a 75° head-up tilt. In particular, we seek to answer which of the parameters included in the analysis has the greatest impact on mean arterial pressure and heart rate.

We address this question by repeatedly simulating tilt experiments while varying each of the parameters by a certain percentage of their nominal values. In Fig. 2.6, we report the changes in mean arterial pressure and heart rate from their respective supine baselines in response to a 4 min head-up tilt to 75° for varying levels of total blood volume. We note that head-up tilt usually results in a slight *increase* in mean arterial pressure measured at heart level, with a concomitant increase in heart rate. Figure 2.6 reflects this fact as the baseline simulation (0 % decrement in total blood volume, or 70 ml/kg of body weight) shows an increase in mean arterial pressure of about 4 mm Hg and an increase of approximately 20 beats/min in heart rate. As blood volume is progressive reduced, the gentle rise in mean arterial pressure is diminished, but generally maintained up to volume decrements of 5 %. Beyond that, the system fails to maintain mean arterial pressure despite incrementally larger increases in heart rate. The reason for this behavior becomes clear when we consider blood pooling in the dependent vasculature during tilt as a function of hydration status. With increasing degree of hypovolemia, the amount of blood volume pooled in the lower extremities becomes an increasingly larger fraction of distending volume. It therefore becomes progressively more difficult for the cardiovascular system to maintain right atrial pressure, and thus cardiac output, during head-up tilt.

In Fig. 2.7, we display the results of the same analysis for the venous compliance of the legs, the right-ventricular end-diastolic compliance, and the arterial and venous tone feedback gain values (top to bottom). Each of the simulations underlying Fig. 2.7 starts with the same baseline blood volume, which, for future reference, we term the euvolemic baseline state. When comparing the results in Fig. 2.7 with the volume-loss results in Fig. 2.6, it is obvious that deleterious changes in any of the parameters shown in Fig. 2.7 only marginally impact the hemodynamic response to tilt if the volume status if euvolemic. In other words, in the absence of hypovolemia, the body can tolerate significant detrimental changes in any of the other parameters without developing a seriously compromised hemodynamic response to tilt.

Next, we demonstrate that this behavior can change drastically if the baseline volume status is changed. In Fig. 2.8, we vary the four parameters of Fig. 2.7 by the same fractional changes, yet their variation is superimposed on a baseline state that is 5 % hypovolemic compared to the euvolemic baseline states of Figs. 2.6 and 2.7. The results demonstrate that against the backdrop of an otherwise benign reduction in total blood volume, even modest 5 % to 10 % detrimental changes in each of the parameters can significantly impact the hemodynamic response to tilt.

The results of the simulations show that the level of hydration has by far the greatest impact on blood pressure homeostasis during tilt. Furthermore, the impact of changes in other parameters varies significantly with the level of hydration. In the euvolemic state, changes in the four parameters considered in Figs. 2.7 and 2.8 have similar effects on the mean arterial pressure and heart rate responses. In the hypovolemic case, changes in venous tone seem to impact the hemodynamic response to tilt more when compared with the same fractional changes in the other parameters, yet all of the parameters considered significantly influence the heart rate and mean arterial pressure responses to head-up tilt.

The simulations presented in this section demonstrate the importance of blood volume in maintaining mean arterial pressure during orthostatic stress. Changes in the other parameters included in this analysis are largely inconsequential if total blood volume is maintained near euvolemic levels (70 ml/kg). However, if the baseline state is hypovolemic, even relatively modest changes in these parameters can aggravate the cardiovascular system's failure to adapt properly to the upright posture. Reductions in both the arterial resistance gains and the venous tone gains affect mean arterial pressure most; impairment of the venous tone feedback, however, has a stronger effect when the same fractional decrements in the nominal values are considered.

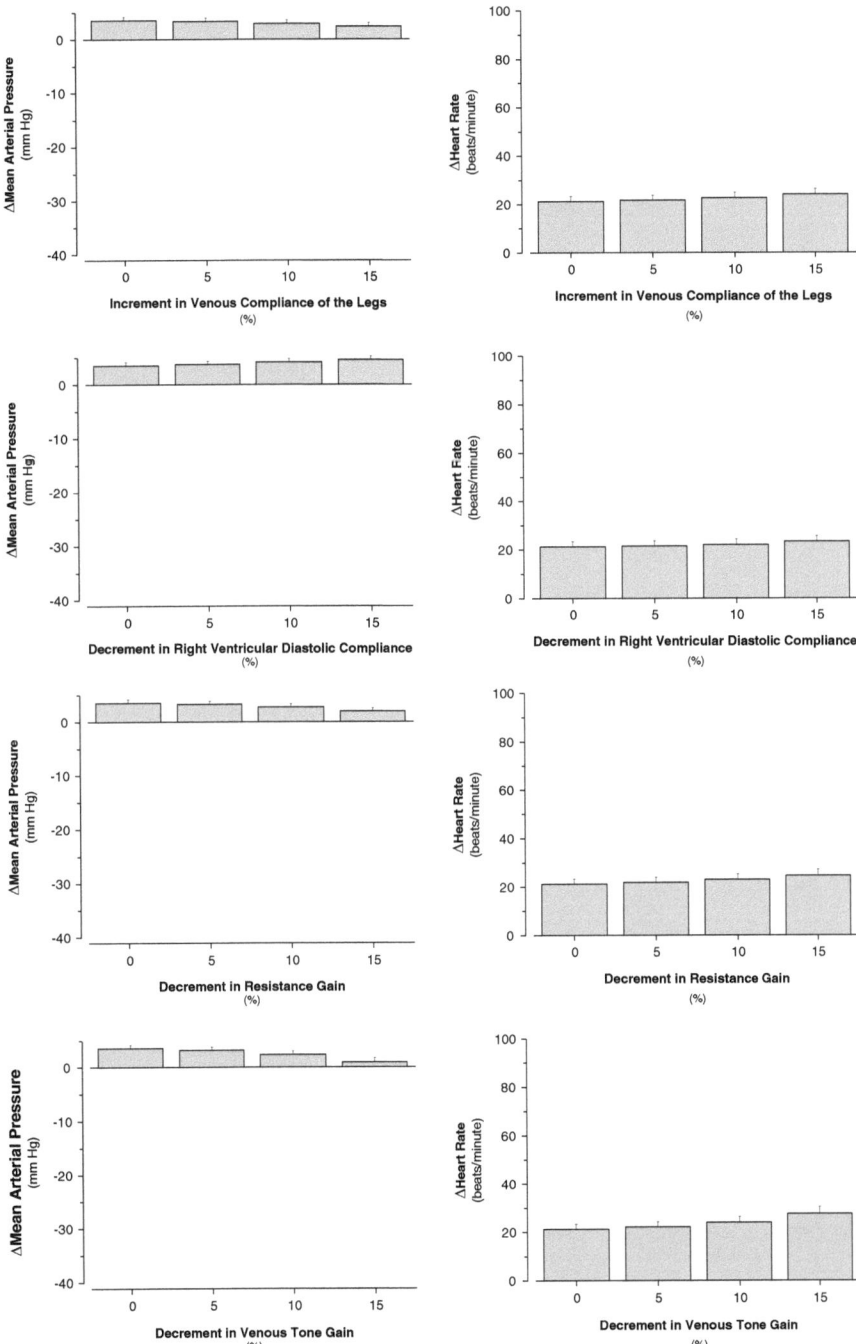

Fig. 2.7 Mean arterial pressure and heart rate changes in response to a 75° head-up tilt under varying parametric conditions. Baseline volume status is euvolemic. Mean response ± SE based on 20 simulations

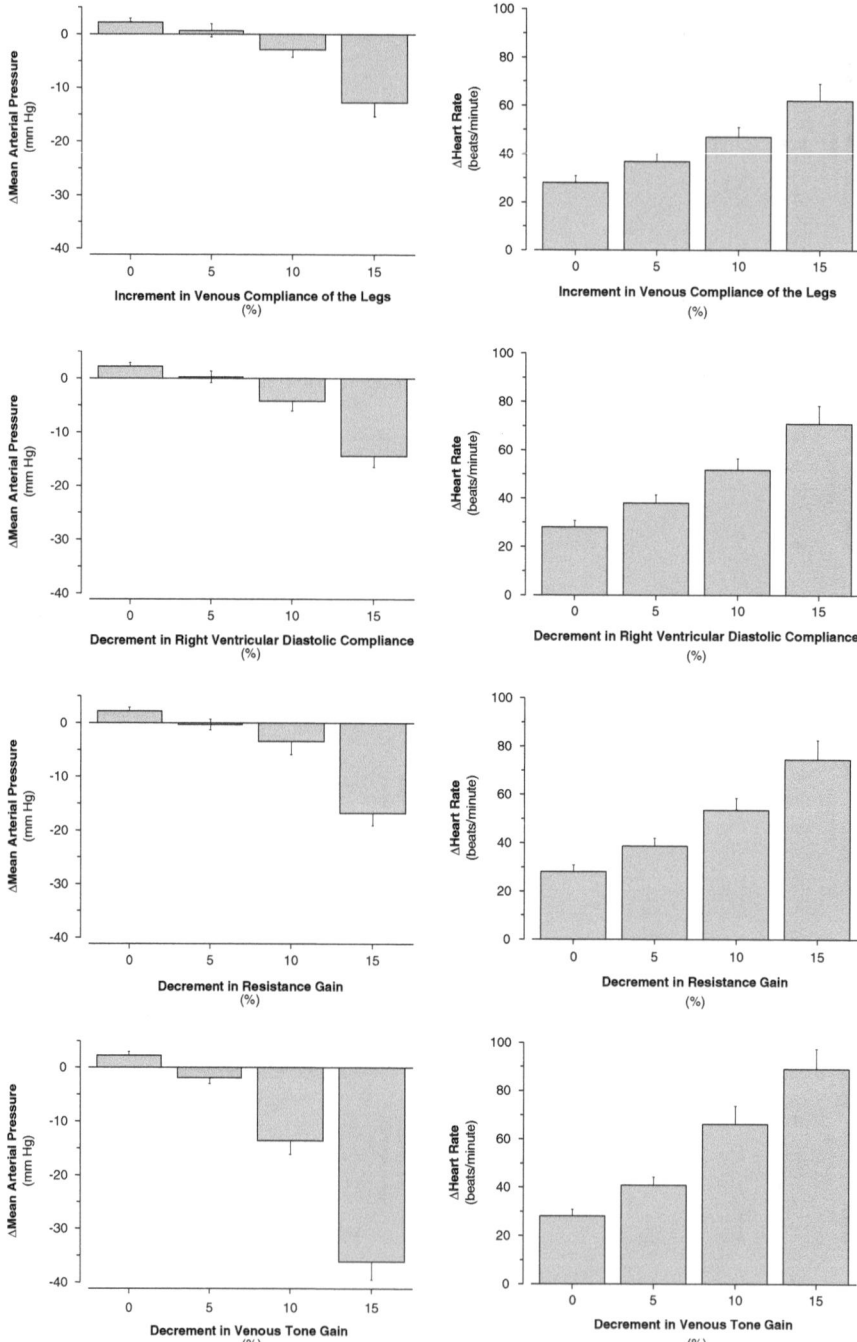

Fig. 2.8 Mean arterial pressure and heart rate changes in response to a 75° head-up tilt under varying parametric conditions. Baseline volume status is 5% hypovolemic. Mean response ± SE based on 20 simulations

Reductions in total blood volume in returning astronauts have been well established, though the magnitude of the hypovolemia is highly variable. The work by Waters and co-workers suggest a mean overall reduction of about 6 % in male pre-syncopal astronauts [47]. Our simulations give credence to the view that hypovolemia can be the principal mechanism of post-spaceflight orthostatic intolerance, yet at an average level of 6 % might not be solely responsible for the clinical picture of the syndrome. Of particular importance, therefore, are perturbations to the α-sympathetically mediated reflex pathways of arteriolar resistance and venous tone, both of which depend on the peripheral neurotransmitter norepinephrine. Ramsdell et al. demonstrated that an exogenous α-adrenergic agonist, when given to subjects undergoing orthostatic stress testing at the conclusion of a prolonged bedrest protocol, significantly reduces the incidence of syncope or pre-syncopal symptoms [37]. Furthermore, Meck et al. showed that those astronauts who become pre-syncopal on landing day fail to mount a significant vasoconstrictor response due to an inappropriately low release of norepinephrine in the upright posture [30]. Finally, Platts and co-workers demonstrated that the same α-agonist used by Ramsdell was beneficial in alleviating post-spaceflight orthostatic intolerance in a female astronaut who had a prior history of orthostatic hypotension after previous missions [36]. In the context of our model, failure to release norepinephrine at the smooth muscle synapses is interpreted as a reduction in the gain from arterial blood pressure and right atrial pressure to venous tone and arterial resistance.

Together, these experimental findings corroborate the results from our simulation studies, namely that a critical combination of mechanisms might have to be invoked to explain the phenomenon of post-spaceflight orthostatic intolerance, and that apart from hypovolemia, detrimental changes to the feedback pathway to vascular smooth muscle (venous tone and arteriolar resistance) play a dominant role.

2.4 Model Identification

Mechanistic physiological models tend to have relatively high spatial and temporal resolution for their ultimate application and therefore involve large numbers of parameters and exhibit rich dynamic behavior, spanning several time scales. Such a modeling approach, in which models are built from detailed analysis of the underlying physiological or biological processes, is commonly referred to as *forward modeling*. Often, the purpose of modeling is to use the resultant model in conjunction with experimental data to estimate the values of some model parameters. Such a strategy falls under the general umbrella of *inverse methods* in which attributes of a system are estimated based on measurements that are only indirectly related to these attributes [46]. The link between the attributes and the data is the mathematical model.

In such parameter estimation problems, the parameters of a model are tuned such that a measure of error between the model output and a corresponding set of observations is minimized. Methods for solving this minimization problem depend quite naturally on the error criterion and on the structure of the chosen model. Furthermore, the quality of the resultant parameter estimates also depends on the fidelity of the available data.

In this volume, the contributions by Attarian and co-workers (Chap. 4), Banks and co-workers (Chap. 3), Ottesen et al. (Chap. 10), and Hartung and Turi (Chap. 6) focus on specific aspects of inverse modeling. Attarian focuses on the topic of Kalman filtering for both parameter and state estimation, while Banks' chapter provides an overview of statistical methods to address the parameter estimation problem, particularly the issue of subset selection which we will touch upon later in this section, too. Ottesen highlights the application of parameter estimation methods to the clinical problem of estimating cerebrovascular variables. Hartung and Turi apply parameter estimation to the identification of the respiratory control system. The contribution by Bruce (Chap. 7) similarly focuses on identifying respiratory control loops.

Our own interest in parameter estimation initially developed in the context of cardiovascular applications, where the number of signals available from physiological experiments are usually small in number. Quite often, the only signals recorded are beat-to-beat mean arterial pressure and heart rate during rest or a particular stress-test intervention to challenge the cardiovascular system. The models developed, however, can be quite detailed in their representation of cardiovascular physiology. Thus a disparity exists between the high resolution and rich dynamic behavior of the model and the low resolution and limited dynamic behavior of the measurements. In the parameter estimation setting, this disparity leads to a very sensitive—or *ill-conditioned*—estimation problem, in which small changes to the input data can lead to very large and obviously undesirable changes to the resultant parameter estimates. Since treatment of ill-conditioning is not commonly found in the physiological modeling literature, we present in some detail one possible approach, namely subset selection, to overcome this problem. Our exposition in the remainder of this section is largely based on [17].

2.4.1 Non-linear Least-Squares Estimation

If the model outputs are linear functions of the model parameters, and if the error measure is chosen to be the sum of squares of the prediction error on each output, the resultant minimization problem is the well-known *linear least squares problem*. If the model output is a *non-linear* function of the parameters, the minimization problem is usually solved iteratively through a sequence of linear least squares problems that involve the gradient or higher-order derivatives of the cost function.

Let $\Phi(\hat{\mathbf{y}}(\theta), \mathbf{y})$ denote a non-negative measure of error, or cost function, between model output $\hat{\mathbf{y}}(\theta) \in R^n$ and experimental data $\mathbf{y} \in R^n$. Through the model output, Φ is an implicit function of the vector of model parameters $\theta \in R^m$ and can therefore be considered a function from R^m to the non-negative real line, $\Phi = \Phi(\theta) : R^m \to R^+ \cup \{0\}$, for a given experimental data set. Since we aim to minimize this measure of error iteratively, we seek a computational scheme that provides us with a new parameter vector θ_1 such that $\Phi(\theta_1) < \Phi(\theta_0)$ given an initial best guess θ_0. Once such a θ_1 has been identified, it can assume the role of θ_0, and we can repeat the computational scheme in an effort to reduce the measure of error even further.

Let us assume that a second-order Taylor series approximation Ψ of $\Phi(\theta)$ is a good approximation of the local behavior of the cost function for small perturbations $\Delta\theta = \theta - \theta_0$ around the initial parameter estimate. [$\Psi(\theta)$ is the best second-order approximation to the surface defined by $\Phi(\theta)$ around θ_0.] $\Psi(\theta)$ is given by

$$\Psi(\theta) = \Phi(\theta_0) + \left[\frac{\partial\Phi}{\partial\theta}\right]_{\theta_0} \Delta\theta + \frac{1}{2}\,\Delta\theta^T\left[\frac{\partial^2\Phi}{\partial\theta^2}\right]_{\theta_0}\Delta\theta,$$

where $[\partial\Phi/\partial\theta]_{\theta_0}$ and $[\partial^2\Phi/\partial\theta^2]_{\theta_0}$ are the appropriate matrices of the first- and second-order derivatives evaluated at the current best guess of the parameter vector as indicated by the subscript θ_0. To find its minimum, we equate to zero the gradient of $\Psi(\theta)$:

$$\frac{\partial}{\partial\theta}\Psi(\theta) = \left[\frac{\partial\Phi}{\partial\theta}\right]_{\theta_0} + \left[\frac{\partial^2\Phi}{\partial\theta^2}\right]_{\theta_0}\Delta\theta = 0,$$

which leads to the following condition for the stationary point θ_1:

$$\left[\frac{\partial^2\Phi}{\partial\theta^2}\right]_{\theta_0}(\theta_1 - \theta_0) = -\left[\frac{\partial\Phi}{\partial\theta}\right]_{\theta_0}.$$

If the inverse of the second-order derivative matrix exists, the stationary point is given by:

$$\theta_1 = \theta_0 - \left[\frac{\partial^2\Phi}{\partial\theta^2}\right]_{\theta_0}^{-1}\cdot\left[\frac{\partial\Phi}{\partial\theta}\right]_{\theta_0}. \qquad (2.1)$$

It can be shown that $\Phi(\theta_1) < \Phi(\theta_0)$ if and only if the matrix of second-order derivatives is positive definite [2].

Note that we have not yet specified the cost function Φ. The results obtained so far only require it to be twice differentiable. Let us assume now that we are aiming to minimize the square of the residual error $\mathbf{r}(\theta) = \hat{\mathbf{y}}(\theta) - \mathbf{y}$ between model output and actual measurements, that is

$$\Phi(\theta) = \frac{1}{2}\cdot\|\mathbf{r}(\theta)\|^2 = \frac{1}{2}\cdot\|\hat{\mathbf{y}}(\theta) - \mathbf{y}\|^2 = \frac{1}{2}\cdot\mathbf{r}^T\mathbf{r} \to \min,$$

where the factor of $1/2$ has been included for convenience, and the superscript ^{T} denotes transposition. The gradient of this cost function is given by:

$$\left[\frac{\partial\Phi}{\partial\theta}\right] = \mathbf{J}^T\cdot\mathbf{r}(\theta) \quad\text{where}\quad J_{ij} = \frac{\partial r_i(\theta)}{\partial\theta_j} = \frac{\partial\hat{y}_i(\theta)}{\partial\theta_j}.$$

$\mathbf{J} \in R^{n \times m}$ denotes the Jacobian matrix of the error vector with respect to the parameter vector. Similarly, we can compute the elements of the Hessian matrix $\mathbf{H} \in R^{m \times m}$ of second derivatives of the cost function:

$$H_{ij} = \frac{\partial^2 \Phi}{\partial \theta_i \partial \theta_j} = (\mathbf{J}^\mathsf{T} \mathbf{J})_{ij} + \sum_{l=1}^{n} r_l \cdot \frac{\partial^2 r_l}{\partial \theta_i \partial \theta_j}. \tag{2.2}$$

Note that for small residuals, the Hessian can be approximated by

$$H_{ij} = \frac{\partial^2 \Phi}{\partial \theta_i \partial \theta_j} \approx (\mathbf{J}^\mathsf{T} \mathbf{J})_{ij}, \tag{2.3}$$

since the second term involves the elements of the vector of residuals directly. This approximation is known as the Gauss–Newton approximation to the Hessian.

Inserting the expressions for the derivatives into Eq. (2.1), we obtain the iterative parameter updates of the Newton method:

$$\mathbf{H} \cdot (\theta_{i+1} - \theta_i) = -\mathbf{J}^\mathsf{T} \cdot \mathbf{r} \tag{2.4}$$

and the Gauss–Newton approximation thereof:

$$\mathbf{J}^\mathsf{T} \mathbf{J} \cdot (\theta_{i+1} - \theta_i) = -\mathbf{J}^\mathsf{T} \cdot \mathbf{r}. \tag{2.5}$$

Let $\mathbf{R}$ denote either the full Hessian $\mathbf{H}$ or its Gauss–Newton approximation $\mathbf{J}^\mathsf{T} \mathbf{J}$. If $\mathbf{R}$ has full column rank, Eqs. (2.4) and (2.5) can be solved exactly or in a least squares sense, depending on whether or not $\mathbf{J}^\mathsf{T} \cdot \mathbf{r}$ is in the column space of $\mathbf{R}$. In either case, however, the solution is unique, and efficient algorithms exist to solve the set of linear equations numerically [14]. If, on the other hand, $\mathbf{R}$ is rank-deficient, then the m columns of $\mathbf{R}$ actually contain less than m linearly independent vectors. As a consequence, $\mathbf{R}$ is semi-definite with at least one of its eigenvalues at zero. The following argument by Burth and co-workers [6] illustrates the problem of a singular matrix $\mathbf{R}$ in the context of parameter estimation. Assume $\mathbf{R}$ has a single eigenvalue at zero with some associated eigenvector ϑ. Within the limits of our second-order approximation, ϑ can be added to any step direction without affecting the error criterion, since

$$\mathbf{R} \cdot (\theta_{i+1} - \theta_i + \vartheta) = \mathbf{R} \cdot (\theta_{i+1} - \theta_i) + \mathbf{R} \cdot \vartheta = \mathbf{R} \cdot (\theta_{i+1} - \theta_i) = -\mathbf{J}^\mathsf{T} \cdot \mathbf{r}.$$

This implies that we can arbitrarily change parameter values along the direction of ϑ without affecting the error criterion. Such indeterminacy suggests that the parameters of the model cannot be estimated uniquely from the given measurements; the estimation problem is said to be ill-conditioned.

Frequently, the matrix $\mathbf{R}$ is not exactly rank deficient but quite nearly so, in the sense that the largest eigenvalue is orders of magnitude larger than the smallest eigenvalue. Closeness to singularity is measured by the condition number $\kappa(\mathbf{R})$ which, for real, symmetric matrices, is given by the ratio of the largest to the smallest eigenvalue.

Burth reviews several related consequences of a large condition number for estimation problems [6]. We simply note here that a hallmark of an ill-conditioned problem is its extreme sensitivity to small perturbations in either the experimental data or the elements of the matrix $\mathbf{R}$. In the following section we discuss subset selection as an approach to overcome this ill-conditioning.

2.4.2 Subset Selection

Subset selection aims to determine which parameter axes lie closest to the singular directions of the Hessian matrix [45]. Changes in the corresponding parameters do relatively little to change the fit of the model to the data, and therefore these parameters are hard to estimate reliably. Once these ill-conditioned parameter axes are identified, one can fix the associated parameters at prior values throughout the estimation process, thus improving the conditioning of the resultant reduced-order estimation problem. Fixing values of the ill-conditioned parameters has the effect of introducing some bias error into the model, but by removing these parameters from the estimated set, we improve the reliability with which the remaining parameters are estimated.

Subset selection is most powerful if the eigenvalue spectrum of the Hessian matrix exhibits a large gap between ρ large eigenvalues and $m - \rho$ small ones. Such a situation suggests that the Hessian matrix has numerical rank ρ and that $m - \rho$ appropriately chosen parameters should be fixed. Equations (2.4) and (2.5) then only involve reduced-order Hessian and Jacobian matrices, which we will denote by $\mathbf{H}_\rho$ and $\mathbf{J}_\rho$, respectively.

The following subset selection algorithm for non-linear least squares estimation is based on the work of Vélez-Reyes [45] and is essentially an extension of a subset selection algorithm for the linear least squares problem [13, 14]:

1. Given an initial estimate θ_0, compute the Hessian $\mathbf{H}(\theta_0)$ and its eigenvalue decomposition $\mathbf{H} = \mathbf{V}\Lambda\mathbf{V}^\mathsf{T}$.
2. Determine ρ and an ordering of the eigenvalues in the decomposition such that the first ρ eigenvalues of $\mathbf{H}$ are much larger than the remaining $m - \rho$.
3. Partition the matrix of eigenvectors according to $\mathbf{V} = [\mathbf{V}_\rho \ \mathbf{V}_{m-\rho}]$.
4. Determine a permutation matrix $\mathbf{P}$ by constructing a QR decomposition with column pivoting [14, p. 248] for $\mathbf{V}_\rho^\mathsf{T}$, i.e., determine $\mathbf{P}$ such that

$$\mathbf{V}_\rho^\mathsf{T} \cdot \mathbf{P} = \mathbf{Q} \cdot \mathbf{R},$$

where $\mathbf{Q}$ is an orthogonal matrix and the first ρ columns of $\mathbf{R}$ form an upper triangular matrix.

5. Use $\mathbf{P}$ to re-order the parameter vector θ according to $\tilde{\theta} = \mathbf{P}^\mathsf{T}\theta$.
6. Make the partition $\tilde{\theta} = [\tilde{\theta}_\rho^\mathsf{T} \ \tilde{\theta}_{m-\rho}^\mathsf{T}]^\mathsf{T}$, where $\tilde{\theta}_\rho^\mathsf{T}$ contains the first ρ elements of $\tilde{\theta}$.
   Fix $\tilde{\theta}_{m-\rho}$ at a prior estimate $\hat{\tilde{\theta}}_{m-\rho}$.
7. Compute the new estimate of the parameter vector $\hat{\tilde{\theta}}$ by solving the reduced-order minimization problem

$$\hat{\tilde{\theta}} = \arg\min_{\tilde{\theta}} \Phi(\tilde{\theta}) \quad \text{subject to} \quad \hat{\tilde{\theta}}_{m-\rho} = \tilde{\theta}_{m-\rho}.$$

The eigenvalue decomposition in the first step is a pre-requisite for the determination of the numerical rank of $\mathbf{H}$ and the subset selection step. The rank determination in step 2 is based on reasonably sized gaps in the eigenvalue spectrum. Such gaps might not exist, and in those cases subset selection might only be of limited help in overcoming ill-conditioning [13]. Usually several gaps of differing sizes can be identified and one has to choose between including more parameters and keeping the condition number of the reduced order Hessian $\mathbf{H}_\rho$ small. The former choice usually increases the reduced-order model's ability to represent experimental data, while the latter leads to more reliable estimation of the remaining parameters. The numerical rank estimate tells us how many parameters to include in our analysis. Step 4, the actual subset selection step, determines *which* parameters to include. This information is encoded in the permutation matrix $\mathbf{P}$. Step 5 reorders the parameter vector θ such that the ρ dominant parameters move to the top of the vector. Steps 6 and 7 describe the reduced-order estimation step.

We applied the subset selection algorithm outlined above to the problem of estimating cardiovascular parameters from the transient hemodynamic response to standing up detailed in Sect. 2.2.

Figure 2.9 shows the eigenvalue spectrum $\{\lambda_i\}$ of the approximate Hessian matrix $\mathbf{H}(\theta_0) = \mathbf{J}^\mathsf{T}\mathbf{J}$. Since the spectrum covers almost eight orders of magnitude, we plot the eigenvalues on a logarithmic scale. The corresponding condition number of the full-order Hessian matrix is $\kappa(\mathbf{H}) = 3.4 \cdot 10^7$, indicating substantial ill-conditioning if attempts were made to solve the full-order estimation problem. Figure 2.9 also shows the gap structure $\{\lambda_i/\lambda_{i+1}\}$ of the eigenvalue spectrum. It is evident from the two figures that no single dominantly large gap exists that would suggest an obvious choice for the rank estimate. However, the gap structure does show three breakpoints that are sufficiently removed from the remainder of the spectrum. They correspond to the rank estimates $\rho_1 = 3$, $\rho_2 = 1$, and $\rho_3 = 4$, where ρ_1 corresponds to the largest gap, ρ_2 to the second largest and so on.

We compared the rank estimate $\rho = 4$ against an arbitrarily chosen one, $\rho = 15$, to demonstrate the effect of ill-conditioning on the parameter estimates. Table 2.1 shows the results of this exploration and demonstrates that including more parameters in the estimation scheme than warranted by the subset selection criterion yields unreliable estimates even for those parameters that by themselves might be identifiable. A more detailed discussion of the methodology, evaluation strategy, and results of subset selection in the context of cardiovascular parameter estimation can be found in [17].

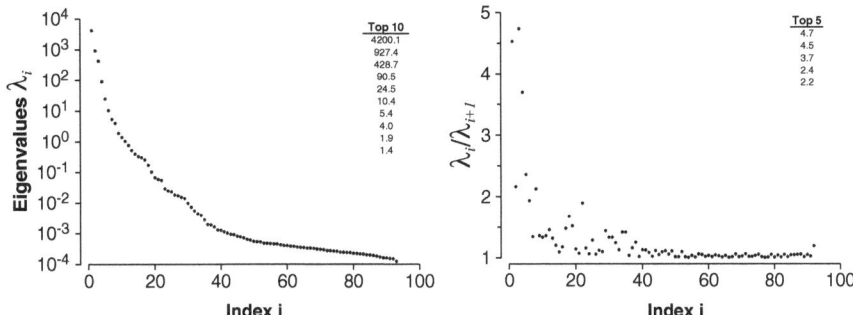

Fig. 2.9 Eigenvalue spectrum of the Hessian matrix (*left*) and gaps λ_i/λ_{i+1} of the eigenvalue spectrum

Table 2.1 Mean relative errors of estimated parameters with respect to their true values. Numeric values are given in %. Parameters are: Setpoint of the arterial blood pressure (1), right-ventricular end-diastolic elastance (2), total blood volume (3), and nominal heart rate (4)

		Parameter index			
	$\kappa(\mathbf{H}_\rho)$	1	2	3	4
$\rho = 4$	55.3	(-0.9 ± 0.3)	(0.1 ± 0.2)	(0.4 ± 0.1)	(0.7 ± 0.4)
$\rho = 15$	$8.0\cdot10^4$	(-21.1 ± 2.5)	(4.6 ± 4.2)	(-5.6 ± 1.2)	(32.1 ± 6.8)

2.5 The Case for Structured Model-Order Reduction

When developing physiological models, researchers commonly do not strive to build a minimal model capable of representing particular experimental observations. Rather, they usually represent all components of the physiological system that may contribute significantly to particular observations. Consequently, the resultant models can become quite sizable, representing physiology at a variety of time and space scales. Such models typically embody or reflect the underlying physical or mechanistic understanding we have about the system, as well as structural features such as the delineation of subsystems and their interconnections. Often, these models have been built up over decades of study and reflect the cumulative knowledge and contributions of many researchers in a field.

However, increasing the complexity of a model can significantly work against its usefulness in many respects, as simulation times are increased and it becomes difficult to understand in a fundamental way what parts of a model are actually being exercised and how. In addition, a significant feature of such models is the uncertainty associated with many or most of the parameters of the model. The data that one can collect from the associated physiological system is rarely rich enough to allow reliable identification of all model parameters (see Sect. 2.4), yet there are good reasons not to be satisfied with direct black-box identification of a reduced-order model [1]. The challenge then is to develop meaningful reduced-order models that reflect the detailed, hard-won knowledge one has about the system, while being better suited to identification, simulation, and control design than the original large model. Currently, no such tools for *structured* or *gray-box* model-order reduction exist, though the need to develop such tools is quite obvious: only if we are able to identify those components of a model that contribute significantly to a particular simulated response do we actually increase our understanding of the system under study, and only if we are able to quantify the effects of parametric uncertainties can we assess the model's range of validity and can we suggest experiments that might help reduce prediction uncertainties.

2.6   Conclusions

Physiology has always been an integrative and quantitative science in which mathematical analyses and modeling featured very prominently alongside experimentation to illuminate experimental results and to test our understanding of physiology in an exact, quantitative framework. The need to continue in this tradition is ever more urgent as the recent explosion in biological, biomedical, and clinical data necessitates novel approaches to the integration, analysis, and interpretation of such overwhelmingly copious amounts of data. In this introductory chapter, we have touched upon some key themes of mathematical modeling, namely simulation, exploration of hypotheses, parameter estimation and model-order reduction. Many of the contributions in this volume elaborate on one or the other of these themes in their application of modeling in physiology or biology.

Acknowledgements This work was partially supported by the United States National Aeronautics and Space Administration through the Cooperative Agreement NCC-52 with the National Space Biomedical Research Institute, and through grant R01 001659 from the National Institute of Biomedical Imaging and Bioengineering of the US National Institutes of Health.

References

1. Antoulas, A.: Approximation of Large-Scale Dynamical Systems. Advances in Design and Control. SIAM, Philadelphia, PA (2005)
2. Bard, Y.: Nonlinear Parameter Estimation. Academic, New York (1974)
3. Bassingthwaighte, J.: Strategies for the physiome project. Ann. Biomed. Eng. **28**, 1043–1058 (2000)
4. Boyers, D., Cuthbertson, J., Luetscher, J.: Simulation of the human cardiovascular system: A model with normal response to change in posture, blood loss, transfusion, and autonomic blockade. Simulation **18**, 197–205 (1972)
5. Brunberg, A., Heinke, S., Spillner, J., Autschbach, R., Abel, D., Leonhardt, S.: Modeling and simulation of the cardiovascular system: A review of applications, methods, and potentials. Biomed. Tech. (Berl.) **54**(5), 233–244 (2009)

6. Burth, M., Verghese, G., Vélez-Reyes, M.: Subset selection for improved parameter estimation in on-line identification of a synchronous generator. IEEE Trans. Power Syst. **14**(1), 218–225 (1999)
7. Consensus Committee of the American Autonomic Society and the American Academy of Neurology: Consensus statement on the definition of orthostatic hypotension, pure autonomic failure and multiple system atrophy. Clin. Auton. Res. **6**, 125–126 (1996)
8. Crampin, E., Halstead, M., Hunter, P., Nielsen, P., Noble, D., Smith, N., Tawhai, M.: Computational physiology and the physiome project. Exp. Physiol. **89**(1), 1–26 (2004)
9. Croston, R., Fitzjerrell, D.: Cardiovascular model for the simulation of exercise, lower body negative pressure, and tilt table experiments. Proceedings of the Fifth Annual Pittsburgh Conference on Modeling and Simulation, pp. 471–476 (1974)
10. Croston, R., Rummel, J., Kay, F.: Computer model of cardiovascular control system response to exercise. J. Dyn. Syst. Meas. Control **95**, 301–307 (1973)
11. Dickinson, C.: A digital computer model of the effects of gravitational stress upon the heart and venous system. Med. Biol. Eng. **7**, 267–275 (1969)
12. Frank, O.: Die Grundform des arteriellen Pulses. Erste Abhandlung. Mathematische Analyse. Z. Biol. **37**, 483–526 (1899)
13. Golub, G., Klema, V., Stewart, G.: Rank degeneracy and least squares problems. Technical Report TR-456, Department of Computer Science, University of Maryland, College Park, MD (1976)
14. Golub, G., van Loan, C.: Matrix Computations, 3rd edn. Johns Hopkins University Press, Baltimore, MD (1996)
15. Green, J., Miller, N.: A model describing the response of the circulatory sytem to acceleration stress. Ann. Biomed. Eng. **1**(4), 455–467 (1973)
16. Guyton, A., Coleman, T.: Quantitative analysis of the pathophysiology of hypertension. Circ. Res. **24**(Suppl. I), I1–I19 (1969)
17. Heldt, T.: Computational models of cardiovascular response to orthostatic stress. Doctoral dissertation, Harvard-MIT Division of Health Sciences and Technology, Massachusetts Institute of Technology, Cambridge, MA (2004). Http://dspace.mit.edu/handle/1721.1/28761
18. Heldt, T., Mark, R.: Scaling cardiovascular parameters for population simulations. Comput. Cardiol. **31**, 133–136 (2004). Available at http://www.cinc.org/archives/2004/133.pdf
19. Heldt, T., Mark, R.: Understanding post-spaceflight orthostatic intolerance – A simulation study. Comput. Cardiol. **32**, 631–634 (2005). Available at http://www.cinc.org/archives/2005/0631.pdf
20. Heldt, T., Mukkamala, R., Moody, G., Mark, R.: CVSim: An open-source cardiovascular simulator for teaching and research. Open Pacing Electrophysiol. Ther. J. **3**, 45–54 (2010)
21. Heldt, T., Oefinger, M., Hoshiyama, M., Mark, R.: Circulatory response to passive and active changes in posture. Comput. Cardiol. **29**, 263–266 (2003). Available at http://www.cinc.org/archives/2003/263.pdf
22. Heldt, T., Shim, E., Kamm, R., Mark, R.: Computational modeling of cardiovascular response to orthostatic stress. J. Appl. Physiol. **92**(3), 1239–1254 (2002)
23. Heldt, T., Verghese, G., Long, W., Szolovits, P., Mark, R.: Integrating data, models, and reasoning in critical care. In: Proceedings of the 28th IEEE EMBC International Conference, pp. 350–353. IEEE Engineering in Medicine and Biology Society (2006)
24. van Heusden, K., Gisolf, J., Stok, W., Dijkstra, S., Karemaker, J.: Mathematical modeling of gravitational effects on the circulation: Importance of the time course of venous pooling and blood volume changes in the lungs. Am. J. Physiol. **291**(5), H2152–H2165 (2006)
25. Hodgkin, A., Huxley, A.: A quantitative description of membrane current and its application to conduction and excitation in nerve. J. Physiol. **117**, 500–544 (1952)
26. Hunter, P., Borg, T.: Integration from proteins: The physiome project. Nat. Rev. Mol. Cell Biol. **4**, 237–243 (2003)
27. Jaron, D., Moore, T., Bai, J.: Cardiovascular response to acceleration stress: A computer simulation. Proc. IEEE **76**(6), 700–707 (1988)

28. Jaron, D., Moore, T., Chu, C.L.: A cardiovascular model for studying impairment of cerebral function during +Gz stress. Aviat. Space Environ. Med. **55**(1), 24–31 (1984)

29. Leonard, J., Leach, C., Rummel, J.: Computer simulations of postural change, water immersion, and bedrest: An integrative approach for understanding the spaceflight response. Physiologist **22**(6), S31–S32 (1979)

30. Meck, J., Waters, W., Ziegler, M., deBlock, H., Mills, P., Robertson, D., Huang, P.: Mechanisms of post-spaceflight orthostatic hypotension: α_1-adrenergic receptor responses before flight and central autonomic dysregulation post-flight. Am. J. Physiol. **286**(4), H1486–H1495 (2004)

31. Melchior, F., Srinivasan, R., Clère, J.: Mathematical modeling of the human response to LBNP. Physiologist **35**(1 Suppl.), S204–S205 (1992)

32. Melchior, F., Srinivasan, R., Thullier, P., Clère, J.: Simulation of cardiovascular response to lower body negative pressure from 0 to −40 mmHg. J. Appl. Physiol. **77**(2), 630–640 (1994)

33. Noble, D.: The surprising heart: A review of recent progress in cardiac electrophysiology. J. Physiol. **353**, 1–50 (1984)

34. Olufsen, M., Ottesen, J., Tran, H., Ellwein, L., Lipsitz, L., Novak, V.: Blood pressure and blood flow variation during postural change from sitting to standing: Model development and validation. J. Appl. Physiol. **99**(4), 1523–1537 (2005)

35. Peterson, K., Ozawa, E., Pantalos, G., Sharp, M.: Numerical simulation of the influence of gravity and posture on cardiac performance. Ann. Biomed. Eng. **30**(2), 247–259 (2002)

36. Platts, S., Ziegler, M., Waters, W., Mitchell, B., Meck, J.: Midodrine prescribed to improve recurrent post-spaceflight orthostatic hypotension. Aviat. Space Environ. Med. **75**(6), 554–556 (2004)

37. Ramsdell, C., Mullen, T., G.H., S., Rostoft, S., Sheynberg, N., Aljuri, N., Maa M. amd Mukkamala, R., Sherman, D., Toska, K., Yelle, J., Bloomfield, D., Williams, G., Cohen, R.: Midodrine prevents orthostatic intolerance associated with simulated microgravity. J. Appl. Physiol. **90**(6), 2245–2248 (2001)

38. Simanonok, K., Srinivasan R.S. Myrick, E., Blomkalns, A., Charles, J.: A comprehensive guyton model analysis of physiologic responses to preadapting the blood volume as a countermeasure to fluid shifts. J. Clin. Pharmacol. **34**(5), 440–453 (1994)

39. Simanonok, K., Srinivasan, R., Charles, J.: A computer simulation study of preadaptation of the circulation by removal of different blood volumes to counteract fluid shifts. Physiologist **35**(1 Suppl.), S111–S112 (1992)

40. Smith, J., Hughes, C., Ptacin, M., Barney, J., Tristani, F., Ebert, T.J.: The effect of age on hemodynamic response to graded postural stress in normal men. J. Gerontol. **42**(4), 406–411 (1987)

41. Snyder, M., Rideout, V.: Computer simulation studies of the venous circulation. IEEE Trans. Biomed. Eng. **BME-16**(4), 325–334 (1969)

42. Srinivasan, R., Simanonok, K., Charles, J.: Computer simulation analysis of the effects of countermeasures for reentry orthostatic intolerance. Physiologist **35**(1 Suppl.), S165–S168 (1992)

43. Sud, V., Srinivasan, R., Charles, J., Bungo, M.: Effects of lower body negative pressure on blood flow with applications to the human cardiovascular system. Med. Biol. Eng. Comput. **31**(6), 569–575 (1993)

44. Summers, R., Martin, D., Meck, J., Coleman, T.: Computer systems analysis of spaceflight induced changes in left ventricular mass. Comput. Biol. Med. **37**(3), 358–363 (2007)

45. Vélez-Reyes, M.: Decomposed algorithms for parameter estimation. Doctoral dissertation, Massachusetts Institute of Technology, Cambridge, MA (1992)

46. Vogel, C.: Computational Methods for Inverse Problems. Frontiers in Applied Mathematics. SIAM, Philadelphia (2002)

47. Waters, W., Ziegler, M., Meck, J.: Postspaceflight orthostatic hypotension occurs mostly in women and is predicted by low vascular resistance. J. Appl. Physiol. **2**(92), 596–594 (2002)

48. White, R., Blomqvist, C.: Central venous pressure and cardiac function during spaceflight. J. Appl. Physiol. **85**(2), 738–746 (1998)

49. White, R., Fitzjerrell, D., Croston, R.: Cardiovascular modelling: Simulating the human response to exercise, lower body negative pressure, zero gravity and clinical conditions. In: Advances in Cardiovascular Physics, vol. 5 (Part 1), pp. 195–229. Karger, Basel (1983)

50. White, R., Fitzjerrell, D., Croston, R.: Fundamentals of lumped compartmental modelling of the cardiovascular system. In: Advances in Cardiovascular Physics, vol. 5 (Part 1), pp. 162–184. Karger, Basel (1983)

51. White, R., Leonard, J., Srinivasan, R., Charles, J.: Mathematical modeling of acute and chronic cardiovascular changes during extended duration orbiter (EDO) flights. Acta Astronaut. **23**, 41–51 (1991)

52. Zamanian, A.: Modeling and simulating the human cardiovascular response to acceleration. SM thesis, Department of Electrical Engineering and Computer Science, Massachusetts Institute of Technology, Cambridge, MA (2007)

Chapter 3
Parameter Selection Methods in Inverse Problem Formulation

H.T. Banks, Ariel Cintrón-Arias, and Franz Kappel

Abstract We discuss methods for *a priori* selection of parameters to be estimated in inverse problem formulations (such as Maximum Likelihood, Ordinary and Generalized Least Squares) for dynamical systems with numerous state variables and an even larger number of parameters. We illustrate the ideas with an in-host model for HIV dynamics which has been successfully validated with clinical data and used for prediction and a model for the reaction of the cardiovascular system to an ergometric workload.

H.T. Banks (✉)
Center for Research in Scientific Computation and Center for Quantitative Sciences
in Biomedicine, Department of Mathematics, North Carolina State University, Raleigh,
NC 27695-8212
e-mail: htbanks@ncsu.edu

A. Cintrón-Arias
Center for Research in Scientific Computation and Center for Quantitative Sciences
in Biomedicine, Department of Mathematics, North Carolina State University, Raleigh,
NC 27695-8212

Department of Mathematics and Statistics, East Tennessee State University, Johnson City,
TN 37614-0663
e-mail: cintronarias@mail.etsu.edu

F. Kappel
Center for Research in Scientific Computation and Center for Quantitative Sciences
in Biomedicine, Department of Mathematics, North Carolina State University, Raleigh,
NC 27695-8212

Institute for Mathematics and Scientific Computation, University of Graz, A8010 Graz, Austria
e-mail: franz.kappel@uni-graz.at

J.J. Batzel et al. (eds.), *Mathematical Modeling and Validation in Physiology*,
Lecture Notes in Mathematics 2064, DOI 10.1007/978-3-642-32882-4_3,
© Springer-Verlag Berlin Heidelberg 2013

3.1 Introduction

There are many topics of great importance and interest in the areas of modeling and inverse problems which are properly viewed as essential in the use of mathematics and statistics in scientific inquiries. A brief, noninclusive list of topics include the use of traditional sensitivity functions (TSF) and generalized sensitivity functions (GSF) in experimental design (what type and how much data is needed, where/when to take observations) [9–11, 16, 56], choice of mathematical models and their parameterizations (verification, validation, model selection and model comparison techniques) [8, 12, 13, 17, 21–24, 41], choice of statistical models (observation process and sampling errors, residual plots for statistical model verification, use of asymptotic theory and bootstrapping for computation of standard errors, confidence intervals) [8, 14, 30, 31, 54, 55], choice of cost functionals (maximum likelihood estimation, ordinary least squares, weighted least squares, generalized least squares, etc.) [8, 30], as well as parameter identifiability and selectivity. There is extensive literature on each of these topics and many have been treated in surveys in one form or another ([30] is an excellent monograph with many references on the statistically related topics) or in earlier lecture notes [8].

We discuss here an enduring major problem: selection of those model parameters which can be readily and reliably (with quantifiable uncertainty bounds) estimated in an inverse problem formulation. This is especially important in many areas of biological modeling where often one has large dynamical systems (many state variables), an even larger number of unknown parameters to be estimated and a paucity of longitudinal time observations or data points. As biological and physiological models (at the cellular, biochemical pathway or whole organism level) become more sophisticated (motivated by increasingly detailed understanding—or lack thereof—of mechanisms), it is becoming quite common to have large systems (10–20 or more differential equations), with a plethora of parameters (25–100) but only a limited number (50–100 or fewer) of data points per individual organism. For example, we find models for the cardiovascular system [16, Chap. 1] (where the model has 10 state variables and 22 parameters) and [50, Chap. 6] (where the model has 22 states and 55 parameters), immunology [48] (8 states, 24 parameters), metabolic pathways [32] (8 states, 35 parameters) and HIV progression [7, 42]. Fortunately, there is a growing recent effort among scientists to develop quantitative methods based on sensitivity, information matrices and other statistical constructs (see, for example, [9–11, 25, 27, 37, 38, 59]) to aid in identification or parameter estimation formulations. We discuss here one approach using sensitivity matrices and asymptotic standard errors as a basis for our developments. To illustrate our discussions, we will use two models from the biological sciences: (a) a recently developed in-host model for HIV dynamics which has been successfully validated with clinical data and used for prediction [3, 7]; (b) a global non-pulsatile model for the cardiovascular system which has been validated with data from bicycle ergometer tests [16, 44].

The topic of system and parameter identifiability is actually an old one. In the context of parameter determination from system observations or output it is at least 40 years old and has received much attention in the peak years of linear system and control theory in the investigation of observability, controllability and detectability [6, 18, 19, 35, 39, 43, 46, 52, 53]. These early investigations and results were focused primarily on engineering applications, although much interest in other areas (e.g., oceanography, biology) has prompted more recent inquiries for both linear and nonlinear dynamical systems [5, 15, 29, 34, 40, 47, 57, 59–61].

3.2 Statistical Models for the Observation Process

One has errors in any data collection process and the presence of these errors is reflected in any parameter estimation result one might obtain. To understand and treat this, one usually specifies a *statistical model* for the observation process in addition to the *mathematical model* representing the dynamics. To illustrate ideas here we use ordinary least squares (OLS) consistent with an error model for absolute error in the observations. For a discussion of other frameworks (maximum likelihood in the case of known error distributions, generalized least squares appropriate for relative error models) see [8].

In order to be more specific we assume that the dynamics of the system is modeled by a system of ordinary differential equations:

$$\dot{x}(t) = g(t, x(t), \theta), \ t \geq t_0, \quad x(t_0) = x_0(\theta), \tag{3.1}$$

$$z(t) = h(t, x(t), \theta), \quad t \geq t_0, \ \theta \in \mathscr{A}, \tag{3.2}$$

where $G \subset \mathbb{R}^n$ and $\mathscr{A} \subset \mathbb{R}^p$ are open sets and $g : [t_0, \infty) \times G \times \mathscr{A} \to \mathbb{R}^n$, $x_0 : \mathscr{A} \to \mathbb{R}^n$ and $h : [t_0, \infty) \times G \times \mathscr{A} \to \mathbb{R}$ are sufficiently smooth functions. The set $\mathscr{A}$ is called the set of *admissible parameters* and $z(\cdot)$ is the *measurable output* of the system, which for simplicity we assume to be scalar. Let $x(t) = x(t; \theta)$ denote the solution of (3.1) for given $\theta \in \mathscr{A}$ and set

$$f(t, \theta) = h(t, x(t; \theta), \theta), \quad t \geq t_0, \ \theta \in \mathscr{A}.$$

Then

$$z(t) = f(t; \theta), \quad t \geq t_0, \ \theta \in \mathscr{A}, \tag{3.3}$$

is the output model corresponding to the model (3.1), (3.2). It is clear that an output model of the form (3.3) can also originate from dynamical models, where instead of the ODE-system (3.1) we may have a system of delay equations or some partial differential equation. In order to describe the observation process we assume there exists a vector $\theta_0 \in \mathscr{A}$, referred to as the *true or nominal parameter vector*, for which the output $z(t) = f(t, \theta_0)$ describes the output of the real system exactly. At given *sampling times*

$$t_0 \leq t_1 < \cdots < t_N,$$

we have measurements y_j for the output of the real system, $j = 1, \ldots, N$. It is also reasonably assumed that each of the N longitudinal measurements y_j is affected by random deviations ϵ_j from the true underlying output. That is, we assume that the measurements are given by

$$y_j = f(t_j; \theta_0) + \epsilon_j, \quad j = 1, \ldots, N. \tag{3.4}$$

The measurement errors ϵ_j are assumed to be realizations of random variables $\mathcal{E}_j$ satisfying the following assumptions:

(i) The errors $\mathcal{E}_j$ have mean zero, $\mathsf{E}(\mathcal{E}_j) = 0$.
(ii) The errors $\mathcal{E}_j$ have finite common variance, $\mathrm{var}(\mathcal{E}_j) = \sigma_0^2 < \infty$.
(iii) the errors $\mathcal{E}_j$ are independent (i.e., $\mathrm{cov}(\mathcal{E}_j, \mathcal{E}_k) = 0$ whenever $j \neq k$) and identically distributed.

According to (3.4) the measurements y_j are realizations of random variables Y_j, the *observations* at the sampling times t_j. Then the statistical model for the scalar observation process is

$$Y_j = f(t_j; \theta_0) + \mathcal{E}_j, \quad j = 1, \ldots, N. \tag{3.5}$$

Assumptions (i)–(iii) imply that the mean of the observations is equal to the model output for the nominal parameter vector, $\mathsf{E}(Y_j) = f(t_j; \theta_0)$, and the variance in the observations is constant in time, $\mathrm{var}(Y_j) = \sigma_0^2$, $j = 1, \ldots, N$.

For given measurements $y = (y_1, \ldots, y_N)^\mathsf{T}$ the estimate $\hat{\theta}_{\mathrm{OLS}}$ for θ_0 is obtained by minimizing

$$J(y, \theta) = \sum_{j=1}^{N} (y_j - f(t_j; \theta))^2 = |y - F(\theta)|^2 = |F(\theta) - F(\theta_0) - \epsilon|^2, \tag{3.6}$$

where we have set

$$F(\theta) = \big(f(t_1; \theta), \ldots, f(t_N; \theta)\big)^\mathsf{T}, \quad \epsilon = \big(\epsilon_1, \ldots, \epsilon_N\big)^\mathsf{T},$$

and $|\cdot|$ is the Euclidean norm in $\mathbb{R}^N$. The estimate $\hat{\theta}_{\mathrm{OLS}}$ is a realization of a random variable, the *least squares estimator* Θ_{OLS}. In order to indicate the dependence on N we shall write $\hat{\theta}_{\mathrm{OLS}}^N$ and Θ_{OLS}^N when needed. From [54] we find that under a number of regularity and sampling conditions, as $N \to \infty$, Θ_{OLS}^N is approximately distributed according to a multivariate normal distribution, i.e.,

$$\Theta_{\mathrm{OLS}}^N \sim \mathcal{N}_p\left(\theta_0, \Sigma_0^N\right), \tag{3.7}$$

where $\Sigma_0^N = \sigma_0^2 (N \Omega_0)^{-1} \in \mathbb{R}^{p \times p}$ and

$$\Omega_0 = \lim_{N \to \infty} \frac{1}{N} \chi^N(\theta_0)^\mathsf{T} \chi^N(\theta_0).$$

The $N \times p$ matrix $\chi^N(\theta)$ is known as the *sensitivity matrix* of the system, and is defined as

$$\chi^N(\theta_0) = \left(\frac{\partial f(t_i; \theta_0)}{\partial \theta_j}\right)_{1 \leq i \leq N, \, 1 \leq j \leq p} = \frac{\partial F}{\partial \theta}(\theta_0) = \nabla_\theta F(\theta_0). \qquad (3.8)$$

Asymptotic theory [8, 30, 54] requires existence and non-singularity of Ω_0. The $p \times p$ matrix Σ_0^N is the covariance matrix of the estimator Θ^N.

If the output model (3.3) corresponds to the model (3.1), (3.2) then the derivatives of f with respect to the parameters are given by

$$\frac{\partial f}{\partial \theta_j}(t, \theta) = \frac{\partial h}{\partial x}(t, x(t; \theta), \theta)\frac{\partial x}{\partial \theta_j}(t; \theta) + \frac{\partial h}{\partial \theta_j}(t, x(t; \theta), \theta), \quad j = 1, \ldots, p,$$

$$(3.9)$$

where $w(t; \theta) = \left((\partial x/\partial \theta_1)(t; \theta), \ldots, (\partial x/\partial \theta_p)(t; \theta)\right) \in \mathbb{R}^{N \times p}$ is obtained by solving

$$\dot{x}(t; \theta) = g(t, x(t; \theta), \theta), \quad x(t_0; \theta) = x_0(\theta),$$

$$\dot{w}(t, \theta) = \frac{\partial g}{\partial x}(t, x(t; \theta), \theta)w(t; \theta) + \frac{\partial g}{\partial \theta}(t, x(t; \theta), \theta), \quad w(t_0; \theta) = \frac{\partial x_0}{\partial \theta}(\theta),$$

$$(3.10)$$

from $t = t_0$ to $t = t_N$. One could alternatively obtain the sensitivity matrix using difference quotients (usually less accurately) or by using automated differentiation software (for additional details on sensitivity matrix calculations, see [8, 9, 27, 28, 33, 36]).

The estimate $\hat{\theta}_{\text{OLS}} = \hat{\theta}_{\text{OLS}}^N$ is a realization of the estimator Θ_{OLS}, and is calculated using a realization $\{y_i\}_{i=1}^N$ of the observation process $\{Y_i\}_{i=1}^N$, while minimizing (3.6) over θ. Moreover, the estimate $\hat{\theta}_{\text{OLS}}$ is used in the calculation of the sampling distribution for the parameters. The generally unknown error variance σ_0^2 is approximated by

$$\hat{\sigma}_{\text{OLS}}^2 = \frac{1}{N - p} \sum_{j=1}^N (y_j - f(t_j; \hat{\theta}_{\text{OLS}}^N))^2, \qquad (3.11)$$

while the covariance matrix Σ_0^N is approximated by

$$\hat{\Sigma}_{\text{OLS}}^N = \hat{\sigma}_{\text{OLS}}^2 \left(\chi^N(\hat{\theta}_{\text{OLS}})^{\mathsf{T}} \chi^N(\hat{\theta}_{\text{OLS}})\right)^{-1}. \qquad (3.12)$$

As discussed in [8, 30, 54] an approximate for the sampling distribution of the estimator is given by

$$\Theta_{\text{OLS}} = \Theta_{\text{OLS}}^N \sim \mathcal{N}_p(\theta_0, \Sigma_0^N) \approx \mathcal{N}_p(\hat{\theta}_{\text{OLS}}^N, \hat{\Sigma}_{\text{OLS}}^N). \qquad (3.13)$$

Asymptotic standard errors can be used to quantify uncertainty in the estimation, and they are calculated by taking the square roots of the diagonal elements of the covariance matrix $\hat{\Sigma}_{\mathrm{OLS}}^N$, i.e.,

$$\mathrm{SE}_k(\hat{\theta}_{\mathrm{OLS}}^N) = \sqrt{(\hat{\Sigma}_{\mathrm{OLS}}^N)_{k,k}}, \quad k = 1, \ldots, p. \tag{3.14}$$

Before describing the algorithm in detail and illustrating its use, we provide some motivation underlying the use of the sensitivity matrix $\chi(\theta_0)$ of (3.8) and the Fisher information matrix $\mathcal{F}(\theta_0) = (1/\sigma_0^2)\chi^{\mathsf{T}}(\theta_0)\chi(\theta_0)$. Both of these matrices play a fundamental role in the development of the approximate asymptotic distributional theory resulting in (3.13) and (3.14). Since a prominent measure of the ability to estimate a parameter is related to its associated standard errors in estimation, it is worthwhile to briefly outline the underlying approximation ideas in the asymptotic distributional theory.

Ordinary least squares problems involve choosing $\Theta = \Theta_{\mathrm{OLS}}$ to minimize the difference between observations Y and model output $F(\theta)$, i.e., minimize $|Y - F(\theta)|$. However the approximate asymptotic distributional theory (e.g., see [54, Chap. 12]) which is *exact* for model outputs linear in the parameters, employs a fundamental linearization in the parameters in a neighborhood of the hypothesized "true" parameters θ_0. Replacing the model output with a first order linearization about θ_0, we then may seek to minimize for θ in the approximate functional

$$\left| Y - F(\theta_0) - \nabla_\theta F(\theta_0)(\theta - \theta_0) \right|.$$

If we use the statistical model $Y = F(\theta_0) + \mathcal{E}$ and let $\delta\theta = \theta - \theta_0$, we thus wish to minimize

$$\left| \mathcal{E} - \chi(\theta_0)\delta\theta \right|,$$

where $\chi(\theta_0) = \nabla_\theta F(\theta_0)$ is the $N \times p$ sensitivity matrix defined in (3.8). This is a standard optimization problem [45, Sect. 6.11] whose solution can be given using the pseudo inverse $\chi(\theta_0)^\dagger$ defined in terms of minimal norm solutions of the optimization problem and satisfying $\chi(\theta_0)^\dagger = (\chi(\theta_0)^{\mathsf{T}}\chi(\theta_0))^\dagger \chi(\theta_0)^{\mathsf{T}} = \sigma_0^2 \mathcal{F}(\theta_0)^\dagger \chi(\theta_0)^{\mathsf{T}}$. The solution is

$$\delta\Theta = \chi(\theta_0)^\dagger \mathcal{E}$$

or

$$\Theta_{\mathrm{LIN}} = \theta_0 + \chi(\theta_0)^\dagger \mathcal{E} = \theta_0 + \sigma_0^2 \mathcal{F}(\theta_0)^\dagger \chi(\theta_0)^{\mathsf{T}} \mathcal{E}.$$

If $\mathcal{F}(\theta_0)$ is invertible, then the solution (to first order) of the OLS problem is

$$\Theta_{\mathrm{OLS}} \approx \Theta_{\mathrm{LIN}} = \theta_0 + \sigma_0^2 \mathcal{F}(\theta_0)^{-1} \chi(\theta_0)^{\mathsf{T}} \mathcal{E}. \tag{3.15}$$

This approximation, for which the asymptotic distributional theory is exact, can be a reasonable one for use in developing an approximate nonlinear asymptotic theory if $\delta\Theta$ is small, i.e., if the OLS estimated parameter is close to θ_0.

From these calculations, we see that the rank of $\chi(\theta_0)$ and the conditioning (or ill-conditioning) of $\mathscr{F}(\theta_0)$ play a significant role in solving OLS inverse problems as well as in any asymptotic standard error formulations based on this linearization. Observe that the error (or noise) $\mathscr{E}$ in the data will in general be amplified as the ill-conditioning of $\mathscr{F}$ increases. We further note that the $N \times p$ sensitivity matrix $\chi(\theta_0)$ is of full rank p if and only if the $p \times p$ Fisher matrix $\mathscr{F}(\theta_0)$ has rank p, or equivalently, is nonsingular. These underlying considerations have motivated a number of efforts (e.g., see [9–11]) on understanding the conditioning of the Fisher matrix as a function of the number N and longitudinal locations $\{t_j\}_{j=1}^{N}$ of data points as a key indicator for well-formulated inverse problems and as a tool in optimal design, especially with respect to computation of uncertainty (standard errors, confidence intervals) for parameter estimates.

In view of the considerations above (which are very *local* in nature—both the sensitivity matrix and the Fisher information matrix are taken at the nominal vector θ_0), one should be pessimistic about using these quantities to obtain any *nonlocal* selection methods or criteria for estimation. Indeed, for nonlinear complex systems, it is easy to argue that questions related to some type of global parameter identifiability are not fruitful questions to pursue.

3.3 Subset Selection Algorithm

The focus of our presentation here is how one chooses *a priori* (i.e., *before* any inverse problem calculations are carried out) which parameters can be readily estimated with a typical longitudinal data set. We illustrate an algorithm, developed recently in [27], to select parameter vectors that can be estimated from a given data set using an ordinary least squares inverse problem formulation (similar ideas apply if one is using a relative error statistical model and generalized least squares formulations). Let $q \in \mathbb{R}^{p_0}$ be the parameter vectors being at our disposal for parameter estimation and denote by $q_0 \in \mathbb{R}^{p_0}$ the vector of the corresponding nominal values. Given a number $p < p_0$ of parameters we wish to identify, the algorithm searches all possible choices of p different parameters among the p_0 parameters and selects the one which is identifiable (i.e., the corresponding sensitivity matrix has full rank p) and minimizes a given uncertainty quantification (e.g. by means of asymptotic standard errors). Prior knowledge of a nominal set of values for all parameters along with the observation times for data (but not the values of the observations) will be required for our algorithm. For $p < p_0$ we set

$$\mathscr{S}_p = \{\theta \in \mathbb{R}^p \mid \theta \text{ is a sub-vector of } q \in \mathbb{R}^{p_0}\},$$

i.e., $\theta \in \mathscr{S}_p$ is given as $\theta = (q_{i_1}, \ldots, q_{i_p})^\mathsf{T}$ for some $1 \le i_1 < \cdots < i_p \le p_0$. The corresponding nominal vector is $\theta_0 = ((q_0)_{i_1}, \ldots, (q_0)_{i_p})^\mathsf{T}$.

As we have stated above, to apply the parameter subset selection algorithm we require prior knowledge of nominal variance and nominal parameter values. These nominal values of σ_0 and θ_0 are needed to calculate the sensitivity matrix, the Fisher matrix and the corresponding covariance matrix defined in (3.12). For our illustration presented in Sect. 3.5, we use the variance and parameter estimates obtained in previous investigations of the models as nominal values. In problems for which no prior estimation has been carried out, one must use knowledge of the observation process error and some knowledge of viable parameter values that might be reasonable with the model under investigation.

The uncertainty quantification we shall use is based on the considerations given in the previous section. Let $\theta \in \mathbb{R}^p$ be given. As an approximation to the covariance matrix of the estimator for θ we take

$$\Sigma(\theta_0) = \sigma_0^2 (\chi(\theta_0)^\mathsf{T} \chi(\theta_0))^{-1} = \mathscr{F}(\theta_0)^{-1}.$$

We introduce the *coefficients of variation* for θ

$$\nu_i(\theta_0) = \frac{(\Sigma(\theta_0)_{i,i})^{1/2}}{(\theta_0)_i}, \quad i = 1, \ldots, p, \tag{3.16}$$

and take as a uncertainty quantification for the estimates of θ the *selection score* given by the Euclidean norm in $\theta \in \mathbb{R}^p$ of $\nu(\theta_0)$, i.e.,

$$\alpha(\theta_0) = |\nu(\theta_0)|,$$

where $\nu(\theta_0) = (\nu_1(\theta_0), \ldots, \nu_p(\theta_0))^\mathsf{T}$. The components of the vector $\nu(\theta_0)$ are the ratios of each standard error for a parameter to the corresponding nominal parameter value. These ratios are dimensionless numbers warranting comparison even when parameters have considerably different scales and units (e.g., in case of the HIV-model N_T is on the order of 10^1, while k_1 is on the order of 10^{-6}, whereas in case of the CVS-model we have c_ℓ on the order 10^{-2} and $A_{\text{pesk}}^{\text{exer}}$ on the order 10^2). A selection score $\alpha(\theta_0)$ near zero indicates lower uncertainty possibilities in the estimation, while large values of $\alpha(\theta_0)$ suggest that one could expect to find substantial uncertainty in at least some of the components of the estimates in any parameter estimation attempt.

Let $\mathscr{F}_0$ be the Fisher information matrix corresponding to the parameter vectors $q \in \mathbb{R}^{p_0}$ and $\mathscr{F}_p$ the Fisher information matrix corresponding to the parameter vectors in $\theta \in \mathscr{S}_p$. Then rank $\mathscr{F}_0(q_0) = p_0$ implies that rank $\mathscr{F}_p(\theta_0) = p$ for any $\theta \in \mathscr{S}_p$, $p < p_0$, i.e., if $\mathscr{F}_0(q_0)$ is non-singular then also all $\mathscr{F}_p(\theta_0)$ are non-singular for all $p < p_0$ and all θ_0 corresponding to a $\theta \in \mathscr{S}_p$. Moreover, if rank $\mathscr{F}_0(q_0) = p_1$ with $p_1 < p_0$, then rank $\mathscr{F}(\theta_0) < p$ for all $p_1 < p < p_0$ and all $\theta \in \mathscr{S}_p$.

On the basis of the considerations given above our algorithm proceeds as follows:

Selection Algorithm. *Given $p < p_0$ the algorithm considers all possible choices of indices $i_1, \ldots, i_p$ with $1 \leq i_1 < \cdots < i_p \leq p_0$ in lexicographical ordering starting with the first choice $(i_1^{(1)}, \ldots, i_p^{(1)}) = (1, \ldots, p)$ and completes the following steps:*

Initializing step: Set $\mathrm{ind}^{\mathrm{sel}} = (1, \ldots, p)$ and $\alpha^{\mathrm{sel}} = \infty$.

Step k: For the choice $(i_1^{(k)}, \ldots, i_p^{(k)})$ compute $r = \mathrm{rank}\,\mathscr{F}\big((q_0)_{i_1^{(k)}}, \ldots, (q_0)_{i_p^{(k)}}\big)$.

> *If $r < p$, go to Step $k + 1$.*
> *If $r = p$, compute $\alpha_k = \alpha\big((q_0)_{i_1^{(k)}}, \ldots, (q_0)_{i_p^{(k)}}\big)$.*
> > *If $\alpha_k \geq \alpha^{\mathrm{sel}}$, go to Step $k + 1$.*
> > *If $\alpha_k < \alpha^{\mathrm{sel}}$, set $\mathrm{ind}^{\mathrm{sel}} = (i_1^{(k)}, \ldots, i_p^{(k)})$, $\alpha^{\mathrm{sel}} = \alpha_k$ and go to Step $k + 1$.*

Following the initializing step the algorithm performs $\binom{p_0}{p}$ steps and provides the index vector $\mathrm{ind}^{\mathrm{sel}} = \big(i_1^*, \ldots, i_p^*\big)$ which gives the sub-vector $\theta^* = \big(q_{i_1^*}, \ldots, q_{i_p^*}\big)^{\mathsf{T}}$ such that the selection score $\alpha\big((q_0)_{i_1^*}, \ldots, (q_0)_{i_p^*}\big)$ is minimal among all possible choices of sub-vectors in $\mathscr{S}_p$. If rank $\mathscr{F}_{p_0} = p_0$ then the rank test in Step k can be cancelled, of course.

3.4 Models

In the following, we shall illustrate the parameter selection ideas with results obtained by use of the subset selection algorithm described in the previous section for two specific models. These models have a moderate number of parameters to be identified yet are sufficiently complex to make a trial-error approach unfeasible.

3.4.1 A Mathematical Model for HIV Progression with Treatment Interruption

As our first illustrative example we use one of the many dynamic models for HIV progression found in an extensive literature (e.g., see [1–4, 7, 20, 26, 49, 51, 58] and the many references therein). For our example model, the dynamics of in-host HIV is described by the interactions between uninfected and infected type 1 target cells (T_1 and T_1^*) (CD4$^+$ T-cells), uninfected and infected type 2 target cells (T_2 and T_2^*)

(such as macrophages or memory cells, etc.), infectious free virus V_I, and immune response E (cytotoxic T-lymphocytes CD8$^+$) to the infection. This model, which was developed and studied in [1, 3] and later extended in subsequent efforts (e.g., see [7]), is based on one suggested in [26], but includes an immune response compartment and dynamics as in [20]. The model equations are given by

$$\dot{T}_1 = \lambda_1 - d_1 T_1 - \left(1 - \bar{\epsilon}_1(t)\right) k_1 V_I T_1,$$

$$\dot{T}_1^* = \left(1 - \bar{\epsilon}_1(t)\right) k_1 V_I T_1 - \delta T_1^* - m_1 E T_1^*,$$

$$\dot{T}_2 = \lambda_2 - d_2 T_2 - \left(1 - f\bar{\epsilon}_1(t)\right) k_2 V_I T_2,$$

$$\dot{T}_2^* = \left(1 - f\bar{\epsilon}_1(t)\right) k_2 V_I T_2 - \delta T_2^* - m_2 E T_2^*,$$

$$\dot{V}_I = \left(1 - \bar{\epsilon}_2(t)\right) 10^3 N_T \delta (T_1^* + T_2^*) - c V_I$$

$$\qquad - \left(1 - \bar{\epsilon}_1(t)\right) 10^3 k_1 T_1 V_I - \left(1 - f\bar{\epsilon}_1(t)\right) 10^3 k_2 T_2 V_I,$$

$$\dot{E} = \lambda_E + \frac{b_E\left(T_1^* + T_2^*\right)}{T_1^* + T_2^* + K_b} E - \frac{d_E\left(T_1^* + T_2^*\right)}{T_1^* + T_2^* + K_d} E - \delta_E E,$$

(3.17)

together with an initial condition vector

$$\left(T_1(0), T_1^*(0), T_2(0), T_2^*(0), V_I(0), E(0)\right)^{\mathsf{T}}.$$

The differences in infection rates and treatment efficacy help create a low, but non-zero, infected cell steady state for T_2^*, which is compatible with the idea that macrophages or memory cells may be an important source of virus after T-cell depletion. The populations of uninfected target cells T_1 and T_2 may have different source rates λ_i and natural death rates d_i. The time-dependent treatment factors $\bar{\epsilon}_1(t) = \epsilon_1 u(t)$ and $\bar{\epsilon}_2(t) = \epsilon_2 u(t)$ represent the effective treatment impact of a reverse transcriptase inhibitor (RTI) (that blocks new infections) and a protease inhibitor (PI) (which causes infected cells to produce non-infectious virus), respectively. The RTI is potentially more effective in type 1 target cells (T_1 and T_1^*) than in type 2 target cells (T_2 and T_2^*), where the efficacy is $f\bar{\epsilon}_1$, with $f \in [0, 1]$. The relative effectiveness of RTIs is modeled by ϵ_1 and that of PIs by ϵ_2, while the time-dependent treatment function $0 \leq u(t) \leq 1$ represents therapy levels, with $u(t) = 0$ for fully off and $u(t) = 1$, for fully on. Although HIV treatment is nearly always administered as combination therapy, the model allows the possibility of mono-therapy, even for a limited period of time, implemented by considering separate treatment functions $u_1(t), u_2(t)$ in the treatment factors.

As in [1, 3], for our numerical investigations we consider a log-transformed and reduced version of the model. This transformation is frequently used in the HIV modeling literature because of the large differences in orders of magnitude in state values in the model and the data and to guarantee non-negative state values as well as because of certain probabilistic considerations (for further discussions see [3]). This results in the nonlinear system of differential equations

$$\dot{x}_1 = \frac{10^{-x_1}}{\ln(10)}\left(\lambda_1 - d_1 10^{x_1} - \left(1 - \bar{\varepsilon}_1(t)\right)k_1 10^{x_5}10^{x_1}\right),$$

$$\dot{x}_2 = \frac{10^{-x_2}}{\ln(10)}\left(\left(1 - \bar{\varepsilon}_1(t)\right)k_1 10^{x_5}10^{x_1} - \delta 10^{x_2} - m_1 10^{x_6}10^{x_2}\right),$$

$$\dot{x}_3 = \frac{10^{-x_3}}{\ln(10)}\left(\lambda_2 - d_2 10^{x_3} - \left(1 - f\bar{\varepsilon}_1(t)\right)k_2 10^{x_5}10^{x_3}\right),$$

$$\dot{x}_4 = \frac{10^{-x_4}}{\ln(10)}\left(\left(1 - f\bar{\varepsilon}_1(t)\right)k_2 10^{x_5}10^{x_3} - \delta 10^{x_4} - m_2 10^{x_6}10^{x_4}\right),$$

$$\dot{x}_5 = \frac{10^{-x_5}}{\ln(10)}\left(\left(1 - \bar{\varepsilon}_2(t)\right)10^3 N_T \delta\left(10^{x_2} + 10^{x_4}\right) - c10^{x_5}\right.$$

$$\left. - \left(1 - \bar{\varepsilon}_1(t)\right)10^3 k_1 10^{x_1}10^{x_5} - \left(1 - f\bar{\varepsilon}_1(t)\right)10^3 k_2 10^{x_3}10^{x_5}\right),$$

$$\dot{x}_6 = \frac{10^{-x_6}}{\ln(10)}\left(\lambda_E + \frac{b_E\left(10^{x_2} + 10^{x_4}\right)}{10^{x_2} + 10^{x_4} + K_b}10^{x_6}\right.$$

$$\left. - \frac{d_E\left(10^{x_2} + 10^{x_4}\right)}{10^{x_2} + 10^{x_4} + K_d}10^{x_6} - \delta_E 10^{x_6}\right),$$

$$(3.18)$$

where the changes of variables are defined by

$$T_1 = 10^{x_1},\ T_1^* = 10^{x_2},\ T_2 = 10^{x_3},\ T_2^* = 10^{x_4},\ V_I = 10^{x_5},\ E = 10^{x_6}.$$

The initial conditions for equations (3.18) are denoted by $x_i(t_0) = x_i^0$, $i = 1, \ldots, 6$. We note that this model has six state variables and the following 20 (in general, unknown) system parameters in the right-hand sides of equations (3.18)

$$\lambda_1, d_1, \epsilon_1, k_1, \lambda_2, d_2, f, k_2, \delta, m_1, m_2, \ldots$$

$$\ldots \epsilon_2, N_T, c, \lambda_E, b_E, K_b, d_E, K_d, \delta_E.$$

We may also consider the initial conditions as unknowns and thus we have 26 unknown parameters which we collect in the parameter vector θ,

$$\theta = \left(x_1^0, x_2^0, x_3^0, x_4^0, x_5^0, x_6^0, \lambda_1, d_1, \epsilon_1, k_1, \lambda_2, d_2, f, k_2, \delta, m_1, m_2, \ldots\right.$$

$$\left. \ldots \epsilon_2, N_T, c, \lambda_E, b_E, K_b, d_E, K_d, \delta_E\right)^{\mathsf{T}}.$$

A list of the parameters in the model equations along with their units is given below in Table 3.1.

As reported in [1, 3], data to be used with this model in inverse or parameter estimation problems typically consisted of observations for $T_1 + T_1^*$ and V over some extended time period. For the purpose of this paper we are only using the data for $T_1 + T_1^*$ in case of patient # 4 which we depict in Fig. 3.1 together with

Table 3.1 Parameters in the equations of the HIV model

Parameter	Units	Description
λ_1	cells/(ml day)	Production rate of type 1 target cells
d_1	1/day	Death rate of type 1 target cells
ϵ_1	—	Treatment efficacy of type 1 target cells
k_1	ml/(virion day)	Infection rate of type 1 target cells
λ_2	cells/(ml day)	Production rate of type 2 target cells
d_2	1/day	Death rate of type 2 target cells
f	—	Reduction of treatment efficacy for type 2 target cells
k_2	ml/(virion day)	Infection rate of type 2 target cells
δ	1/day	Death rate of infected cells
m_1	ml/(cell day)	Immune-induced clearance rate for type 1 target cells
m_2	ml/(cell day)	Immune-induced clearance rate for type 2 target cells
ϵ_2	—	Treatment efficacy for type 2 target cells
N_T	virions/cell	Virions produced per infected cell
c	1/day	Natural death rate of viruses
λ_E	cells/(ml day)	Production rate for immune effectors
b_E	1/day	Maximum birth rate for immune effectors
K_b	cells/ml	Saturation constant for immune effector birth
d_E	1/day	Maximum death rate for immune effectors
K_d	cells/ml	Saturation constant for immune effector death
δ_E	1/day	Natural death rate for immune effectors

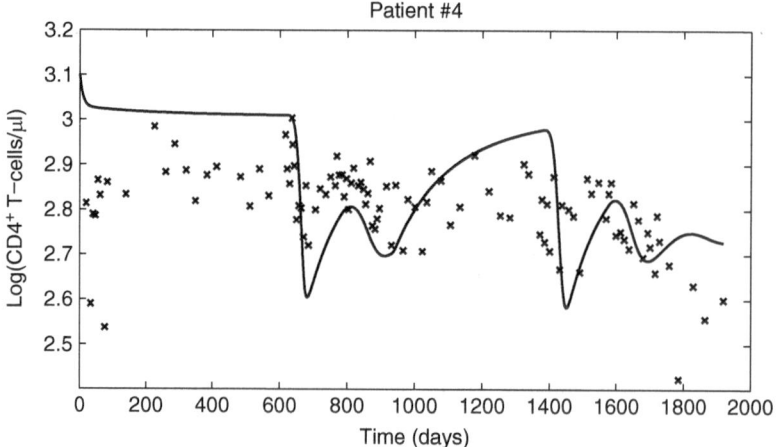

Fig. 3.1 Log-scaled data $\{y_j\}$ for CD4$^+$ T-cells of Patient #4 (*crosses*), and model output $z(t)$ (*solid curve*) evaluated at the parameter estimates obtained in [1, 3]

the corresponding model output for the parameters identified in [1, 3]. Thus our observations are

$$f(t_i; \theta_0) = \log_{10}\left(10^{x_1(t_i;\theta_0)} + 10^{x_2(t_i;\theta_0)}\right), \qquad (3.19)$$

where the nominal parameter vector θ_0 is given at the beginning of Sect. 3.5.1.

While the inverse problem we are considering in this paper for the HIV-model is relatively "small" compared to many of those found in the literature, it still represents a nontrivial estimation challenge and is more than sufficient to illustrate the ideas and methodology we discuss in this presentation. Other difficult aspects (censored data requiring use of the Expectation Maximization algorithm as well as use of residual plots in attempts to validate the correctness of choice of corresponding statistical models introduced and discussed in Sect. 3.2) of such inverse problems are discussed in the review chapter [8] and will not be pursued here.

3.4.2 A Model for the Reaction of the Cardiovascular System to an Ergometric Workload

As a second example to illustrate our methods we chose a model for cardiovascular function. The model was developed in order to describe the reaction of the cardiovascular system to a constant ergometric workload of moderate size. The building blocks of the model are the left ventricle, the arterial and venous systemic compartments representing the systemic circuit as well as the right ventricle, arterial and venous pulmonary compartments representing the pulmonary circuit. The model is non-pulsatile and includes the baroreceptor loop as the essential control loop in the situation to be modeled. The feedback control which steers the system from the equilibrium corresponding to rest to the equilibrium corresponding to the imposed workload is obtained by solving a linear quadratic regulator problem. Furthermore the model includes a sub-model describing the so called autoregulation process, i.e., the control of local blood flow in response to local metabolic demands. The model equations are given as (for details see [44], [16, Chap. 1]):

$$\dot{P}_{as} = \frac{1}{c_{as}}\left(Q_\ell - \frac{1}{R_s}(P_{as} - P_{vs})\right),$$

$$\dot{P}_{vs} = \frac{1}{c_{vs}}\left(\frac{1}{R_s}(P_{as} - P_{vs}) - Q_r\right),$$

$$\dot{P}_{ap} = \frac{1}{c_{ap}}\left(Q_r - \frac{1}{R_p}(P_{ap} - P_{vp}(P_{as}, P_{vs}, P_{ap}))\right),$$

$$\dot{S}_\ell = \sigma_\ell,$$

$$\dot{\sigma}_\ell = -\alpha_\ell S_\ell - \gamma_\ell \sigma_\ell + \beta_\ell H, \tag{3.20}$$

$$\dot{S}_r = \sigma_r,$$

$$\dot{\sigma}_r = -\alpha_r S_r - \gamma_r \sigma_r + \beta_r H,$$

$$\dot{R}_s = \frac{1}{K}\left(A_{pesk}\left(\frac{P_{as} - P_{vs}}{R_s}C_{a,O_2} - M_0 - \rho W\right) - (P_{as} - P_{vs})\right),$$

$$\dot{H} = u(t)$$

with

$$P_{vp} = P_{vp}(P_{as}, P_{vs}, P_{ap}) := \frac{1}{c_{vp}}(V_{tot} - c_{as}P_{as} - c_{vs}P_{vs} - c_{ap}P_{ap}),$$

$$Q_\ell = H\frac{c_\ell P_{vp}(P_{as}, P_{vs}, P_{ap})a_\ell(H)S_\ell}{a_\ell(H)P_{as} + k_\ell(H)S_\ell}, \tag{3.21}$$

$$Q_r = H\frac{c_r P_{vs}a_r(H)S_r}{a_r(H)P_{ap} + k_r(H)S_r},$$

where

$$k_\ell(H) = e^{-(c_\ell R_\ell)^{-1}t_d(H)} \quad \text{and} \quad a_\ell(H) = 1 - k_\ell(H),$$

$$k_r(H) = e^{-(c_r R_r)^{-1}t_d(H)} \quad \text{and} \quad a_r(H) = 1 - k_r(H). \tag{3.22}$$

For the duration t_d of the diastole we use Bazett's formula (duration of the systole $=$ $\kappa/H^{1/2}$) which implies

$$t_d = t_d(H) = \frac{1}{H^{1/2}}\left(\frac{1}{H^{1/2}} - \kappa\right). \tag{3.23}$$

Introducing $x = (P_{as}, P_{vs}, P_{ap}, S_\ell, \sigma_\ell, S_r, \sigma_r, R_s, H)^\mathsf{T} \in \mathbb{R}^9$ system (3.20) can be written as

$$\dot{x}(t) = f(x(t), W, \theta, u(t)),$$

where $W = W^{rest} = 0$ (Watt) for $t \le 0$ and $W = W^{exer} = 75$ (Watt) for $t > 0$. Moreover, θ is the parameter vector of the system. We distinguish two values for each of the parameters R_p and A_{pesk}, one for the resting situation and one for the exercise situation. Consequently we have the parameters R_p^{rest}, A_{pesk}^{rest} and R_p^{exer} and A_{pesk}^{exer} instead of R_p and A_{pesk}. The initial value for system (3.20) is the equilibrium x^{rest}, which is computed from $f(x^{rest}, W^{rest}, \theta, 0) = 0$. Analogously x^{exer} is the equilibrium corresponding to the constant workload W^{exer} (satisfying $f(x^{exer}, W^{exer}, \theta, 0) = 0$).

Let $B = (0, \ldots, 0, 1)^\mathsf{T} \in \mathbb{R}^9$, $C = (1, 0, \ldots, 0) \in \mathbb{R}^{1\times 9}$ and denote by $A(\theta) = \frac{\partial f(x, W^{exer}, \theta, 0)}{\partial x}\big|_{x=x^{exer}}$, the Jacobian of the right-hand side of system (3.20) at the equilibrium x^{exer}. The control $u(t)$ is obtained as the solution of the linear-quadratic regulator problem for the linear system

$$\dot{\xi}(t) = A(\theta)\xi(t) + Bu(t), \quad \xi(0) = x^{rest} - x^{exer}, \tag{3.24}$$

where we have set $\xi(t) = x(t) - x^{exer}$, and the quadratic cost functional

$$J(u) = \int_0^\infty \left(q_{as}^2(P_{as}(t) - P_{as}^{exer})^2 + u(t)^2\right) dt. \tag{3.25}$$

Table 3.2 Parameters of the CVS-model

Parameter	Units	Description
c_{as}	l/mmHg	Compliance of the arterial systemic compartment
c_{vs}	l/mmHg	Compliance of the venous systemic compartment
c_{ap}	l/mmHg	Compliance of the arterial pulmonary compartment
c_{vp}	l/mmHg	Compliance of the venous pulmonary compartment
c_ℓ	l/mmHg	Compliance of the relaxed left ventricle
c_r	l/mmHg	Compliance of the relaxed right ventricle
V_{tot}	l	Total blood volume
R_p	mmHg min/l	Resistance in the peripheral region of the pulmonary circuit
R_ℓ	mmHg min/l	Inflow resistance of the left ventricle
R_r	mmHg min/l	Inflow resistance of the right ventricle
κ	$min^{1/2}$	Coefficient in Bazett's formula (see (3.23))
C_{a,O_2}	1	O_2-concentration in arterial systemic blood
K	1	Constant in the formula for the biochemical energy flow, $M_b = -K\, dC_{v,O_2}/dt$
A_{pesk}	mmHg min/l	Constant in the formula relating peripheral systemic resistance and venous O_2 concentration ($R_s = A_{pesk}C_{v,O_2}$)
M_0	l/min	Metabolic rate in the systemic tissue region corresponding to zero workload
ρ	l/(min Watt)	Coefficient of W in the differential equation for R_s
q_{as}	$min^{-2}(mmHg)^{-1}$	Weighting factor for P_{as} in the cost functional (3.25)
α_ℓ	min^{-2}	Coefficient of S_ℓ in the differential equation for S_ℓ
α_r	min^{-2}	Coefficient of S_r in the differential equation for S_r
β_ℓ	mmHg/min	Coefficient of H in the differential equation for S_ℓ
β_r	mmHg/min	Coefficient of H in the differential equation for S_r
γ_ℓ	min^{-1}	Coefficient of $\dot{S}_\ell$ in the differential equation for S_ℓ
γ_r	min^{-1}	Coefficient of $\dot{S}_r$ in the differential equation for S_r

This functional reflects the fact that the baroreceptor loop, which is the basic control loop for the situation we are considering here, generates the control using the arterial systemic pressure $P_{as}(t)$.

According to the theory of the linear-quadratic control problem $u(t)$ is given by

$$u(t) = -B^\mathsf{T} X \xi(t) = -B^\mathsf{T} X(x(t) - x^{exer}), \quad t \geq 0, \qquad (3.26)$$

where $X = X(\theta)$ is the solution of the Riccati matrix equation $XA(\theta) + A(\theta)^\mathsf{T} X - XBB^\mathsf{T}X + C^\mathsf{T}C = 0$. The feedback control (3.26) is also a stabilizing control for system (3.20), i.e., we have $\lim_{t\to\infty} x(t) = x^{exer}$ provided $\|x(0) - x^{exer}\|_2$ is sufficiently small.

The parameter vector of the system is

$$q = \big(c_\ell, c_r, c_{as}, c_{vs}, c_{ap}, c_{vp}, R_\ell, R_r, \alpha_\ell, \alpha_r, \beta_\ell, \beta_r, \gamma_\ell, \ldots$$

$$\ldots \gamma_r, K, \kappa, M_0, \rho, C_{a,O_2}, q_{as}, V_{tot}, R_p^{rest}, A_{pesk}^{rest}, R_p^{exer}, A_{pesk}^{exer}\big)^\mathsf{T} \in \mathbb{R}^{25}. \quad (3.27)$$

Tables 3.2 and 3.3 contain the descriptions and units for the parameters q and the state variables x, respectively, of the system.

Table 3.3 The state variables and other variables of the CVS-model

Variable	Unit	Description
P_{as}	mmHg	Pressure in the arterial systemic compartment
P_{vs}	mmHg	Pressure in the venous systemic compartment
P_{ap}	mmHg	Pressure in the arterial pulmonary compartment
P_{vp}	mmHg	Pressure in the venous pulmonary compartment
S_ℓ	mmHg	Contractility of the left ventricle
σ_ℓ	mmHg/min	Time derivative of S_ℓ
S_r	mmHg	Contractility of the right ventricle
σ_r	mmHg/min	Time derivative of S_r
R_s	mmHg min/l	Peripheral resistance in the systemic circuit
H	min^{-1}	Heart rate
Q_ℓ	l/min	Cardiac output of the left ventricle
Q_r	l/min	Cardiac output of the right ventricle
W	Watt	Workload imposed on the test person

3.5 Results and Discussion

In this section we discuss some of our findings when we applied the Selection Algorithm presented in Sect. 3.3 to the HIV-model and the CVS-model.

3.5.1 The HIV-Model

As the nominal parameter vector we take the estimates obtained in [1, 3] for Patient # 4. More precisely, we assume that the error variance is $\sigma_0^2 = 0.11$, and that the nominal parameter values (for description and units see Table 3.1) are given as:

$$x_1^0 = \log_{10}(1.202\text{e}+3), \quad x_2^0 = \log_{10}(6.165\text{e}+1), \quad x_3^0 = \log_{10}(1.755\text{e}+1),$$

$$x_4^0 = \log_{10}(6.096\text{e}-1), \quad x_5^0 = \log_{10}(9.964\text{e}+5), \quad x_6^0 = \log_{10}(1.883\text{e}-1),$$

$$\lambda_1 = 4.633, \qquad\qquad d_1 = 4.533 \times 10^{-3}, \qquad \epsilon_1 = 6.017 \times 10^{-1},$$

$$k_1 = 1.976 \times 10^{-6}, \qquad \lambda_2 = 1.001 \times 10^{-1}, \qquad d_2 = 2.211 \times 10^{-2},$$

$$f = 5.3915 \times 10^{-1}, \qquad k_2 = 5.529 \times 10^{-4}, \qquad \delta = 1.865 \times 10^{-1},$$

$$m_1 = 2.439 \times 10^{-2}, \qquad m_2 = 1.3099 \times 10^{-2}, \qquad \epsilon_2 = 5.043 \times 10^{-1},$$

$$N_T = 1.904 \times 10^1, \qquad c = 1.936 \times 10^1, \qquad \lambda_E = 9.909 \times 10^{-3},$$

$$b_E = 9.785 \times 10^{-2}, \qquad K_b = 3.909 \times 10^{-1}, \qquad d_E = 1.021 \times 10^{-1},$$

$$K_d = 8.379 \times 10^{-1}, \qquad \delta_E = 7.030 \times 10^{-2}.$$

In Fig. 3.1 above we depicted the log-scaled longitudinal observations $\{y_i\}$ on the number of CD4$^+$ T-cells and the model output $z(t_j) = f(t_j, \theta_0)$, $j = 1, \ldots, N$, evaluated at the nominal parameter vector and given in (3.19).

It is assumed that the following parameters are always fixed at the values given above:

$$x_3^0, \ x_4^0, \ \text{and} \ x_6^0 \ .$$

As reported in [1, 3], simulation and relative sensitivity studies revealed that the validated models are not sensitive to these initial conditions and following [3] we therefore fix them. In other words, the parameters to be selected for estimation will always constitute a sub-vector of

$$q = \left(x_1^0, x_2^0, x_5^0, \lambda_1, d_1, \epsilon_1, k_1, \lambda_2, d_2, f, k_2, \delta, \ldots \right.$$
$$\left. \ldots m_1, m_2, \epsilon_2, N_T, c, \lambda_E, b_E, K_b, d_E, K_d, \delta_E \right) \in \mathbb{R}^{23}. \quad (3.28)$$

Let $\mathscr{F}_{23}$ denote the Fisher information matrix of system (3.18) for the 23 parameters of q as given in (3.28) at their nominal values q_0 as given above. Then we have

$$\text{cond} \, \mathscr{F}_{23}(q_0) = 1.712 \times 10^{24},$$

i.e., $\mathscr{F}_{23}(q_0)$ is non-singular, but ill-conditioned. Since $\mathscr{F}_{23}(q_0)$ is non-singular, the Fisher information matrix for any sub-vector θ of q at the corresponding nominal parameter values is also non-singular. Consequently the regularity check for the Fisher information matrix in the subset selection algorithm can be deleted in case of the HIV-model.

In [1, 3], the authors estimate the parameter vector

$$\theta = \left(x_1^0, x_2^0, x_5^0, \lambda_1, d_1, \epsilon_1, k_1, \epsilon_2, N_T, c, b_E \right)^\mathsf{T} \in \mathbb{R}^{11}. \quad (3.29)$$

The selection score for this parameter vector is $\alpha(\theta_0) = 4.611 \times 10^3$. In Table 3.4 we display, for the five selections of sub-vector $\theta \in \mathbb{R}^{11}$ of q with the minimal selection scores, the condition numbers of the corresponding Fisher information matrices and the selection scores. In addition we also show the selection $\theta \in \mathbb{R}^{11}$ with the maximal selection score. As we can see, the selection score values range from 2.881×10^1 to 2.340×10^5 for the $\binom{23}{11} = 1{,}352{,}078$ different parameter vectors in $\mathbb{R}^{11}$ which can be selected from the 23 parameters in (3.28).

As we also can see from Table 3.4 that the selection algorithm chooses most of the parameters in the vector (3.29). For instance, the sub-vector $(x_5^0, \lambda_1, d_1, \epsilon_1, \epsilon_2)$ of (3.29) appears in every one of the top five parameter vectors displayed in Table 3.4. In fact the top five parameter vectors have the sub-vector $(x_5^0, \lambda_1, d_1, \epsilon_1, \lambda_2, d_2, k_2, \delta, \epsilon_2) \in \mathbb{R}^9$ in common and differ only by one or two of the parameters x_1^0, x_2^0, k_1, c, and N_T. Use of the subset selection algorithm discussed here (had it been available) might have proved valuable in the efforts reported in [1, 3].

Table 3.4 The top five parameter vectors obtained with subset selection algorithm for $p = 11$. For each parameter vector θ the condition number $\kappa(\mathscr{F}(\theta_0))$ of the Fisher information matrix and the selection score $\alpha(\theta_0)$ are displayed. The next two lines show $\kappa(\mathscr{F}(\theta_0))$ and $\alpha(\theta_0)$ for the sub-vector $\theta \in \mathbb{R}^9$ which is common to the top five parameter vectors and for the optimal parameter vector in $\mathbb{R}^9$. The last line presents the sub-vector in $\mathbb{R}^{11}$ with the largest selection score

Parameter vector θ	$\kappa(\mathscr{F}(\theta_0))$	$\alpha(\theta_0)$
$(x_1^0, x_5^0, \lambda_1, d_1, \epsilon_1, \lambda_2, d_2, k_2, \delta, \epsilon_2, N_T)$	9.841×10^{10}	2.881×10^1
$(x_1^0, x_5^0, \lambda_1, d_1, \epsilon_1, \lambda_2, d_2, k_2, \delta, \epsilon_2, c)$	9.845×10^{10}	2.883×10^1
$(x_1^0, x_5^0, \lambda_1, d_1, \epsilon_1, k_1, \lambda_2, d_2, k_2, \delta, \epsilon_2)$	4.388×10^{16}	2.896×10^1
$(x_2^0, x_5^0, \lambda_1, d_1, \epsilon_1, \lambda_2, d_2, k_2, \delta, \epsilon_2, N_T)$	9.235×10^{10}	2.904×10^1
$(x_2^0, x_5^0, \lambda_1, d_1, \epsilon_1, \lambda_2, d_2, k_2, \delta, \epsilon_2, c)$	9.241×10^{10}	2.906×10^1
$(x_5^0, \lambda_1, d_1, \epsilon_1, \lambda_2, d_2, k_2, \delta, \epsilon_2)$	9.083×10^{10}	2.193×10^1
$(x_1^0, x_5^0, \lambda_1, d_1, k_1, d_2, k_2, \delta, \epsilon_2)$	4.335×10^{15}	1.050×10^1
$(d_2, k_2, \delta, m_2, N_T, \lambda_E, b_E, K_b, d_E, K_d, \delta_E)$	7.247×10^{17}	2.340×10^5

In Fig. 3.2a, we depict the minimal selection score as a function of the number of parameters. Table 3.5 contains the values of the corresponding selection scores. Figure 3.2b is a semilog plot of Fig. 3.2a, i.e., it displays the logarithm of the selection score as a function of the number of parameters. Figure 3.2b, suggests that parameter vectors with more than 13 parameters might be expected to have large uncertainty when estimated from observations, because the minimal selection score is already larger than 100. Figure 3.2b also depicts the regression line, which fits the logarithm of the selection score. From this linear regression we conclude the selection score $\alpha_{\min}(p)$ grows exponentially with the number p of parameters to be estimated. More precisely, we find

$$\alpha_{\min}(p) \approx 0.01133 e^{0.728p}, \quad p = 1, \ldots, 22. \tag{3.30}$$

Computing the minimal selection scores for $p = 1, \ldots, 22$ requires to consider all possible choices of sub-vectors of (3.28) in $\mathbb{R}^p$, $p = 1, \ldots, 22$, i.e., to consider $\sum_{i=1}^{22} \binom{23}{i} = 8{,}388{,}605$ cases. If we determine the regression line only using the minimal selection scores for $p = 1, 2, 3, 20, 21, 22$ we obtain

$$\alpha_{\min}(p) \approx 0.01441 e^{0.710p}, \quad p = 1, \ldots, 22. \tag{3.31}$$

Computing the minimal selection scores needed for (3.31) requires to consider only $\binom{23}{1} + \binom{23}{2} + \binom{23}{3} + \binom{23}{20} + \binom{23}{21} + \binom{23}{22} = 2(\binom{23}{1} + \binom{23}{2} + \binom{23}{3}) = 4{,}094$ cases. In Fig. 3.2a, we show the curves given by (3.30) and (3.31), whereas in Fig. 3.2b the corresponding regression lines are depicted. Table 3.6 shows the time it takes on a laptop computer with an Intel© Core™2 Duo processor using a MATLAB programm to compute $\alpha_{\min}(p)$, $p = 1, \ldots, 22$, once the Fisher information matrix $\mathscr{F}_{23}(q_0)$ for the nominal parameter vector q_0 has been computed. In order to reduce computational efforts it is important to observe that for any selection θ of

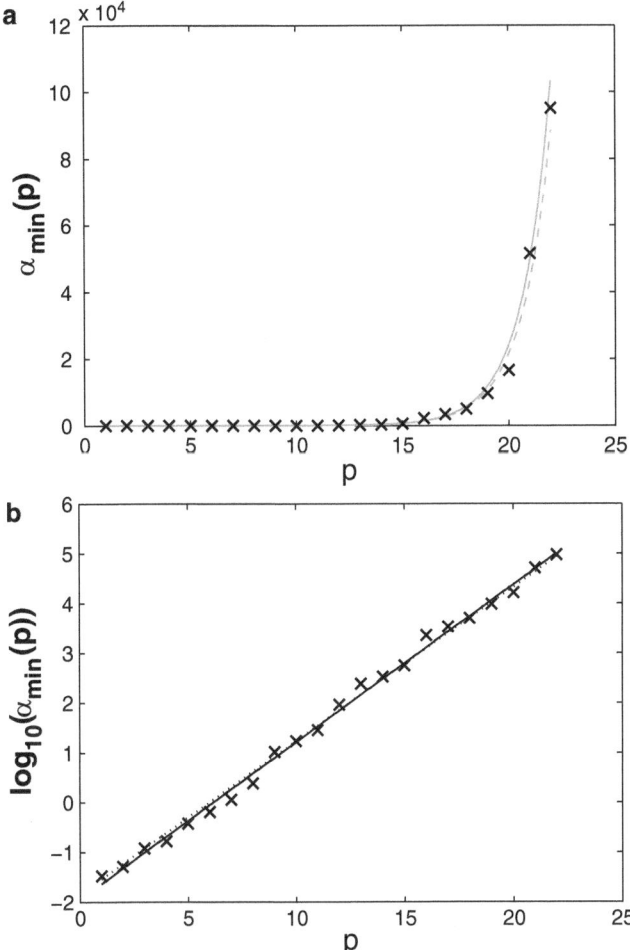

Fig. 3.2 (**a**) Minimal selection scores (*crosses*) and exponential approximations (3.30) (*grey solid line*) respectively (3.31) (*grey dashed line*) versus the number of parameters p. (**b**) Logarithm of minimal selection scores (*crosses*) and regression lines corresponding to (3.30) (*gray solid line*) respectively to (3.31) (*grey dashed line*) versus number of parameters p

Table 3.5 Minimal selection scores $\alpha_{\min}(p)$ for sub-vectors in $\mathbb{R}^p$ of (3.28), $p = 1, 2, \ldots, 22$

p	1	2	3	4	5	6	7	8
$\alpha_{\min}(p)$	0.0340	0.0523	0.1203	0.1679	9.3796	0.6511	1.1375	2.4166

p	9	10	11	12	13	14	15	16
$\alpha_{\min}(p)$	10.503	17.482	28.81	92.91	243.34	336.77	566.86	2,274.3

p	17	18	19	20	21	22
$\alpha_{\min}(p)$	3,372.4	5,047.9	9,664.4	16,585	51,631	95,128

Table 3.6 Time for computing $\theta \in \mathbb{R}^p$ with the minimal selection score on a laptop computer, once the Fisher information matrix $\mathscr{F}_{23}(q_0)$ for the HIV-model has been computed

p	1	2	3	4	5	6	7	8
time (s)	0.054	0.015	0.075	0.472	1.66	5.52	14.05	29.81

p	9	10	11	12	13	14	15	16
time (s)	53.49	80.02	100.4	105.8	96.74	74.01	47.15	24.09

p	17	18	19	20	21	22
time (s)	10.69	3.64	1.02	0.211	0.034	0.0041

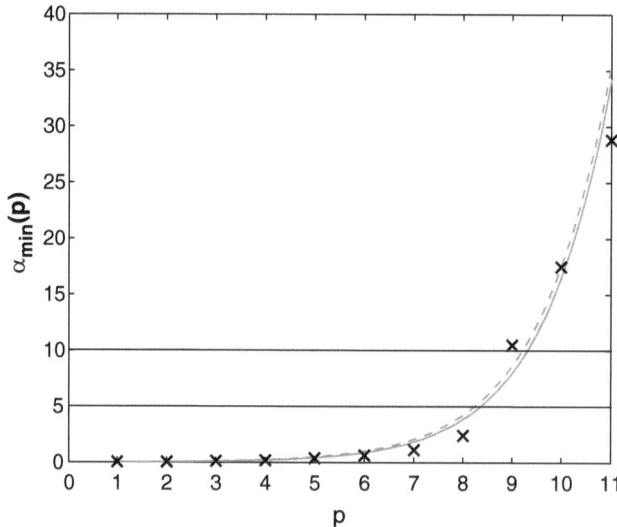

Fig. 3.3 Minimal selection scores (*crosses*) and exponential approximations (3.30) (*grey solid line*) respectively (3.31) (*grey dashed line*) for $p = 1, \dots, 11$

p parameters from the set of p_0 parameters the corresponding Fisher information matrix $\mathscr{F}(\theta_0)$ (θ_0 being the vector of nominal values of the components of θ) is obtained from $\mathscr{F}_{23}(q_0)$ simply by selecting the rows and columns corresponding to the components of θ.

If the condition number of the Fisher information matrix corresponding to the original parameter set is becoming very large, then the approximation of $\alpha_{\min}(p)$ by an exponential function might be possible only for $p = 1, \dots, p^*$ with $p^* < p_0$. See the results for the CVS-model in Sect. 3.5.2.

Figure 3.3 is the same as Fig. 3.2a, but for $p = 1, \dots, 11$. The curves (3.30) respectively (3.31) can be used to determine p such that the selection score $\alpha_{\min}(p)$ is smaller than a given upper bound. From Fig. 3.3 we can see that in order to have

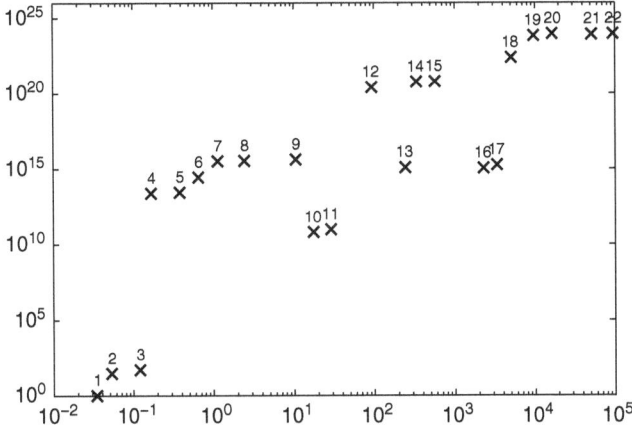

Fig. 3.4 Condition number $\kappa(\mathscr{F}_p(\theta_0))$ versus minimal selection score $\alpha_{\min}(p)$ for the IIIV-model, where $\theta \in \mathbb{R}^p$ is the sub-vector of (3.28) with the minimal selection score, $p = 1, \ldots, 22$. Both axes are in logarithmic scale

Table 3.7 The top five parameter vectors $\theta \in \mathbb{R}^{11}$ selected according the criterion of minimal condition number for the Fisher information matrix

Parameter vector θ	$\kappa(\mathscr{F}(\theta_0))$	$\alpha(\theta_0)$
$(x_1^0, x_2^0, \lambda_1, \epsilon_1, f, m_1, m_2, N_T, \lambda_E, d_E, \delta_E)$	1.735×10^6	1.908×10^3
$(x_1^0, x_2^0, \lambda_1, \epsilon_1, f, m_1, m_2, N_T, \lambda_E, b_E, \delta_E)$	1.738×10^6	1.897×10^3
$(x_1^0, x_2^0, \lambda_1, \epsilon_1, f, m_1, m_2, c, \lambda_E, d_E, \delta_E)$	1.744×10^6	1.908×10^3
$(x_1^0, x_2^0, \lambda_1, \epsilon_1, f, m_1, m_2, c, \lambda_E, b_E, \delta_E)$	1.747×10^6	1.896×10^3
$(x_2^0, x_5^0, \lambda_1, \epsilon_1, f, m_1, m_2, c, \lambda_E, d_E, \delta_E)$	1.788×10^6	1.910×10^3

$\alpha_{\min}(p) < 5$ we should choose $p \le 8$, which is correct according to each of the two curves (see Table 3.5). In order to have $\alpha_{\min} < 10$ the curves suggest $p \le 9$, which is not quite correct, because according to Table 3.5 we have $\alpha_{\min}(9) = 10.5$, so that we should choose $p \le 8$. In Fig. 3.4, we graph (in logarithmic scales) the condition number $\kappa(\mathscr{F}_p(\theta_0))$ of the corresponding Fisher information matrix versus the smallest selection score $\alpha_{\min}(p)$ for $p = 1, \ldots, 22$. It is clear from Fig. 3.4 that the condition numbers for the Fisher information matrix corresponding to the selected parameter vector $\theta \in \mathbb{R}^p$ does not show a monotone behavior with respect to p. We could also determine the selection of parameters according to the criterion of minimal condition number for the corresponding Fisher information matrix. In Table 3.7, we present the best five selections $\theta \in \mathbb{R}^{11}$ according to this criterion together with the condition numbers of the Fisher information matrix and the corresponding selection scores.

In Table 3.8 we examine the effect that removing parameters from an estimation has in uncertainty quantification. The coefficient of variation (CV) is shown for each parameter (see (3.16)). Five cases are considered:

Table 3.8 Coefficient of variation (CV) of the parameter vectors for the HIV-model as listed above

Parameter	CV for $\theta^{(18)}$	$\theta^{(5,1)}$	$\theta^{(5,2)}$	$\theta^{(5,3)}$	$\theta^{(5,4)}$
x_1^0	—	—	4.09×10^{-2}	—	—
x_2^0	3.80	—	—	—	—
x_5^0	1.58×10^1	3.43×10^{-1}	—	—	—
λ_1	8.19×10^{-1}	3.56×10^{-1}	1.13×10^{-1}	—	—
d_1	9.39×10^{-1}	3.94×10^{-1}	—	—	—
ϵ_1	1.26×10^2	—	—	8.49	—
k_1	7.67×10^2	8.17×10^{-2}	9.57×10^{-2}	—	—
λ_2	4.74×10^1	—	—	9.99	—
d_2	4.62×10^1	—	—	—	—
f	2.51×10^2	—	—	—	—
k_2	7.53×10^2	—	—	—	—
δ	—	—	3.29×10^{-1}	—	—
m_1	2.29×10^3	—	—	9.85×10^2	2.10×10^1
ϵ_2	1.63×10^2	1.11×10^{-1}	9.33×10^{-2}	6.34	—
c	7.74×10^2	—	—	—	—
λ_E	2.18×10^3	—	—	9.83×10^2	—
b_E	—	—	—	—	1.22×10^4
K_b	2.55×10^3	—	—	—	4.62×10^3
d_E	4.56×10^2	—	—	—	1.16×10^4
K_d	1.98×10^3	—	—	—	4.40×10^3
δ_E	1.72×10^3	—	—	—	—
$\alpha(\cdot)$	5.05×10^3	6.47×10^{-1}	3.75×10^{-1}	1.39×10^3	1.80×10^4

(i) The parameter vector

$$\theta^{(18)} = \left(x_2^0, x_5^0, \lambda_1, d_1, \epsilon_1, k_1, \lambda_2, d_2, f, k_2, m_1, \epsilon_2, c, \lambda_E, d_E, K_d, \delta_E\right)^{\mathsf{T}},$$

which is the sub-vector in $\mathbb{R}^{18}$ with the minimal selection score.

(ii) The parameter vector

$$\theta^{(5,1)} = \left(x_5^0, \lambda_1, d_1, k_1, \epsilon_2\right)^{\mathsf{T}},$$

which is the sub-vector of $\theta^{(18)}$ in $\mathbb{R}^5$ with the minimal selection score.

(iii) The parameter vector

$$\theta^{(5,2)} = \left(x_1^0, \lambda_1, k_1, \delta, \epsilon_2\right)^{\mathsf{T}},$$

which is the sub-vector in $\mathbb{R}^5$ of q as given by (3.28) with the minimal selection score.

(iv) The parameter vector

$$\theta^{(5,3)} = (\epsilon_1, \lambda_2, m_1, \epsilon_2, \lambda_E)^{\mathsf{T}},$$

which is the sub-vector of $\theta^{(18)}$ in $\mathbb{R}^5$ with the maximal selection score.

(v) The parameter vector

$$\theta^{(5,4)} = (m_1, b_E, K_b, d_E, K_d)^{\mathsf{T}},$$

which is the sub-vector in $\mathbb{R}^5$ of q as given by (3.28) with the maximal selection score.

We see that here are substantial improvements in uncertainty quantification when considering $\theta^{(5,1)}$ instead of $\theta^{(18)}$. However, just taking a considerably lower dimensional sub-vector of $\theta^{(18)}$ in $\mathbb{R}^5$ does not lead necessarily to a drastic improvement of the estimate.

3.5.2 The CVS-Model

For this model we take as the nominal parameters the following estimates obtained in [44] using data obtained at bicycle ergometer tests (for a description and units see Table 3.2):

$$c_\ell = 0.02305, \qquad c_r = 0.04413, \qquad c_{as} = 0.01016,$$

$$c_{vs} = 0.6500, \qquad c_{ap} = 0.03608, \qquad c_{vp} = 0.1408,$$

$$R_\ell = 0.2671, \qquad R_r = 0.04150, \qquad \alpha_\ell = 30.5587,$$

$$\alpha_r = 28.6785, \qquad \beta_\ell = 25.0652, \qquad \beta_r = 1.4132,$$

$$\gamma_\ell = -1.6744, \qquad \gamma_r = -1.8607, \qquad K = 16.0376,$$

$$\kappa = 0.05164, \qquad M_0 = 0.35, \qquad \rho = 0.011,$$

$$C_{a,O_2} = 0.2, \qquad q_{as} = 163.047, \qquad V_{tot} = 5.0582,$$

$$R_p^{rest} = 1.5446, \qquad A_{pesk}^{rest} = 177.682, \qquad R_p^{exer} = 0.3165,$$

$$A_{pesk}^{exer} = 254.325.$$

For the estimates given above measurements for P_{as}, H and Q_ℓ were available. The sampling times t_j for the measurements of P_{as} and H were uniformly distributed on the time interval from 0 to 10 min with $t_{j+1} - t_j = 2$ s, i.e., 601 measurements for P_{as} and H. The measurements for Q_ℓ were obtained by Doppler echo-cardiography and consequently were much less frequent (only 20 measurements) and also irregularly distributed. In the following we shall consider P_{as} as the only

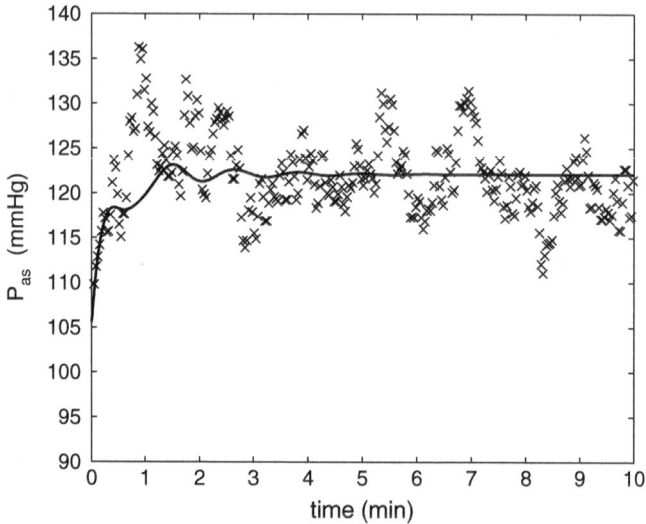

Fig. 3.5 The P_{as}-component of the solution of system (3.20) with the nominal parameter values and the P_{as}-measurements

Table 3.9 The equilibria x^{rest} and x^{exer} for the CVS-model

Variable	P_{as}	P_{vs}	P_{ap}	P_{vp}	S_ℓ	σ_ℓ	S_r	σ_r	R_s	H
x^{rest}	105.595	4.277	12.474	5.367	64.675	0	3.886	0	22.020	78.85
x^{exer}	122.115	3.595	10.441	7.844	88.092	0	5.293	0	14.445	107.4

measured output of the system, i.e., we have $f(t; \theta) = P_{as}(t; \theta)$. The variance of the measurement errors was roughly estimated to be $\sigma_0^2 = 10$ (Fig. 3.5).

The equilibria x^{rest} and x^{exer} are given in Table 3.9 (for units see Table 3.2).

As for the HIV-model also in case of the CVS-model the Fisher information matrix $\mathscr{F}_{25}(q_0)$ at the nominal values of the parameters (3.27) is non-singular, but highly ill-conditioned:

$$\text{cond } \mathscr{F}_{25}(q_0) = 2.5755 \times 10^{31}.$$

Therefore we can also delete the regularity test for the Fisher information matrix in the selection algorithm in case of the CVS-model. In Fig. 3.7, we show the minimal selection scores for $p = 3, \ldots, 20$, whereas in Table 3.10 we list the minimal selection scores $\alpha_{min}(p)$, $p = 1, \ldots, 24$. Table 3.10 clearly shows the effect of the extreme ill-conditioning of the Fisher information matrix $\mathscr{F}_{25}$. We see an extreme increase of the selection score from $3 \cdot 10^5$ at $p = 21$ to $2.4 \cdot 10^{31}$ at $p = 22$. This is also reflected in Fig. 3.6 which clearly shows that there is no reasonable approximation of $\log(\alpha_{min}(p))$, $p = 1, \ldots, 24$, by a regression line. However, a regression line makes sense for $p = 1, \ldots, 21$ as can be see from Fig. 3.6. In

Table 3.10 Minimal selection scores $\alpha_{\min}(p)$, $p = 1, \ldots, 24$, for the CVS-model

p	1	2	3	4	5	6
$\alpha_{\min}(p)$	9.397×10^{-4}	2.042×10^{-2}	5.796×10^{-2}	1.096×10^{-1}	2.315×10^{-1}	3.299×10^{-1}

p	7	8	9	10	11	12
$\alpha_{\min}(p)$	4.340×10^{-1}	6.251×10^{-1}	8.018×10^{-1}	1.084	1.723	3.238

p	13	14	15	16	17	18
$\alpha_{\min}(p)$	6.336	17.13	37.31	59.37	2.224×10^2	4.301×10^2

p	19	20	21	22	23	24
$\alpha_{\min}(p)$	1.015×10^3	1.265×10^5	3.038×10^5	2.396×10^{31}	1.834×10^{34}	2.459×10^{108}

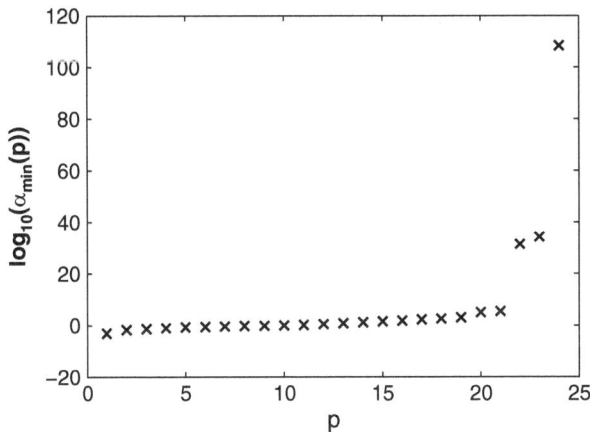

Fig. 3.6 Logarithm of the minimal selection scores, $p = 1, \ldots, 24$, for the CVS-model

Fig. 3.7, we depict the minimal selection scores $\alpha_{\min}(p)$, $p = 2, \ldots, 19$, together with the approximating exponential functions

$$\alpha_{\min}(p) \approx 0.00670e^{0.580p}, \quad p = 1, \ldots, 19, \tag{3.32}$$

obtained from the regression line for $p = 1, \ldots, 19$ (solid grey line) and

$$\alpha_{\min}(p) \approx 0.00799e^{0.610p}, \quad p = 1, \ldots, 19, \tag{3.33}$$

obtained for $p = 2, 3, 4, 17, 18, 19$ (dashed grey line). Table 3.11 for the CVS-model is the analogue to Table 3.6 (for the HIV-model) concerning computing times.

In Table 3.12, we present the sub-vector $\theta \in \mathbb{R}^{10}$ of q given by (3.27) with the minimal selection score together with the coefficients of variation. In Fig. 3.8, we depict the classical sensitivities for the chosen sub-vector $\theta \in \mathbb{R}^{10}$ (black solid lines) and the sensitivities for the other 15 parameters (grey dashed lines). We see that the algorithm chooses not only the parameters with large sensitivities, but also

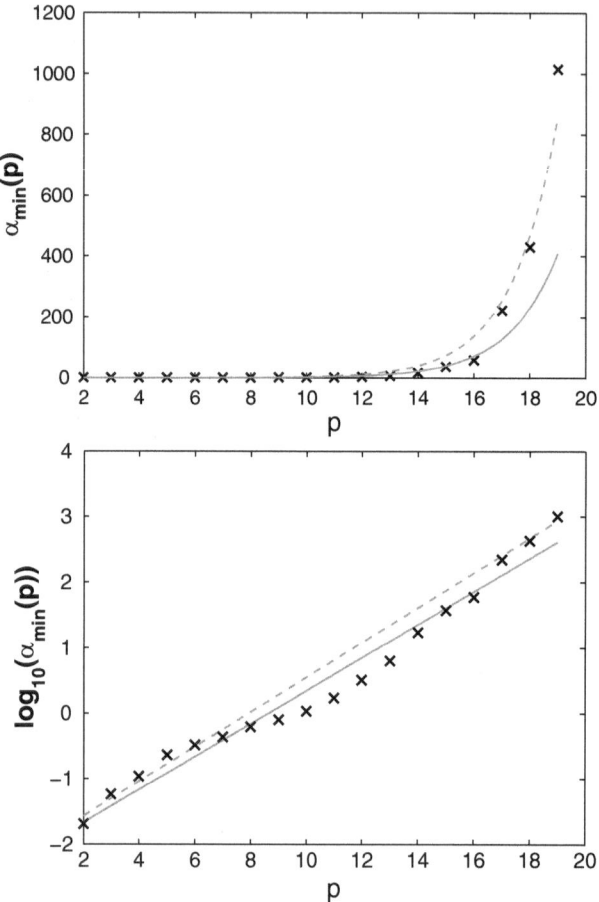

Fig. 3.7 Minimal selection scores $\alpha_{min}(p)$ (*upper panel*) and $\log_{10}\left(\alpha_{min}(p)\right)$ (*lower panel*), $p = 2,\ldots, 19$, for the CVS-model

Table 3.11 Time for computing $\theta \in \mathbb{R}^p$ with the minimal selection score on a laptop computer, once the Fisher information matrix $\mathscr{F}_{25}(q_0)$ for the CVS-model has been computed

p	1	2	3	4	5	6	7	8	9
time (s)	0.0038	0.018	0.104	0.564	2.364	7.848	21.24	46.60	88.31

p	10	11	12	13	14	15	16	17	18
time (s)	139.7	188.3	211.1	208.9	172.9	123.1	71.41	36.10	14.92

p	19	20	21	22	23	24
time (s)	5.137	1.382	0.329	0.093	0.0645	0.0005

Table 3.12 Coefficients of variance for the sub-vector $\theta \in \mathbb{R}^{10}$ with the minimal selection score

$\theta \in \mathbb{R}^{10}$	α_ℓ	α_r	β_ℓ	γ_ℓ	γ_r	ρ	C_{a,O_2}	q_{as}	V_{tot}	R_p^{rest}
CV	0.086	0.183	0.435	0.353	0.559	0.166	0.235	0.485	0.368	0.240
$\alpha_{min}(10)$	1.086									

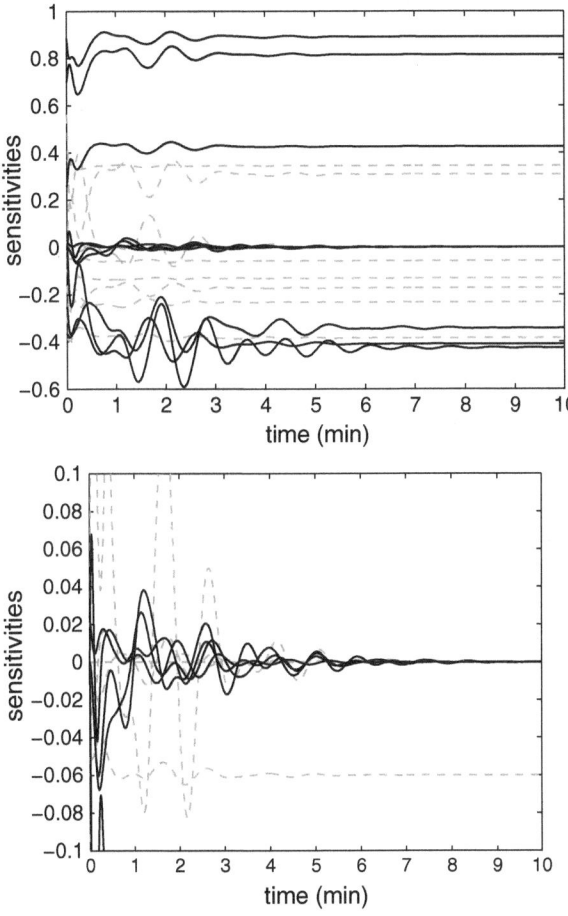

Fig. 3.8 Classical sensitivities of P_{as} with respect to the parameters of the selection $\theta = (\alpha_\ell, \alpha_r, \beta_\ell, \gamma_\ell, \gamma_r, \rho, C_{a,O_2}, q_{as}, V_{tot}, R_p^{rest})^\top \in \mathbb{R}^{10}$ which provides the minimal selection score (*black solid lines*) and classical sensitivities of Pas with respect to the remaining 15 parameters (*dashed grey lines*). The *lower panel* is a blow-up of the part of the *upper panel* with sensitivities between -0.1 and 0.1

parameters with rather small sensitivities. In this context one has to observe that the subset selection algorithm chooses parameters which can be jointly identified with a given accuracy (in terms of asymptotic standard errors), whereas classical sensitivities characterize the sensitivity of the measurable output with respect to a single parameter.

3.6 Concluding Remarks

As we have noted, inverse problems for complex system models containing a large number of parameters are difficult. There is great need for quantitative methods to assist in posing inverse problems that will be well formulated in the sense of the ability to provide parameter estimates with quantifiable small uncertainty estimates. We have introduced and illustrated use of such an algorithm that requires prior local information about ranges of admissible parameter values and initial values of interest along with information on the error in the observation process to be used with the inverse problem. These are needed in order to implement the sensitivity/Fisher matrix based algorithm.

Because sensitivity of a model with respect to a parameter is fundamentally related to the ability to estimate the parameter, and because sensitivity is a local concept, we observe that the pursuit of a global algorithm to use in formulating parameter estimation or inverse problems is most likely a quest that will go unfulfilled.

Acknowledgements This research was supported in part by Grant Number R01AI071915-07 from the National Institute of Allergy and Infectious Diseases and in part by the Air Force Office of Scientific Research under grant number FA9550-09-1-0226. The content is solely the responsibility of the authors and does not necessarily represent the official views of the NIAID, the NIH or the AFOSR.

References

1. Adams, B.M.: Non-parametric parameter estimation and clinical data fitting with a model of HIV infection. Ph.D. thesis, North Carolina State University, Raleigh, NC (2005)
2. Adams, B.M., Banks, H.T., Davidian, M., Kwon, H., Tran, H.T., Wynne, S.N., Rosenberg, E.S.: HIV dynamics: Modeling, data analysis, and optimal treatment protocols. J. Comp. Appl. Math. **184**, 10–49 (2005)
3. Adams, B.M., Banks, H.T., Davidian, M., Rosenberg, E.S.: Model fitting and prediction with HIV treatment interruption data; CRSC-TR05-40, NCSU, October 2005. Bull. Math. Biol. **69**, 563–584 (2007)
4. Adams, B.M., Banks, H.T., Tran, H.T., Kwon, H.: Dynamic multidrug therapies for HIV: Optimal and STI control approaches. Math. Biosci. Eng. **1**, 223–241 (2004)
5. Anh, D.T., Bonnet, M.P., Vachaud, G., Minh, C.V., Prieur, N., Duc, L.V., Anh, L.L.: Biochemical modeling of the Nhue River (Hanoi, Vietnam): Practical identifiability analysis and parameter estimation. Ecol. Model. **193**, 182–204 (2006)
6. Astrom, K.J., Eykhoff, P.: System identification – A survey. Automatica **7**, 123–162 (1971)
7. Banks, H.T., Davidian, M., Hu, S., Kepler, G.M., Rosenberg, E.S.: Modeling HIV immune response and validation with clinical data; CRSC-TR07-09, March 2007. J Biol. Dyn. **2**, 357–385 (2008)
8. Banks, H.T., Davidian, M., Samuels, J.R., Sutton, K.L.: An inverse problem statistical methodology summary; CRSC-TR08-1, NCSU, January 2008. In: Chowell G., et al., (eds.) Mathematical and Statistical Estimation Approaches in Epidemiology, pp. 249–302. Springer, New York (2009)

9. Banks, H.T., Dediu, S., Ernstberger, S.E.: Sensitivity functions and their uses in inverse problems; CRSC-TR07-12, NCSU, July, 2007. J. Inverse Ill Posed Probl. **15**, 683–708 (2007)
10. Banks, H.T., Dediu, S., Ernstberger, S.L., Kappel, F.: Generalized sensitivities and optimal experimental design; CRSC-TR08-12, NCSU, September 2008, revised, November 2009. J. Inverse Ill Posed Probl. **18**, 25–83 (2010)
11. Banks, H.T., Ernstberger, S.L., Grove, S.L.: Standard errors and confidence intervals in inverse problems: Sensitivity and associated pitfalls. J. Inverse Ill Posed Probl. **15**, 1–18 (2007)
12. Banks, H.T., Fitzpatrick, B.G.: Inverse problems for distributed systems: statistical tests and ANOVA; LCDS/CCS Rep. 88–16, July 1988, Brown University. In: Proc. International Symposium on Math. Approaches to Envir. and Ecol. Problems. Springer Lecture Notes in Biomathematics, vol. 81, pp. 262–273. Springer, New York (1989)
13. Banks, H.T., Fitzpatrick, B.G.: Statistical methods for model comparison in parameter estimation problems for distributed systems; CAMS Tech. Rep. 89-4, September 1989, University of Southern California. J. Math. Biol. **28**, 501–527 (1990)
14. Banks, H.T., Holm, K., Robbins, D.: Standard error computations for uncertainty quantification in inverse problems: Asymptotic theory vs. bootstrapping; CRSC-TR09-13, NCSU, June 2009, revised August 2009; Math. Comput. Model. **52**, 1610–1625 (2010)
15. Banks, H.T., Samuels Jr., J.R.: Detection of cardiac occlusions using viscoelastic wave propagation; CRSC-TR08-23, December 2008. Adv. Appl. Math. Mech. **1**, 1–28 (2009)
16. Batzel, J.J., Kappel, F., Schneditz, D., Tran, H.T.: Cardiovascular and Respiratory Systems: Modeling, Analysis and Control. Frontiers in Applied Mathematics, vol. FR34. SIAM, Philadelphia (2006)
17. Bedrick, E.J., Tsai, C.L.: Model selection for multivariate regression in small samples. Biometrics **50**, 226–231 (1994)
18. Bellman, R., Astrom, K.M.: On structural identifiability. Math. Biosci. **7**, 329–339 (1970)
19. Bellman, R., Kalaba, R.: Quasilinearization and Nonlinear Boundry Value Problems. American Elsevier, New York (1965)
20. Bonhoeffer, S., Rembiszewski, M., Ortiz, G.M., Nixon, D.F.: Risks and benefits of structured antiretroviral drug therapy interruptions in HIV-1 infection. AIDS **14**, 2313–2322 (2000)
21. Bozdogan, H.: Model selection and Akaike's Information Criterion (AIC): The general theory and its analytical extensions. Psychometrika **52**, 345–370 (1987)
22. Bozdogan, H.: Akaike's Information Criterion and recent developments in information complexity. J. Math. Psychol. **44**, 62–91 (2000)
23. Burnham, K.P., Anderson, D.R.: Model Selection and Multimodel Inference: A Practical Information-Theoretic Approach. Springer, Berlin, Heidlberg, New York (2002)
24. Burnham, K.P., Anderson, D.R.: Multimodel inference: Understanding AIC and BIC in model selection. Socio. Meth. Res. **33**, 261–304 (2004)
25. Burth, M., Verghese, G.C., Vélez-Reyes, M.: Subset selection for improved parameter estimation in on-line identification of a synchronous generator. IEEE T. Power Syst. **14**, 218–225 (1999)
26. Callaway, D.S., Perelson, A.S.: HIV-1 infection and low steady state viral loads. Bull. Math. Biol. **64**, 29–64 (2001)
27. Cintrón-Arias, A., Banks, H.T., Capaldi, A., Lloyd, A.L.: A sensitivity matrix based methodology for inverse problem formulation; CRSC-TR09, NCSU, April 2009. J. Inverse Ill Posed Probl. **17**, 545–564 (2009)
28. Cintrón-Arias, A., Castillo-Chávez, C., Bettencourt, L.M.A., Lloyd, A.L., Banks, H.T.: The estimation of the effective reproductive number from disease outbreak data; CRSC-TR08-08, NCSU, April 2008. Math. Biosci. Eng. **6**, 261–283 (2009)
29. Cobelli, C., DiStefano 3rd, J.J.: Parameter and structural identifiability concepts and ambiguities: a critical review and analysis. Am. J. Physiol. **239**, R7–R24 (1980)
30. Davidian, M., Giltinan, D.M.: Nonlinear Models for Repeated Measurement Data. Chapman & Hall, Boca Raton (1995)

31. Efron, B., Tibshirani, R.J.: An Introduction to the Bootstrap. Chapman & Hall/CRC, Boca Raton (1998)
32. Engl, H.W., Flamm, C., Kügler, P., Lu, J., Müller, S., Schuster, P.: Inverse problems in system biology. Inverse Probl. **25**, 123,014 (51pp.) (2009)
33. Eslami, M.: Theory of Sensitivity in Dynamic Systems: An Introduction. Springer, New York (1994)
34. Evans, N.D., White, L.J., Chapman, M.J., Godfrey, K.R., Chappell, M.J.: The structural identifiability of the susceptible infected recovered model with seasonal forcing. Math. Biosci. **194**, 175–197 (2005)
35. Eykhoff, P.: System Identification: Parameter and State Estimation. Wiley, New York (1974)
36. Fink, M.: myAD: fast automatic differentiation code in MATLAB. http://gosh.gmxhome.de/ (2006)
37. Fink, M., Attarian, A., Tran, H.T.: Subset selection for parameter estimation in an HIV model. Proc. Appl. Math. Mech.**7**, 11212,501–11221,502 (2008)
38. Fink, M., Batzel, J.J., Tran, H.T.: A respiratory system model: Parameter estimation and sensitivity analysis. Cardiovasc. Eng. **8**, 120–134 (2008)
39. Glover, K., Willems, J.C.: Parametrizations of linear dynamical systems: Canonical forms and identifiability. IEEE Trans. Automat. Contr. **AC-19**, 640–645 (1974)
40. Holmberg, A.: On the practical identifiability of microbial growth models incorporating Michaelis-Menten type nonlinearities. Math. Biosci. **62**, 23–43 (1982)
41. Hurvich, C.M., Tsai, C.L.: Regression and time series model selection in small samples. Biometrika **76**, 297–307 (1989)
42. Jones, L.E., Perelson, A.S.: Opportunistic infection as a cause of transient viremia in chronically infected HIV patients under treatment with HAART. Bull. Math. Biol. **67**, 1227–1251 (2005)
43. Kalman, R.E.: Mathematical description of linear dynamical systems. SIAM J. Contr. **1**, 152–192 (1963)
44. Kappel, F., Peer, R.O.: A mathematical model for fundamental regulation processes in the cardiovascular system. J. Math. Biol. **31**, 611–631 (1993)
45. Luenberger, D.G.: Optimization by Vector Space Methods. Wiley, New York (1969)
46. Mehra, A.K., Lainiotis, D.G.: System Identification. Academic, New York (1976)
47. Navon, I.M.: Practical and theoretical aspects of adjoint parameter estimation and identifiability in meteorology and oceanograph. Dynam. Atmosp. Oceans **27**, 55–79 (1997)
48. Nelson, P., Smith, N., Cuipe, S., Zou, W., Omenn, G.S., Pietropaolo, M.: Modeling dynamic changes in type 1 diabetes progression: Quantifying β-cell variation after the appearance of islet-specific autoimmune responses. Math. Biosci. Eng. **6**, 753–778 (2009)
49. Nowak, M.A., Bangham, C.R.M.: Population dynamics of immune responses to persistent viruses. Science **272**, 74–79 (1996)
50. Ottesen, J.T., Olufsen, M.S., Larsen, J.K.: Applied Mathematical Models in Human Physiology, vol. MM09. SIAM, Philadelphia (2004)
51. Perelson, A.S., Nelson, P.W.: Mathematical analysis of HIV-1 dynamics in vivo. SIAM Rev. **41**, 3–44 (1999)
52. Reid, J.G.: Structural identifiability in linear time-invariant systems. IEEE Trans. Automat. Contr. **22**, 242–246 (1977)
53. Sage, A.P., Melsa, J.L.: System Identification. Academic, New York (1971)
54. Seber, G.A.F., Wild, C.J.: Nonlinear Regression. Wiley, Chichester (2003)
55. Shao, J., Tu, D.: The Jackknife and Bootstrap. Springer, New York (1995)
56. Thomaseth, K., Cobelli, C.: Generalized sensitivity functions in physiological system identification. Ann. Biomed. Eng. **27**(5), 607–616 (1999)
57. White, L.J., Evans, N.D., Lam, T.J.G.M., Schukken, Y.H., Medley, G.F., Godfrey, K.R., Chappell, M.J.: The structural identifiability and parameter estimation of a multispecies model for the transmission of mastitis in diary cows. Math. Biosci. **174**, 77–90 (2001)
58. Wodarz, D., Nowak, M.A.: Specific therapy regimes could lead to long-term immunological control of HIV. Proc. Natl. Acad. Sci. **96**, 14,464–14,469 (1999)

59. Wu, H., Zhu, H., Miao, H., Perelson, A.S.: Parameter identifiability and estimation of HIV/AIDS dynamics models. Bull. Math. Biol. **70**, 785–799 (2008)
60. Xia, X., Moog, C.M.: Identifiability of nonlinear systems with application to HIV/AIDS models. IEEE Trans. Automat. Contr. **48**, 330–336 (2003)
61. Yue, H., Brown, M., He, F., Jia, J., Kell, D.B.: Sensitivity analysis and robust experimental design of a signal transduction pathway system. Int. J. Chem. Kinet. **40**, 730–741 (2008)

Chapter 4
Application of the Unscented Kalman Filtering to Parameter Estimation

Adam Attarian, Jerry J. Batzel, Brett Matzuka, and Hien Tran

Abstract Filtering is a methodology used to combine a set of observations with a model to obtain the optimal state. This technique can be extended to estimate the state of the system as well as the unknown model parameters. Estimating the model parameters given a set of data is often referred to as the inverse problem. Filtering provides many benefits to the inverse problem by providing estimates in real time and allowing model errors to be taken into account. Assuming a linear model and Gaussian noises, the optimal filter is the Kalman filter. However, these assumptions rarely hold for many problems of interest, so a number of extensions have been proposed in the literature to deal with nonlinear dynamics. In this chapter, we illustrate the application of one approach to deal with nonlinear model dynamics, the so-called unscented Kalman filter. In addition, we will also show how some of the tools for model validation discussed in other chapters of this volume can be used to improve the estimation process.

A. Attarian · H. Tran (✉)
Center for Research in Scientific Computation and Department of Mathematics, Box 8205,
North Carolina State University, Raleigh, NC 27695
e-mail: arattari@unity.ncsu.edu; tran@math.ncsu.edu

J.J. Batzel
Institute for Mathematics and Scientific Computing, University of Graz, and Institute of
Physiology, Medical University of Graz, A 8010 Austria
e-mail: jerry.batzel@uni-graz.at

B. Matzuka
Biomathematics Graduate Program, Box 8203, North Carolina State University, Raleigh,
NC 27695
e-mail: bjmatzuk@ncsu.edu

J.J. Batzel et al. (eds.), *Mathematical Modeling and Validation in Physiology*,
Lecture Notes in Mathematics 2064, DOI 10.1007/978-3-642-32882-4_4,
© Springer-Verlag Berlin Heidelberg 2013

4.1 Introduction

In modeling biological processes, modelers frequently wish to relate biological parameters characterizing a model, θ, to collected observations making up some data set, y. In this paper, it is assumed that the relationship between θ and y is described by a nonlinear function G

$$G(\theta) = y. \tag{4.1}$$

For example, consider a simple model for the concentration of a drug introduced in a biological system [2]

$$\frac{dx(t)}{dt} = -ax(t) + bu(t),$$

$$y(t) = cx(t),$$

where $x(t)$ denotes the concentration of a drug, $u(t)$ is the input injection of the drug and $y(t)$ the temporal measurement of the drug concentration, which is assumed to be proportional to $x(t)$. Assuming that $x(0) = 0$, the solution, which is computed from the variation of constants formula, is given by

$$y(t) = cb \int_0^t e^{-a(t-s)} u(s)\, ds \equiv G(\theta), \tag{4.2}$$

where $\theta = (a, b, c)$. The forward problem is given θ find y. In this paper, our focus is on the inverse problem, that is, given observations $y(t)$ find the model parameters θ. In general, inverse problems are *ill-posed*. That is, there may be no mathematical model that exactly fits the data. The classical example of an ill-posed inverse problem is the Fredholm integral equation of the first kind

$$\int_0^t g(t, s)\theta(s)\, ds = y(t), \quad c \leq t \leq d,$$

where the kernel g and the function y are given. This problem may be treated as an inverse problem for the unknown function θ. Even in the simplest case where the kernel $g(t, s) \equiv 1$, the inverse problem

$$\int_0^1 \theta(s)\, ds = y(t)$$

has no solution unless the measurement $y(t)$ is a constant. Secondly, even if a solution to the inverse problem

$$\int_0^1 \theta(s)\,ds = c_1,$$

where c_1 is a constant, exists, the solution is not unique. Finally, even if we do not encounter existence or uniqueness issues, *instability* is a fundamental feature of the inverse problem. Consider the following linear system of equations, written in the form $A\theta = y$, where

$$A = \begin{bmatrix} 1 & 1 \\ 1 & 1.0000001 \end{bmatrix}, \quad y = \begin{bmatrix} 2 \\ 2 \end{bmatrix}. \tag{4.3}$$

This simple linear inverse problem has a unique solution $\theta = (2, 0)$. Now, assuming that observations y are corrupted by noises and are now given by

$$y = \begin{bmatrix} 2 \\ 2.0000001 \end{bmatrix}.$$

The unique solution to the linear inverse problem is $\theta = (1, 1)$. Here, small changes in measurements, y, give rise to large changes in θ. This ill-posedness is referred to as *ill-conditioned*. It is important to note that this type of instability in the solution process lies in the mathematical model itself (and not in the particular algorithm used to solve the problem). In fact, this is a well known problem in numerical linear algebra, where the relative error propagation from the data to the solution is controlled by the *condition number* [8]: if Δy is a variation of y and $\Delta\theta$ the corresponding variation of θ, then for any compatible matrix and vector norms

$$\frac{\|\Delta\theta\|}{\|\theta\|} \le \mathrm{cond}(A)\frac{\|\Delta y\|}{\|y\|},$$

where the condition number of the matrix A is defined by

$$\mathrm{cond}(A) = \|A\|\|A^{-1}\|.$$

Since the relative error in the solution, $\|\Delta\theta\|/\|\theta\|$, is less than or equal to the condition number multiplied by the relative error in the data, smaller values of $\mathrm{cond}(A)$ are more desirable. For the matrix A defined by (4.3) and using the 2-norm, the condition number of A is $\mathrm{cond}(A) = 4 \times 10^7$! (not a small number). Clearly, if the inverse problem is ill-conditioned, we cannot trust the solution as much (as solution to the linear system of equations is susceptible to large numerical errors).

Assuming that the inverse problem (4.1) is well-posed (i.e., the solution exists, is unique, and the inverse mapping $y \mapsto \theta$ is continuous), there are several commonly used techniques for parameter estimation. These include the ordinary least squares (OLS), the generalized least squares (GLS), the maximum likelihood estimate (MLE) [1] as well as the Kalman filtering. In this paper, we consider the Unscented

Kalman Filtering (UKF) for the on-line estimation of model parameters. The UKF, which was developed by Julier and Uhlmann [6], is an efficient sampling method for nonlinear filtering. The approach is based on a sigma-point implementation called the *unscented transform* to propagate the means and covariance matrices through nonlinear transformations. The paper also shows how local sensitivity and identifiability analyses can be performed *a priori* to increase the reliability and accuracy of parameter estimates.

4.2   The Unscented Kalman Filter

There exist many well-established methods to solve the parameter estimation problem. These include general minimization procedures, least-squares type techniques as well as Bayesian based approaches. In this paper, we consider the Kalman filter based approach to solve the parameter estimation problem because it possesses several desirable properties, among them: it takes explicitly into account the measurement errors, it takes measurement data into account incrementally (on-line estimation), it is an efficient and simple to implement computational tool, and it can take into account *a priori* information (if available). Kalman filter is useful for estimation of both dynamic systems and stationary problems. It is recursive in the sense that at each step, the updated estimate is found through the previously estimated state and the observation data at that step.

In 1960, R.E. Kalman published his seminal paper describing an efficient recursive solution to the discrete, linear filtering problem from a series of noisy measurements [7]. Since its discovery over 40 years ago, much research has gone into refining its estimation accuracy and into its extensions to highly nonlinear models. Estimation in nonlinear systems is very important because almost all practical systems, from engineering to biological applications, involve nonlinearities of one kind or another. Probably the most widely used estimation algorithm for nonlinear systems is the extended Kalman filter (EKF). In the EKF, a linearization is carried out on the state-space model and the observation model with respect to the state. While the initial state estimation portion of the algorithm is unchanged, these Jacobians are utilized in the initial covariance estimation and in the estimation update for both the covariance and the state. To elaborate on this, we consider the problem of state estimation of a discrete-time nonlinear dynamical system of the form

$$x_{k+1} = f(k, x_k) + w_k, \tag{4.4}$$

$$y_{k+1} = h(k + 1, x_{k+1}) + v_{k+1}, \tag{4.5}$$

where $x \in R^n$ is the state, $y \in R^m$ the output observation, and the random variables w_k and v_k are the model and observation noises, respectively. They are assumed to be uncorrelated and that they are all Gaussian white noise sequences

$$p(w_k) \sim N(0, Q_k),$$
$$p(v_{k+1}) \sim N(0, R_{k+1}).$$

In general, the model noise covariance Q_k and measurement noise covariance R_k matrices are time dependent. The complete set of EKF equations is described below.

Initialize : for $k = 0$

$$\hat{x}_0 = E[x_0]$$
$$P_0 = E\big[(x_0 - E[x_0])(x_0 - E[x_0])^\mathsf{T}\big]$$

EKF Algorithm : for $k = 1, 2, 3, \ldots$

State Estimation

$$\hat{x}_k^- = f(k - 1, \hat{x}_{k-1})$$

Covariance Estimation

$$P_k^- = F_{k,k-1} P_{k-1} F_{k,k-1}^\mathsf{T} + Q_{k-1},$$

where F is the linearization of the model

$$F_{k|k-1} = \left. \frac{\partial f(k - 1, x)}{\partial x} \right|_{x = \hat{x}_{k-1}}$$

Kalman Gain Matrix

$$K = P_k^- H_k^\mathsf{T} \big[H_k P_k^- H_k^\mathsf{T} + R_k \big]^{-1},$$

where H is the linearization of the observation

$$H_k = \left. \frac{\partial h(k, x)}{\partial x} \right|_{x = \hat{x}_k^-}$$

State Estimation Update

$$\hat{x}_k = \hat{x}_k^- + K\big[y_k - h(k, \hat{x}_{k-1}^-) \big]$$

Covariance Estimation Update

$$P_k = P_k^- \big[I - K H_k \big].$$

As with the basic linear discrete Kalman filter, the measurement update corrects the state and covariance estimates with the measurement y_k. An important feature of the EKF is that the Jacobian H_k in the Kalman gain matrix K serves to propagate the relevant component of the measurement information. Although, in general, the EKF preserves the simplicity and computationally efficiency recursive update form of the Kalman filter, it suffers several limitations. To begin, first-order linearization of the nonlinear system can be applied only if the Jacobian matrices exist. Even if Jacobian matrices exist, linearized transformations are only reliable if the error propagation can be well approximated by a linear function on the time scale of the updates. If this condition does not hold, it can introduce large errors in the true posterior mean and covariance, which may lead to suboptimal performance and sometimes divergence of the filter [9].

To overcome the above limitations of the EKF, the unscented transformation (UT) was developed as a method to propagate the mean and covariance information through nonlinear transformation [4, 6]. In particular, the state and errors are again assumed to be Gaussian random variables (GRV). In the EKF, the GRV is propagated analytically through the first-order linearization of the nonlinear system. In the UKF, the state distribution is now specified using a minimal set of carefully chosen sample points. The sample points, when propagated through the nonlinear system, capture the posterior mean and covariance of the GRV to the second order accuracy (Taylor series expansion) for any nonlinearity [5].

We begin the discussion on the UKF by elaborating on the unscented transformation (UT). It is a method used to calculate the statistics of a random variable that undergoes a nonlinear transformation [6]. To be more specific, it is assume that $\mathbf{x}$ is a random variable with dimension L, which has mean $\hat{x}$ and covariance, P_x. This random variable propagates through a nonlinear function, $\mathbf{y} = g(\mathbf{x})$. To calculate the statistics of $\mathbf{y}$, we form a matrix, $\mathscr{X}$, of $2L + 1$ sigma vectors, $\mathscr{X}_i$. This matrix, $\mathscr{X}$ is constructed as follows,

$$\mathscr{X} = \left[\hat{x}, \quad \hat{x} + \left(\sqrt{(L + \lambda)P_x}\right)_i, \quad \hat{x} - \left(\sqrt{(L + \lambda)P_x}\right)_i\right], \qquad (4.6)$$

where $\hat{x}$ is a column vector and $\left(\sqrt{L + \lambda}\,P_x\right)_i$ is the ith row of the matrix square root. The corresponding weights (W_i) for the sigma vectors are as follows,

$$W_0^{(m)} = \frac{\lambda}{(L + \lambda)}, \qquad (4.7)$$

$$W_0^{(c)} = \frac{\lambda}{(L + \lambda)} + (1 - \alpha^2 + \beta), \qquad (4.8)$$

$$W_i^{(m)} = W_i^{(c)} = \frac{1}{2(L + \lambda)}, \qquad i = 1, \ldots, 2L, \qquad (4.9)$$

where $\lambda = \alpha^2(L + \kappa) - L$ is a scaling parameter. The constant α is used to determine the spread of the sigma points around $\hat{x}$, and takes values between $1 \geq \alpha \geq 1 \times 10^{-4}$ depending on the problem under consideration. The parameter κ is a secondary scaling parameter that is optimally set to $3 - L$ [5]. Lastly, β is used to incorporate prior knowledge of the distribution of $\mathbf{x}$ (for Gaussian distributions, $\beta = 2$ is optimal) [9]. The sigma vectors are then propagated through the nonlinear function accordingly,

$$y_i = g(\mathscr{X}_i), \qquad i = 0, \ldots, 2L, \qquad (4.10)$$

with the mean and covariance of $\mathbf{y}$ being approximated using a weighted sample mean and covariance of the posterior sigma points,

$$\hat{y} \approx \sum_{i=0}^{2L} W_i^{(m)} y_i, \tag{4.11}$$

$$P_y \approx \sum_{i=0}^{2L} W_i^{(c)} (y_i - \hat{y})(y_i - \hat{y})^{\mathsf{T}}. \tag{4.12}$$

Extending the UT for use in the Unscented Kalman filter (UKF) is straightforward. For additive Gaussian noises, the UKF equations are shown below, where $\mathscr{X}_{i,k|k-1}$ is the notation for the ith column of the matrix $\mathscr{X}_{k|k-1}$ (it is noted that we used $i = 0$ to denote the first column).

Initialize : for $k = 0$

$$\hat{x}_0 = E[x_0]$$
$$P_0 = E[(x_0 - E[x_0])(x_0 - E[x_0])^{\mathsf{T}}]$$

UKF Algorithm : for $k = 1, 2, 3, \ldots$

Calculate Sigma Points

$$\mathscr{X}_{k-1} = \left[\hat{x}, \ \hat{x} + \left(\sqrt{(L + \lambda)P_{k-1}}\right)_i, \ \hat{x} - \left(\sqrt{(L + \lambda)P_{k-1}}\right)_i \right]$$

time-update equations

$$\mathscr{X}_{k|k-1}^* = f(k - 1, \mathscr{X}_{k-1})$$

$$\hat{x}_k^- = \sum_{i=0}^{2L} W_i^{(m)} \mathscr{X}_{i,k|k-1}^*$$

$$P_k^- = \sum_{i=0}^{2L} W_i^{(c)} \left(\mathscr{X}_{i,k|k-1}^* - \hat{x}_k^-\right)\left(\mathscr{X}_{i,k|k-1}^* - \hat{x}_k^-\right)^{\mathsf{T}} + Q_{k-1}$$

time-update equations for observation: augment sigma points

$$\mathscr{X}_{k|k-1} = \left[\mathscr{X}_{0,k|k-1}^*, \ \mathscr{X}_{0,k|k-1}^* + \left(\sqrt{(L + \lambda)P_k^-}\right)_i, \right.$$

$$\left. \mathscr{X}_{0,k|k-1}^* - \left(\sqrt{(L + \lambda)P_k^-}\right)_i \right]$$

$$Y_{k|k-1} = h(k, \mathscr{X}_{k|k-1})$$

$$\hat{y}_k^- = \sum_{i=0}^{2L} W_i^{(m)} Y_{i,k|k-1}$$

state and measurement-update equations

$$P_{\hat{y}_k \hat{y}_k} = \sum_{i=0}^{2L} W_i^{(c)} \left(Y_{i,k|k-1} - \hat{y}_k^-\right)\left(Y_{i,k|k-1} - \hat{y}_k^-\right)^{\mathsf{T}} + R_k$$

$$P_{x_k y_k} = \sum_{i=0}^{2L} W_i^{(c)} \left(\mathcal{X}_{i,k|k-1} - \hat{x}_k^- \right) \left(Y_{i,k|k-1} - \hat{y}_k^- \right)^{\mathsf{T}}$$

$$K_k = P_{x_k y_k} P_{\hat{y}_k \hat{y}_k}^{-1}$$

$$\hat{x}_k = \hat{x}_k^- + K_k \left(y_k - \hat{y}_k^- \right)$$

$$P_k = P_k^- - K_k P_{\hat{y}_k \hat{y}_k} K_k^{\mathsf{T}}$$

The complexity of the UKF algorithm is of order L^3, which is the same complexity as the EKF. There are multiple variations of this algorithm for implementation purposes, one of which is the square root unscented Kalman filter [4]. In addition, a more general form of the UKF is also available, which does not assume additive noise. In this case, the state is redefined as the concatenation of the original state and the model and measurement noise processes [5].

4.3 Parameter Estimation

The parameter estimation problem (4.1) can be solved by the UKF by writing a new state-space representation,

$$\theta_{k+1} = \theta_k + r_k, \tag{4.13}$$

$$d_k = G(\theta_k) + n_k, \tag{4.14}$$

where θ is the parameter, assumed to be a stationary process with respective noise, r_k, and G is the nonlinear observation function, with respective noise, n_k. It may not be immediately obvious what benefit the UKF will have in parameter estimation, since the state space is linear and stationary. However, since the observation function is nonlinear, the UKF is directly applicable. Modifying the UKF algorithm as presented in Sect. 4.2, we present the equations describing the parameter estimation UKF as follows,

Initialize : for $k = 0$

$$\hat{\theta}_0 = E[\theta_0]$$

$$P_0 = E[(\theta_0 - E[\theta_0])(\theta_0 - E[\theta_0])^{\mathsf{T}}]$$

Parameter estimation UKF algorithm : for $k = 1, 2, 3, \ldots$

time-update equations

$$\hat{\theta}_k^- = \hat{\theta}_{k-1}$$

$$P_{\theta_k}^- = P_{\theta_{k-1}} + R_{k-1}^r$$

$$W_{k|k-1} = \left[\hat{\theta}_k^-, \ \hat{\theta}_k^- + \sqrt{(L+\lambda)P_{\theta_k}^-}_i, \ \hat{\theta}_k^- - \sqrt{(L+\lambda)P_{\theta_k}^-}_i \right]$$

$$D_{k|k-1} = G(W_{k|k-1})$$

$$\hat{d}_k = \sum_{i=0}^{2L} W_i^{(m)} D_{i,k|k-1}$$

parameter and measurement-update equations

$$P_{\hat{d}_k \hat{d}_k} = \sum_{i=0}^{2L} W_i^{(c)} \left(D_{i,k|k-1} - \hat{d}_k \right) \left(D_{i,k|k-1} - \hat{d}_k \right)^{\mathsf{T}} + R_k^e$$

$$P_{\theta_k d_k} = \sum_{i=0}^{2L} W_i^{(c)} \left(W_{i,k|k-1} - \hat{\theta}_k^- \right) \left(D_{i,k|k-1} - \hat{d}_k \right)^{\mathsf{T}}$$

$$K_k = P_{\theta_k d_k} P_{\hat{d}_k \hat{d}_k}^{-1}$$

$$\hat{\theta}_k = \hat{\theta}_k^- + K_k(d_k - \hat{d}_k)$$

$$P_{\theta_k} = P_{\theta_k}^- - K_k P_{\hat{d}_k \hat{d}_k} K_k^{\mathsf{T}}$$

$$R_k^r = \left(\lambda^{-1} - 1 \right) P_{\theta_k}$$

where R^e is the measurement-noise covariance and R^r is the process noise covariance, and λ is the "forgetting factor" as defined in the recursive least-squares algorithm [5]. There are several options that can be used to anneal the process covariance matrix, R^r, we use the forgetting factor approach as it is most appropriate for on-line estimation problems [5].

4.3.1 Dual Estimation

Dual estimation problems consist of estimating both the states, x_k, and the parameters, θ_k, given noisy data, y_k. This can be done using the technique of joint filtering, where one concatenates the states and the parameters into a single augmented "state" vector [10]. We obtain

$$\dot{x} = f(t, x, \theta),$$

$$\dot{\theta} = 0,$$

where x is a vector of states, and θ is a vector of parameters. This method can be utilized with any of the filters we have discussed, but it has some shortcomings. By using joint estimation, you can greatly increase the number of states if you have a substantial number of parameter, which is the case for most biological applications. In addition, the errors from the state are propagated into the parameter, which is subsequently propagated back into the state. This can quickly lead to inaccurate results or divergence of the filter.

Another method is to use a dual filter [10]. This is done by running two filters concurrently; the state filter estimates the state using the current parameter estimate, $\hat{\delta}_k^-$, while the parameter filter estimates the parameters using the current state estimate, $\hat{x}_k^-$. Unifying the UKF state estimation filter with the UKF parameter estimation filter, we can construct a dual UKF. This has its advantages as we are not increasing the number of states for estimation. Also, the errors will not feedback into the next estimate and continue to build up, as can be encountered in the joint estimation method. In the next section, we present two sample problems to illustrate the benefits of this approach.

4.4 Estimation Examples

To illustrate the effectiveness of the dual estimation algorithm, we have performed numerical experiments on two nonlinear dynamical systems. The first example is a second order differential equation describing a "hard"nonlinear spring. The second test problem is a three-dimensional state-space model of HIV dynamics with six parameters. In both examples, simulated data were generated from the model using "true" parameter values.

4.4.1 A Hard Nonlinear Spring Model

In this example, dual UKF is used to estimate both states and parameters in a spring-mass-dashpot system with a "hard" nonlinear spring given by

$$\ddot{x} + \gamma\dot{x} + kx + bx^3 = 0, \tag{4.15}$$

where $\gamma > 0$ is the damping coefficient and $k > 0$ and $b > 0$ are spring constants. We want to estimate the parameters, $\theta = (k, b, \gamma)$, while simultaneously estimating the states $(x, \dot{x})$. The full state, $y = (x, \dot{x})$, is observed. Parameter and state estimation results using the dual UKF are depicted in Figs. 4.1 and 4.2, respectively. For the plots in Fig. 4.1 we also included curves that are plus or minus 3 standard deviations of the estimated parameter means if we ignore the covariances between parameters. All states and parameters appear to be estimated relatively well. The initial noise covariances for both the model and measurement were set to $0.0005I$, where I is a 2×2 identify matrix. It is noted that the initial noise covariances play a crucial role in the accuracy and convergence of the UKF. The final converged parameter estimates are given in the following table:

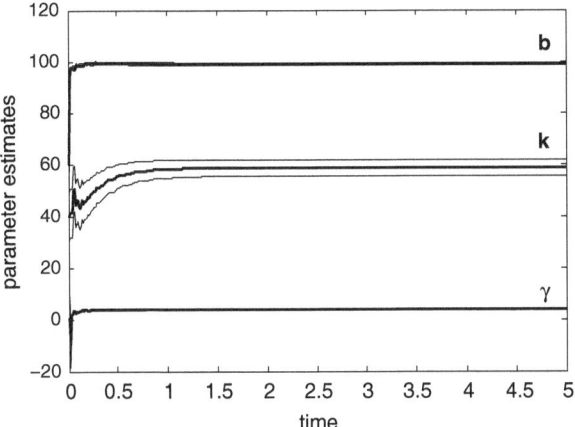

Fig. 4.1 Convergence of parameter estimation for the nonlinear spring model using the dual UKF

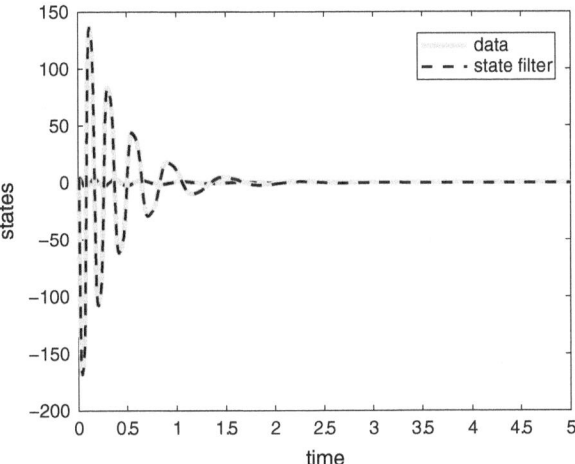

Fig. 4.2 Comparison of the true states (*solid*) versus the dual UKF state estimation for the nonlinear spring model

	k	b	γ
"True model"	60	100	4
"UKF estimate"	59.7	96.985	3.775

4.4.2   An HIV Model

The second numerical experiment is a simple, three-dimensional mathematical model describing HIV dynamics [11]

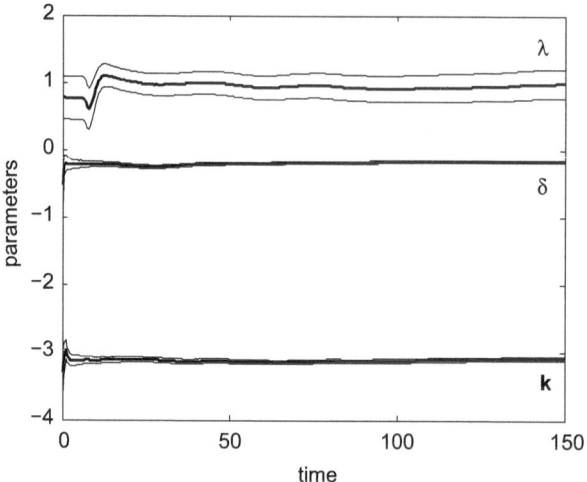

Fig. 4.3 Convergence of parameter estimation for the HIV dynamics using the dual UKF

$$\dot{T} = \lambda - dT - kVT, \qquad (4.16)$$

$$\dot{T^*} = kTV - \delta T^*, \qquad (4.17)$$

$$\dot{V} = N\delta T^* - cV, \qquad (4.18)$$

where λ is the recruitment of uninfected T-cells, T, and d is the per capita death rate of uninfected cells. In the presence of HIV, T-cells become infected with the rate of infection is given by kVT, where k is the infection rate constant. In the second equation, infected cells T^* are produced at a rate of kTV and died at a rate of δT^*. The third equation describes the dynamics of free virions V. The free virions are produced by the infected T-cells at a rate constant N and died at a rate of cV. For observations, all three variables are assumed to be measurable.

To begin, we consider the parameter estimation problem of estimating all six parameters in the model, $\theta = (\lambda, d, k, \delta, N, c)$. However, the UKF algorithm failed to converge. This is not unexpected as it was shown that, using sensitivity analysis and subset selection as described in Chap. 1, only λ, k and δ are identifiable [3]. Therefore, we fixed $d = 0.01$, $N = 100$ and $c = 3$ and estimated $\delta = (\lambda, k, \delta)$ simultaneously with the states using the dual UKF. The estimation results for both the parameters and states are depicted in Figs. 4.3 and 4.4, respectively. In Fig. 4.3, where log-scale is used on the vertical axis, plots of the curves that are plus or minus 3 standard deviations of the estimated parameter means are also given if we ignore the cross covariances. Figures 4.3 and 4.4 illustrate that dual UKF performed very well in estimating both the states and identifiable model parameters. Both figures

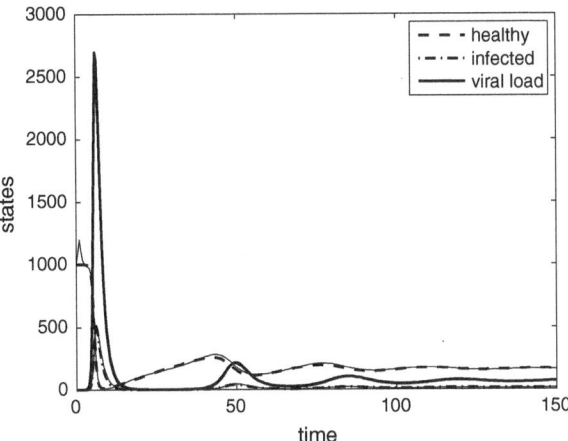

Fig. 4.4 Comparison of the true states (*solid*) versus the dual UKF state estimation for the HIV dynamics

showed that the UKF converges for both the states and parameter estimates. The final converged identifiable parameters are as follows:

	λ	k	δ
"True model"	10	8×10^{-4}	0.7
"UKF estimate"	9.5	8.2×10^{-4}	0.701

It is noted that because the values of identifiable model parameters are in dramatically different scales, varying from 0.7 to 10^{-4}, we transformed them into their log-scales before filtering.

In conclusion, the UKF was shown to be a promising estimation method for the dual estimation problem. In addition, it was demonstrated that selection of parameters, which can be estimated accurately from measurements, is a prerequisite for successful estimation. In particular, local sensitivity and subset selection analyses are important techniques to identify parameter identifiability as well as parameter space reduction. In the future work, the parameter estimation techniques as proposed in this paper will be studied in the context of modeling of biological systems using experimental and/or clinical data.

Acknowledgements The author Hien Tran, gratefully acknowledges partial financial support from the National Institute of Allergy and Infectious Diseases (grant number R01AI071915-07) and from the National Science Foundation (grant number DMS-1022688).

References

1. Banks, H.T., Tran, H.T.: Mathematical and Experimental Modeling of Physical and Biological Processes. Chapman & Hall/CRC, Boca Raton, Florida (2009)
2. Cobelli, C., DiStefano, J.J.: Parameter and structural identifiability concepts and ambiguities: a critical review and analysis. Am. J. Physiol. Regul. Integr. Comp. Physiol. **239**, 7–24 (1980)

3. Fink, M., Attarian, A., Tran, H.T.: Subset selection for parameter estimation in an HIV model. In: Proceedings in Applied Mathematics and Mechanics, vol. 7, pp. 1121,501–1121,502 (2008)
4. Grewal, M.S., Andrews, A.P.: Kalman Filtering: Theory and Practice Using MATLAB. Wiley, Hoboken, New Jersey (2008)
5. Haykin, S.: Kalman Filtering and Neural Networks. Wiley, Toronto, Canada (2001)
6. Julier, S.J., Uhlmann, J.K.: A new extension of the Kalman filter to nonlinear systems. In: Proceedings of of AeroSense: The 11th International Symposium on Aerospace/Defense Sensing, Simulation and Controls, Multi Sensor Fusion, Tracking and Resource Management, pp. 182–193. SPIE, Orlando, Florida (1997)
7. Kalman, R.E.: A new approach to linear filtering and prediction problems. ASME J. Basic Eng. **82**, 34–45 (1960)
8. Trefethen, L., Bau, D.: Numerical Linear Algebra. SIAM, Philadelphia, Pensylvania (1997)
9. Wan, E.A., van der Merwe, R.: The unscented kalman filter for nonlinear estimation. In: Proceedings of of Symposium 2000 on Adaptive Systems for Signal Processing, Communication and Control (AS-SPCC). IEEE, Lake Louise, Alberta, Canada (2000)
10. Wan, E.A., van der Merwe, R., Nelson, A.T.: Dual estimation and the unscented transformation. In: Solla, S., Leen, T., Muller, K.R. (eds.) Advances in Neural Information Processing Systems, vol. 12, pp. 666–672. MIT, Cambridge, Massachusetts (2000)
11. Xia, X.: Estimation of HIV/AIDS parameters. Automatica **39**, 1983–1988 (2003)

Chapter 5
Integrative and Reductionist Approaches to Modeling of Control of Breathing

Chung Tin and Chi-Sang Poon

Abstract Integration and reductionism represent two different modeling approaches in understanding the working mechanism of a system. Traditionally, biological modeling has relied on reductionism, which has the advantage of decomposing the intrinsic complexity of biological systems into smaller subsystems to ease understanding. However, as more information accumulates, it becomes increasingly difficult to integrate these components systematically to explain more complex behavior, particularly when multiple determining factors co-exist. In this chapter, we compare the reductionist and integrative views of biological modeling. The value of integrative modeling is discussed from the perspective of engineering and physics point of view. The need for an integrative approach is illustrated in the context of respiratory control.

5.1 Introduction

Systems biology is a classic discipline which has its root when Norbert Wiener [43] first coined the term "cybernetics". Engineers have a long history of getting inspirations from biology. The wings of Icarus might be considered one of the first "flying machine" models of birds. Although it is just a Greek myth, it shows human's innate appreciation for modeling biological designs. Today, mathematical or computer modeling of biological systems is used to improve our understanding

C. Tin (✉)
Department of Mechanical and Biomedical Engineering, City University of Hong Kong, Hong Kong, China
e-mail: chungtin@cityu.edu.hk

C.-S. Poon
Institute for Medical Engineering and Science, E25-250, Massachusetts Institute of Technology, Cambridge, MA 02139, USA
e-mail: cpoon@mit.edu

J.J. Batzel et al. (eds.), *Mathematical Modeling and Validation in Physiology*, Lecture Notes in Mathematics 2064, DOI 10.1007/978-3-642-32882-4_5, © Springer-Verlag Berlin Heidelberg 2013 of biological phenomena in their full complexity. The traditional approach adopted by biologists for analyzing biological systems is via a strategy of reductionism [1, 12, 27, 33, 35]. A complementary approach is via the method of integrative modeling based on physical principles. This integrative framework will likely facilitate the collaboration of researchers from such diverse disciplines as biology, chemistry, physiology, engineering, computer science, and mathematics to bring out the best of systems biology.

In this chapter, we examine the criteria of a good mechanistic model based on reductionist or integrative approaches. We then discuss the reductionist view and integrated view in the modeling of the respiratory control system. Similar arguments are also applicable to cardiovascular models. (See chapters contributed by Ottesen et al., Heldt et al. and Brown in this volume (Chaps. 10, 2 and 9) for discussion on cardiovascular systems).

5.2   Reductionist View of Biological Modeling

Reductionism is a divide-and-conquer approach to tackle complex phenomena by parsing the problem into smaller, simpler and more tractable components. It has served as an important guiding principle which has proved tremendously useful in understanding many biological problems. However, there are also drawbacks to this approach. Organisms are definitely more than just the sum of their individual parts. The reductionist approach makes several key assumptions in studying biological systems:

1. A singular factor determines each behavior.
   The goal of reductionism is to isolate a single factor to account for each observed behavior. The pharmaceutical community believes that there is a single malfunction for each disease that needs to be cured. Hence, a miracle drug targeting the site of malfunction will solve the problem. Although this is a reasonable assumption for many cases, it does not always apply for more complex situations. For instance, how living habit and genetic factors together affect a person's health [1]? A young immuno-compromised man with pneumococcal pneumonia will get the same antibiotics treatment as an elderly woman with the same infection. A "personalized medicine" approach is more desirable but is infeasible with the reductionist view [41].
2. Response is linearly additive of several factors.
   Reductionism partitions the problem into many pieces, each studied separately. The total response is then viewed as the superposition of these individual effects. This approach is easily executable but it neglects any nonlinear interactions among components.
3. Only static or steady-state response matters.
   Cannon [2] in his book, The Wisdom of the Body first coined the term "homeostasis", describing how our body maintains stability and constancy

robustly in the face of stress. However, the focus on constancy in the steady state ignores the ubiquity of dynamic behaviors like oscillations and chaos, which may also be important for our body to function. For instance, circadian rhythms and heart rate variability are examples of periodic and chaotic behaviors that may be integral to the maintenance of homeostasis.

5.2.1 The Physiome Project and Multiscale Model

The Physiome Project is one of the major efforts to integrate physiology and engineering [10]. It provides a coherent framework to integrate the vast amount of DNA sequences, protein structures and signal transduction pathways data into mathematical models which can facilitate the analysis of complex interaction among these systems of different sizes and time scales. It is an important international effort to apply systems engineering approach to physiological modeling and develop standard languages to facilitate sharing of models and information among different research teams.

The Physiome Project is premised on the notion of "multiscale modeling" in which components on different spatial and temporal scales are integrated into a single model. This is accomplished by combining physiology from molecular level to cell, organ and systems levels. Although the term "multiscale modeling" has been often interpreted as multi-structural modeling across spatiotemporal scales [42], it is increasingly recognized that this modeling approach has its limitations also as it may risk under- or overfitting of parameters and the resulting errors could propagate between models and across spatiotemporal scales (although the benefits of model integration are sometimes considered to outweigh such error costs) [18,29]. Broadly speaking, multiscale modeling could involve any analytical approach that allows increased predictivity (without sacrificing robustness) by extension or generalization of any modeling "scale", including structural or spatiotemporal scale, analytical scale [18] or computational scale [36].

Current models of chemoreflex or exercise hyperpnea are predictive of the ventilatory responses to chemical and metabolic challenges respectively but are generally non-predictive of breathing pattern variables. On the other hand, classical optimization models from the 1950s that consider work of breathing as an objective function for respiratory control have been limited to the prediction of pattern variables such as respiratory frequency but not total ventilation (e.g., [15]). Generalized optimization models that integrate both chemical challenges and work of breathing into an overall control objective [25] can be highly predictive and multiscale (multi-functional scale) in that they can simultaneously predict both total ventilation and breathing pattern responses (including instantaneous inspiratory motor output) to a combination of chemical, mechanical and metabolic challenges (see Eq. (5.5)) as encountered in a variety of physiological and clinical states (on a *multi-disease scale*).

5.3   Engineering and Physics View of Integrative Modeling

The field of physics is traditionally highly reductionistic and multiscale in that it seeks to identify the elemental particles or forces of nature on the smallest spatiotemporal scale that underlie physical phenomena on a macroscale. On the other hand, it is also highly integrative in its quest to discover the physical laws (Newton's law, thermodynamics etc.) that unify diverse physical phenomena from quantum to cosmic scales in order to predict complex behaviors based on first principles. Integrative modeling aims to describe how different system components interacting based on physical principles may give rise to emergent behavior of the system. Elucidation of the integrative mechanisms is important for a conceptual understanding on the working principle of the system behavior, as opposed to an empirical or phenomenological (black box) model.

Engineering and physics approach to integrative modeling has a long history in studying mechanical, fluid, thermal, chemical, electronic and other physical systems. Engineering and physical system equations help to specify the behaviors of individual components and their interactions based on physical laws. The strength of such an integrative modeling approach is that it provides a quantitative description of the interaction of different variables in the system based on the physical mechanisms involved. This leads to an experimentally testable unified theory which can span multiple system scales. Hence, engineers have the right analytical tools for designing complex machines. There are several basic performance measures of any engineering control system: stability, robustness and sensitivity. Stability requires a system's states to remain within a finite space over time with a bounded input. Robustness states that a system can favorably maintain its performance under conditions that it is not designed for. Sensitivity states how small a change can a system detect and respond to.

Engineering systems theory has provided the necessary mathematical tools for us to design and evaluate the performance of control systems, and to understand the underlying mechanisms. During the 1940s, the introduction of frequency-domain methods (e.g., Nyquist, Bode) made it possible for engineers to design linear closed-loop control systems that satisfied some performance requirements. Root-locus method was then developed by Evans in the early 1950s. Nowadays, engineers commonly have to deal with multiple-input-multiple-output (MIMO) and nonlinear systems. Classical control theory, which deals with only linear single-input-single-output (SISO) systems, has become inadequate.

There are many reasons for studying nonlinear systems. First, linear control assumes a small range of operation. Its performance is inevitably compromised or the model may become unstable when the required operation range is large and hence nonlinearity becomes significant. Furthermore, some systems may not be linearizable even within a small working range. These "hard" nonlinearities include Coulomb friction, saturation, dead-zones, backlash and hysteresis. Nonlinear controllers are generally more robust to model uncertainties. Since most systems are inherently nonlinear, a good nonlinear controller model can be paradoxically simpler and more intuitive than a linear model [17, 32].

5.4 Top-Down (Integrative) vs Bottom-Up (Reductionist) Approach to Biological Modeling

Integration and reductionism represent two complementary approaches to study complex biological systems [47]. Reductionism lays the groundwork for biological modeling by providing experimental evidence of discrete system elements at the microscopic scale. This bottom-up approach is an efficient strategy to match observed microscopic evidence with observed macroscopic behavior. For instance, gene-targeting studies attempt to correlate defects of a single gene with abnormalities at the behavioral level. Although this approach will undoubtedly generate useful information for inductive reasoning, the correlation between microscopic and macroscopic events may not be straightforward for complex biological systems [5]. Thus, direct combination of prime data at the microscopic level may not always lead to understanding of biological mechanisms and prediction of behaviors at the macroscopic level. On the other hand, top-down approach translates integrated phenomena at the macroscopic scale into hypotheses about various microscopic system elements and their interactions. Such deductive reasoning helps to formulate a general integrative principle for the system under study.

Both top-down and bottom-up approaches are necessary for studying biological systems which are typically highly complex and multidimensional. Over the past decades, biological research has accumulated so much data that the underlying system is beyond comprehension without formal analytical tools. Lazebnik [12] tells how a biologist might attempt to fix a radio in a different way than does an engineer. Thus, biologists may construct a system diagram of the radio in the form of "all-too-well recognizable diagrams, in which a protein is placed in the middle and connected to everything else with two-way arrows". It is of very slim chance that a radio can be fixed this way. The key difference is that engineers have a formal set of language to describe an electronic device systematically and quantitatively (resistance, capacitance, Kirchhoff's circuit laws, etc). Hence, a trained electronics engineer can unambiguously comprehend a circuit diagram of the radio or any other electronic devices. This shows that reductionism becomes powerless when dealing with multi-dimensional problems.

Yet, top-down approach is necessarily speculative and controversial, especially when the system is complex. Advances in bottom-up approach facilitate top-down investigation. On the other hand, controversies in top-down approach raise scientific inquiries which can lead to potential revolutionary ideas. The process of proving or disproving these ideas will lead to new insights for reducing the search space in the bottom-up approach. Hence, proper combination of top-down and bottom-up approaches is necessary to elucidate complex biological systems. A model is fully validated when bottom-up meets top-down.

5.5   Criteria of a Good Mechanistic Model

Biologists and engineers see modeling in different ways. So then, what is a good model? Pugh and Andersen [27] suggested seven characteristics of a good model. We summarize below four key criteria that we think are most important for a model to be useful.

1. Embody first principles of the system based on physical and chemical laws.

    A good model should be physically realistic and reasonable. It also means the model should incorporate all applicable first principles *quantitatively* and *dynamically*.

2. A minimum model.

    A good model does not have to include every fine details of the system. A model is a tool to provide the intuition to understand the mechanisms underlying a system. An overly complex model may obscure understanding of the system. A *simple* yet sufficient model should be the best model.

3. Summarize existing results with mechanistic insight.

    The sine qua non of a good model is that it is consistent with existing observed phenomena, both qualitatively and quantitatively. A good model should be *general* enough to be consistent with a wide variety of system behaviors. A curve fit to the data is not sufficient; a good model should provide mechanistic insight.

4. Predict behaviors distinct from the data the model is based on.

    A useful model necessarily has to *predict* phenomena beyond those it is based on. If a model has no prediction power, it is simply a curve fitting machine. A good model can predict a wide range of different phenomena that may be surprising or even counterintuitive, such as predicting the Earth is round instead of flat. The prediction will help to guide further experiments to uncover more mechanisms underlying the system.

5.6   Control of Breathing: Reductionism vs Integration

5.6.1   Limitations of Classical Reflex Models: A Case for Sensorimotor Integration

The respiratory control system is a multi-input multi-output nonlinear system [23]. The system comprises a mechanical plant (respiratory mechanics and muscles) and a chemical plant (pulmonary gas exchange) that are controlled by respiratory neurons in the medulla and pons regions of the brainstem. Extensive studies over the past century have accumulated an enormous collection of knowledge about the respiratory system. For instance, Duffin has presented a detailed model of the

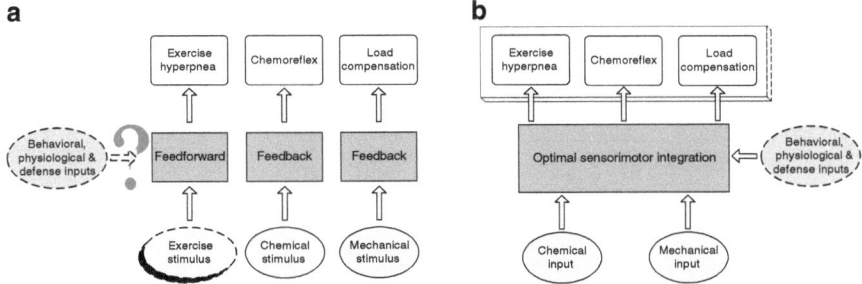

Fig. 5.1 Two views for respiratory control. (**a**) Classical reflex model assumes additive, reducible and superposable characteristics of chemical, mechanical and exercise stimuli. (**b**) The optimization model integrates various afferent-efferent signals in a single model to characterize the complex interactions among these signals (adapted from [26])

chemoreflex control of breathing in this volume (see Chap. 8). Translating these scientific findings into clinical practice becomes critical, as in many other biological research fields [9, 13, 30].

However, the effectiveness of such translational research has been limited by the traditional reductionist paradigm which assumes a simple reflex model of respiratory control [7, 38, 40]. This classical reflex model assumes additive, reducible and superposable characteristics of afferent signals (Fig. 5.1a) from chemical, mechanical and exercise stimuli [4, 14, 31, 38, 39, 48].

In fact, a complex physiological system is more than simply the sum of its parts [35]. There is evidence that various afferent and efferent signals are integrated by the respiratory controller in a sophisticated way to define the overall ventilatory response (Fig. 5.1b). For instance, the available evidence reveals a distinct multiplicative (synergistic) component in the ventilatory response to concomitant exercise and hypercapnia such that CO_2 responsiveness is potentiated during exercise, and vice versa [16, 21, 24]. Ventilatory response to chemical or exercise input is also potentiated by increases in physiological dead space and shunt (e.g., in congestive heart failure patients). Such sensorimotor integration characteristics are ignored in the oversimplified reflex model which mistakenly considers all respiratory inputs as additive to the "exercise stimulus" (Fig. 5.1a). Hence, translational research cannot be effectively conducted without a unified model of respiratory sensorimotor integration that can account for such complex behaviors. The optimization model of ventilatory control first proposed in [19] has proved to hold promise for this purpose.

5.6.2 Optimal Sensorimotor Integration in Respiratory Control

M.J. Purves once queried [28]: "What do we breathe for?".

The primary purpose of breathing is to meet metabolic demands, as evidenced by the apparent respiratory homeostasis during exercise. An implicit objective

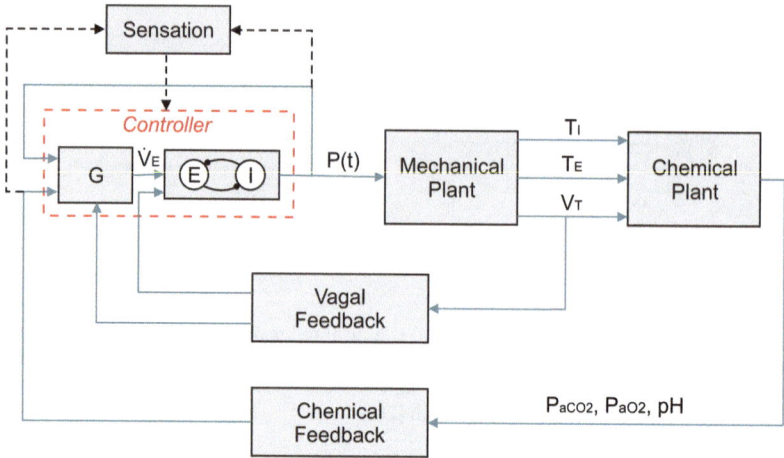

Fig. 5.2 Simplified block diagram of the optimization model of respiratory control. Sensorimotor feedback signals are integrated by an intelligent controller which produces optimal ventilatory drive and breathing pattern that are most cost-effect for meeting metabolic demands, subject to the constraints imposed by the mechanical plant and chemical plant. The functional block "G" represents the transmission gains for sensorimotor integration. *Dyspnea (respiratory discomfort) may involve the processing of similar signals by the higher brain (the somatosensory cortex and limbic system)*

is to maintain chemical homeostasis while another objective is to minimize the energy consumption in the act of breathing. Furthermore, the respiratory system also faces constant challenges of many other causes such as the need for varying behavioral (feeding, smelling, blowing, vocalization, breath-holding, posture, emotion, defecation), physiological (panting, thermal hyperpnea) or defense (coughing, sneezing, emesis, eructation, hiccup) measures. These diverse objectives may not be compatible with one another especially when system stability is considered, and hence maintaining constancy of arterial blood gas level is not always the sole goal. Apparently, the respiratory control system is intelligent enough to maintain a delicate balance between these conflicting objectives [3].

Poon [19,20] first introduced the following cost function to integrate the chemical and mechanical costs of breathing (Fig. 5.2):

$$ J = J_c + J_m = [\alpha(P_{aCO2} - \beta)]^2 + \ln \dot{W}_o. \tag{5.1} $$

The terms J_c, J_m in Eq. (5.1) represent the competing chemical and mechanical costs of breathing (α, β are parameters) in conformance to Steven's power law and Weber–Fechner law of psychophysics respectively [34]. The term $\dot{W}_o$ is a measure of the work rate of breathing subject to the mechanical limitation of the respiratory system.

The optimal total ventilation $\dot{V}_E$ is one that minimizes J (Eq. (5.1)) subject to the gas exchange process and mechanical constraint thus weighing the chemoafferent feedback against the respiratory motor output. The resultant optimal solution simulates the linear $\dot{V}_E - P_{aCO2}$ relationship during CO_2 inhalation and proportional relationship during exercise as follows [19, 20]:

$$\dot{V}_{E0} = 863\alpha^2 (P_{aCO2} - \beta) \frac{\dot{V}_{CO_2}}{(1 - V_D/V_T)}, \tag{5.2}$$

$$\dot{V}_E = \frac{\dot{V}_{E0}}{1 + \dot{V}_{E0}/\dot{V}_{max}}. \tag{5.3}$$

Hence, the controller gain is not constant but may be adjusted to track the metabolic $\dot{V}_{CO_2}$.

Equation (5.1) has been extended [25] to model the integrative control of $\dot{V}_E$ and respiratory pattern, by expressing W_o explicitly in terms of the isometric respiratory driving pressure $P(t)$. The mechanical plant in this case is defined by the following equation of motion:

$$P(t) = \dot{V}(t) R_{rs} + V(t) E_{rs}, \tag{5.4}$$

whereby all ventilatory variables can be derived successively from the $P(t)$ waveform as follows:

$$P(t) \rightarrow \dot{V}(t), V(t) \rightarrow V_T, T_I, T_E \rightarrow \dot{V}_E, \tag{5.5}$$

where R_{rs}, E_{rs} are respectively the total (extrinsic and intrinsic) respiratory resistance and elastance; $\dot{V}(t)$, $V(t)$ are instantaneous respiratory airflow and volume; T_I and T_E are inspiratory and expiratory durations. V_T, T_I and T_E together determine the total ventilation. This integrated model captures both the optimal ventilatory response characteristics of Eq. (5.2), and the corresponding optimal respiratory pattern (Table 5.1).

5.6.3 Hebbian Feedback Covariance Learning Model of Respiratory Motor Control

The optimization model represents an intelligent control system in our breathing. The next question is: what is the brain mechanism that optimizes our breathing? A Hebbian feedback covariance adaptive control paradigm which conforms to the neurophysiological system was proposed based on the Hebbian covariance learning rule (Fig. 5.3). Figure 5.3 is hence an engineering realization of the respiratory optimal controller (Fig. 5.2). The synaptic weight that determines the optimal input–output relationship is computed by correlating the corresponding intrinsic fluctuations that are ubiquitous in physiological signals [45].

Hebbian synaptic plasticity was first postulated by Hebb over 50 years ago as a mechanism of learning and memory [8]. The classical Hebbian model [8] states that

Table 5.1 Glossary

J	Total cost of breathing
J_c	Chemical cost of breathing
J_m	Mechanical cost of breathing
P_{aCO2}	Arterial CO_2 partial pressure
α	Chemoreceptor sensitivity
β	Chemoreceptor response threshold
$\dot{W}_o$	Work rate of breathing
$\dot{V}_E$	Total ventilation
$\dot{V}_{max}$	Maximal ventilatory output
$\dot{V}_{CO_2}$	Metabolic CO_2 production rate
R_{rs}	Total respiratory resistance
E_{rs}	Total respiratory elastance
$\dot{V}(t)$	Instantaneous respiratory volume
$V(t)$	Instantaneous respiratory volume
T_I	Inspiratory duration
T_E	Expiratory duration
$P(t)$	Isometric inspiratory driving pressure measured at functional residual capacity (FRC)

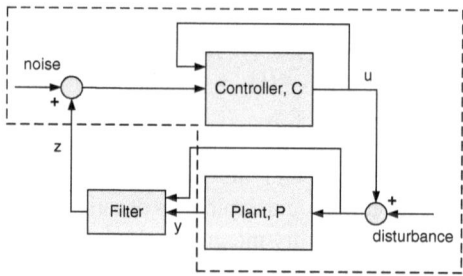

Fig. 5.3 Hebbian feedback covariance control paradigm. The controller gain, C, is adapted based on the covariance between the input (y) and output (u) signals. *Dotted line* indicates the original control scheme for static system. To account for system dynamics, a filter is added to transform the plant output, y into an intermediate variable (adapted from [37])

the strength (or gain) C of a synaptic connection can be modified according to an adaptation law of the form:

$$\frac{dC}{dt} = k(y \cdot u), \tag{5.6}$$

where u, y are mean firing rates of input and output neurons, respectively, and k is an adaptation rate. This synaptic adaptation law has been widely taken as the basis of NMDA receptor-dependent long-term potentiation (LTP) in some hippocampal and neocortical neurons.

However, the inevitable occurrence of runaway instability and irreversible saturation resulting from sustained or random coactivity of interacting neurons limits the implementation of the classical Hebbian model [22, 44, 46]. Moreover, the dependence of the classical Hebbian model on pre- and post-synaptic activities local to the adapting neuron does not lend itself to feedback control applications. To circumvent these difficulties, a stochastic version of the synaptic adaptation law called Hebbian feedback covariance adaptation has been proposed [46]. Instead of pairing the mean input and output activities of the controller neuron, the new adaptation law modifies the synaptic strength by correlating the temporal variations of the pre- and post-synaptic neural activities about their respective mean values. Here, the pre- and post-synaptic activities of the controller neuron correspond to the feedback signal and control signal, respectively. This adaptive control paradigm is analogous to a reinforcement learning system driven by spontaneous, random perturbations in the control and feedback variables (Fig. 5.3). The new adaptation law becomes,

$$\frac{dC}{dt} = k_1(\delta y \cdot \delta u) - k_2 C \cdot g(\delta y, \delta u; y, u), \tag{5.7}$$

where δu and δy are the temporal variations of the pre- and post-synaptic neural activities about their respective mean values and $g(\cdot)$ is some positive definite function, which acts as a decay term to avoid saturation. Depending on the sign of the constant k_1, the algorithm can describe either synaptic LTP (for $k_1 > 0$) or LTD (long-term depression) (for $k_1 < 0$).

The controller gain, C, is updated according to the input/output relationship of the system, without explicit estimation of the plant parameters, hence a direct adaptive control approach. In addition, the adaptive controller can be designed in such a way to optimize a certain objective function (which is essentially the Lyapunov function [32]). By applying the Barbalat's lemma [32], the objective function can be guaranteed to converge to a minimum value at steady state.

The discussions of the algorithm so far have focused on static input–output relationship only. Young and Poon [46] further modified the original algorithm for dynamical systems by introducing a near-term objective function, Q. The Hebbian covariance feedback law is applied to the near-term objective function which, with some suitable transformations of the state variables, will lead to minimization in the long term.

To illustrate, consider a first-order nonlinear system,

$$\dot{y} = q(y, u). \tag{5.8}$$

The steady state solution is obtained as,

$$y_0 = f(u_0). \tag{5.9}$$

An intermediate variable z is defined as a filtered version of the original state variable y,

$$z(y, \dot{y}, u) = f(u), \tag{5.10}$$

such that $z \to y_0$ as $f(u) \to f(u_0)$. Hence, the system becomes static in z, and the same algorithm can be applied on this intermediate (or filtered) state variable with a near-term objective function defined in z and u. In steady state, the near-term objective function converges to the long-term objective function. The same approach is generally applicable to systems of higher order.

The Hebbian feedback covariance control is computationally simple compared to conventional self-tuning or model-reference adaptive control, which generally require a prescribed reference model or desired trajectory. On the other hand, it is also a form of reinforcement learning with an implicit reinforcement signal, namely the covariance of the filtered state and the control signals.

Hebbian feedback covariance control has been successfully applied to the modeling of the respiratory system to predict the optimal adaptation behaviors during exercise and CO_2 inhalation [44, 45]. Moreover, robustness to noise disturbances is also verified with simulations. Further experimental evidence is required to identify the neural correlates of such an algorithm in respiratory control. This Hebbian feedback covariance control has a substantial mathematical background. It shows an example in which reverse engineering is applied to a physiological system toward developing a biologically-inspired engineering control paradigm. Such a paradigm may transcend its biological counterpart to suggest a general engineering control theory that is applicable to any practical system.

5.6.4   Cheyne–Stokes Breathing: Three Different Engineering Control Perspectives

Cheyne–Stokes breathing is of significant clinical concern. It is a respiratory abnormality when a person's breathing wax and wane periodically. It is occurring more frequently in patients with chronic heart failure and left ventricular systolic dysfunction, normals at high altitude, or during sleep. There have been extensive analytical studies about the mechanism of Cheyne–Stokes. Three possible views are discussed below (see also Chap. 7 by Bruce).

5.6.4.1   Cheyne–Stokes Breathing as Linear System Instability

Cheyne–Stokes breathing has been studied on the basis of engineering stability. It has been suggested that Cheyne–Stokes breathing is a result of instability of the respiratory control system based on chemoreflex, as the above mentioned conditions increase the gain and/or phase lag of the system. A vast variety of engineering theory is concerned about the stability of a control system, especially linear time-invariant system. Khoo [11] has performed an extensive analytical study, based on Nyquist's stability criterion, on the dependence of stability of respiratory control system on different parameters and physiological conditions. The analysis determines the local stability of the system about the equilibrium state using a linearized system. An unstable combination of system parameters will diverge the states from the equilibrium, and will be bounded by saturations in chemoreceptors, gas exchange

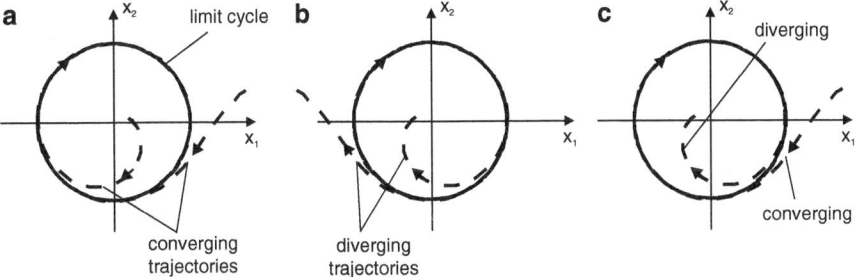

Fig. 5.4 Limit cycle. (**a**) Stable, (**b**) unstable, (**c**) semi-stable (adapted from [32])

process and mechanical limitation of actuators, and hence oscillate. This study represents one of the very successful systematic applications of engineering theories in the study of cardiorespiratory phenomena. It has led to many subsequent studies of Cheyne–Stoke breathing and the mechanisms of possible treatment.

5.6.4.2 Cheyne–Stokes Breathing as Nonlinear Limit Cycle

It should be noted that Cheyne–Stoke breathing can also be interpreted as a stable limit cycle. Limit cycle is an important phenomenon in nonlinear systems, and is distinct from the oscillation in linear systems (Fig. 5.4). A marginally stable linear system may also oscillate. However, the amplitude of a limit cycle is independent of the initial condition, while the oscillation of a marginally stable linear system has its amplitude determined by its initial condition. Furthermore, the oscillation of a marginally stable linear system is not robust and is very sensitive to changes in system parameters. On the other hand, limit cycle may be viewed as a "stable" state. A unique mathematical tool for determining existence and stability of limit cycle is available [32] and may be useful in further study of Cheyne–Stokes breathing.

5.6.4.3 Cheyne–Stokes Breathing as an Optimal Strategy

Ghazanshahi and Khoo [6] have suggested that Cheyne–Stokes breathing may actually represent an optimal breathing strategy at high altitude when the respiratory system is highly stressed. They suggested that Cheyne–Stokes, or more generally periodic breathing, is a cost effective way to breathe when respiratory demand is high. They showed that despite the large fluctuation in blood gas concentration, on average gas transport is more efficient in periodic breathing, and it also saves muscular efforts. The chemical cost is lowered by the large breath to reduce dead space ventilation hence maintaining blood gas homeostasis; while mechanical cost is lowered by the apnea period which partially offset the increased work during the breathing phase.

Acknowledgements Chung Tin is an American Heart Association predoctoral fellow. The research of Chi-Sang Poon was supported by US National Institutes of Health grants HL067966, HL072849, HL079503 and RR028241.

References

1. Ahn, A.C., Tewari, M., Poon, C.S., Phillips, R.S.: The limits of reductionism in medicine: could systems biology offer an alternative? PLoS Med. **3**(6), e208 (2006)
2. Cannon, W.B.: The Wisdom of the Body. W.W. Norton & Company, New York (1932)
3. Cherniack, N.: Potential role of optimization in alveolar hypoventilation and respiratory instability. In: von Euler, C., Lagercrantz, H. (eds.) Neurobiology of the control of breathing. Wenner-Gren International Symposium Series, pp. 45–50. Raven Press, New York (1987)
4. Eldridge, F.L., Morin, D., Romaniuk, J.R., Yamashiro, S., Potts, J.T., Ichiyama, R.M., Bell, H., Phillipson, E.A., Killian, K.J., Jones, N.L., Nattie, E.: Supraspinal locomotor centers do/do not contribute significantly to the hyperpnea of dynamic exercise in humans. J. Appl. Physiol. **100**(5), 1743–1747 (2006)
5. Gerlai, R.: Gene targeting: technical confounds and potential solutions in behavioral brain research. Behav. Brain Res. **125**(1-2), 13–21 (2001)
6. Ghazanshahi, S.D., Khoo, M.C.: Optimal ventilatory patterns in periodic breathing. Ann. Biomed. Eng. **21**(5), 517–30 (1993)
7. Haouzi, P.: Theories on the nature of the coupling between ventilation and gas exchange during exercise. Respir Physiol Neurobiol. **151**(2–3), 267–279 (2006)
8. Hebb, D.O.: The Organization of Behavior. Wiley, New York (1949)
9. Horig, H., Marincola, E., Marincola, F.M.: Obstacles and opportunities in translational research. Nat. Med. **11**(7), 705–708 (2005)
10. Hunter, P.J., Borg, T.K.: Integration from proteins to organs: the physiome project. Nat. Rev. Mol. Cell Biol. **4**(3), 237–243 (2003)
11. Khoo, M.C., Kronauer, R.E., Strohl, K.P., Slutsky, A.S.: Factors inducing periodic breathing in humans: a general model. J. Appl. Physiol. **53**(3), 644–659 (1982)
12. Lazebnik, Y.: Can a biologist fix a radio? – or, what I learned while studying apoptosis. Canc. Cell **2**(3), 179–182 (2002)
13. Mankoff, S.P., Brander, C., Ferrone, S., Marincola, F.M.: Lost in translation: Obstacles to translational medicine. J. Transl. Med. **2**(1), 14 (2004)
14. Mateika, J.H., Duffin, J.: A review of the control of breathing during exercise. Eur. J. Appl. Physiol. Occup. Physiol. **71**(1), 1–27 (1995)
15. Mead, J.: Control of respiratory frequency. J. Appl. Physiol. **15**, 325–336 (1960)
16. Mitchell, G.S., Babb, T.G.: Layers of exercise hyperpnea: modulation and plasticity. Respir. Physiol. Neurobiol. **151**(2-3), 251–266 (2006)
17. Ogata, K.: Modern Control Engineering. Prentice Hall, Englewood Cliffs, New Jersey (1997)
18. Ortoleva, P., Adhangale, P., Cheluvaraja, S., Fontus, M., Shreif, Z.: Deriving principles of microbiology by multiscaling laws of molecular physics. IEEE Eng. Med. Biol. Mag. **28**(2), 70–79 (2009)
19. Poon, C.S.: Optimal control of ventilation in hypoxia, hypercapnia and exercise. In: Whipp, B.J., Wiberg, D.W. (eds.) Modelling and Control of Breathing, pp. 189–196. Elsevier, New York (1983)
20. Poon, C.S.: Ventilatory control in hypercapnia and exercise: optimization hypothesis. J. Appl. Physiol. **62**, 2447–2459 (1987)
21. Poon, C.S.: Effects of inspiratory resistive load on respiratory control in hypercapnia and exercise. J. Appl. Physiol. **66**(5), 2391–2399 (1989)
22. Poon, C.S.: Self-tuning optimal regulation of respiratory motor output by hebbian covariance learning. Neural Netw. **8**, 1–17 (1996)

23. Poon, C.S.: Respiratory models and control. In: Bronzion J.D. (ed.) Biomedical Engineering Handbook, vol. 2, 2nd edn., p. 161. CRC, Boca Raton, Florida (2000)

24. Poon, C.S., Greene, J.G.: Control of exercise hyperpnea during hypercapnia in humans. J. Appl. Physiol. **59**(3), 792–797 (1985)

25. Poon, C.S., Lin, S.L., Knudson, O.B.: Optimization character of inspiratory neural drive. J. Appl. Physiol. **72**(5), 2005–2017 (1992)

26. Poon, C.S., Tin, C., Yu, Y.: Homeostasis of exercise hyperpnea and optimal sensorimotor integration: the internal model paradigm. Respir. Physiol. Neurobiol. **159**(1), 1–13; discussion 14–20 (2007)

27. Pugh E.N. Jr., Andersen, O.S.: Models and mechanistic insight. J. Gen. Physiol. **131**(6), 515–519 (2008)

28. Purves, M.: What do we breathe for? In: von Euler, C.., Lagercrantz, H. (eds.) Central Nervous Control Mechanisms in Breathing (Wenner-Gren Center Int. Symp. Ser.), vol. 32, pp. 7–12. Pergamon, Oxford, UK (1979)

29. Qutub, A., Gabhann, F., Karagiannis, E., Vempati, P., Popel, A.: Multiscale models of angiogenesis. IEEE Eng. Med. Biol. Mag. **28**(2), 14–31 (2009)

30. Saijo, N.: Translational study in cancer research. Intern. Med. **41**(10), 770–773 (2002)

31. Secher, N., Poon, C.S., Ward, S., Whipp, B., Duffin, J.: Supraspinal locomotor centers do/do not contribute significantly to the hyperpnea of dynamic exercise in humans. J. Appl. Physiol. **100**(4), 1417–1418 (2006)

32. Slotine, J.J.E., Li, W.: Applied Nonlinear Control. Prentice-Hall, Englewood Cliffs, NJ (1991)

33. Sorger, P.K.: A reductionist's systems biology: opinion. Curr. Opin. Cell Biol. **17**(1), 9–11 (2005)

34. Stevens, S.: To honor fechner and repeal his law. Science **133**, 80–86 (1961)

35. Strange, K.: The end of "naive reductionism": Rise of systems biology or renaissance of physiology? Am. J. Physiol. Cell Physiol. **288**(5), C968–C974 (2005)

36. Taufer, M., Armen, R., Chen, J., Teller, P., Brooks, C.: Computational multiscale modeling in protein–ligand docking. IEEE Eng. Med. Biol. Mag. **28**(2), 58–69 (2009)

37. Tin, C., Poon, C.S.: Internal models in sensorimotor integration: perspectives from adaptive control theory. J. Neural Eng. **2**(3), S147–S163 (2005)

38. Waldrop, T.G., Iwamoto, G.A., Haouzi, P.: Point:counterpoint: supraspinal locomotor centers do/do not contribute significantly to the hyperpnea of dynamic exercise. J. Appl. Physiol. **100**(3), 1077–1083 (2006)

39. Ward, S.A.: Control of the exercise hyperpnoea in humans: a modeling perspective. Respir. Physiol. **122**(2-3), 149–166 (2000)

40. Wasserman, K., Whipp, B.J., Casaburi, R.: Respiratory control during exercise. In: Cherniack, N.S., Widdicombe, J.G. (eds.) Handbook of Physiology, vol. 2, pp. 595–620. American Physiological Society, Bethesda, MD (1986)

41. Wellstead, P., Bullinger, E., Kalarnatianos, D., Mason, O., Verwoerd, M.: The role of control and system theory in systems biology. Annu. Rev. Contr. **32**(1), 33–47 (2008)

42. White, R., Peng, G., Demir, S.: Multiscale modeling of biomedical, biological, and behavioral systems (part 1). IEEE Eng. Med. Biol. Mag. **28**(2), 12–13 (2009)

43. Wiener, N.: Cybernetics; or, Control and communication in the animal and the machine. Technology Press [Cambridge, MA] (1948)

44. Young, D.L.: Hebbian covariance learning and self-tuning optimal control. Ph.D. thesis, M.I.T (1997)

45. Young, D.L., Poon, C.S.: Hebbian covariance learning. a nexus for respiratory variability, memory, and optimization? Adv. Exp. Med. Biol. **450**, 73–83 (1998)

46. Young, D.L., Poon, C.S.: A hebbian feedback covariance learning paradigm for self-tuning optimal control. IEEE Trans. Syst. Man Cybern. B Cybern. **31**(2), 173–186 (2001)

47. Young, D.L., Poon, C.S.: Soul searching and heart throbbing for biological modeling. Behav. Brain Sci. **24**(06), 1080–1081 (2002)

48. Yu, Y., Poon, C.S.: Critique of "control of arterial Pco2 by somatic afferents". J. Physiol. **572**(3), 897–898 (2006)

Chapter 6
Parameter Identification in a Respiratory Control System Model with Delay

Ferenc Hartung and Janos Turi

Abstract In this paper we study parameter identification issues by computational means for a set of nonlinear delay equations which have been proposed to model the dynamics of a simplified version of the respiratory control system. We design specific inputs for our system to produce "information rich" output data needed to determine values of unknown parameters. We also consider the effects of noisy measurements in the identification process. Several case studies are included.

6.1 Introduction

Mathematical models describing the chemical balance mechanism of the respiratory control system are given in the form of nonlinear, parameter dependent, delay differential equations [3–5]. The physiological features of the respiratory system including a transport delay in the feedback control mechanism is reviewed in Chap. 8.

The analysis of the direct problem (i.e., it is assumed that the values of the parameters are known) corresponding to the model equations shows that the system has a unique equilibrium, and that the stability of this equilibrium depends on the parameter values (see [5] for details). This observation leads naturally to the question of parameter identification in the model equations based on available, but possibly noisy, measurements. In this paper we present a computational procedure, applicable for large classes of functional differential equations [11, 15, 17] which

F. Hartung
Department of Mathematics, University of Pannonia, Veszprém, H-8201, Hungary
e-mail: hartung.ferenc@uni-pannon.hu

J. Turi (✉)
Programs in Mathematical Sciences, University of Texas at Dallas, Richardson, TX 75083, USA
e-mail: turi@utdallas.edu

J.J. Batzel et al. (eds.), *Mathematical Modeling and Validation in Physiology*,
Lecture Notes in Mathematics 2064, DOI 10.1007/978-3-642-32882-4_6,
© Springer-Verlag Berlin Heidelberg 2013

can be used to perform parameter estimation in respiratory control models. We also illustrate how information rich data can enhance the effectiveness of the estimation process. Another issue we study is what are the most promising measurements available for identification purposes (i.e., should one measure gas concentrations or ventilation volumes)?

In Sect. 6.2 we introduce our model equations, in Sect. 6.3 we describe the numerical method we use to run simulations on the model equations. Section 6.4 outlines the parameter estimation process and contains several case studies. In Sect. 6.5 we provide a discussion of our findings.

6.2 Model Equations

We consider following [5] and [3] the system of nonlinear delay equations describing a simple model of the human respiratory control system

$$\dot{x}(t) = a_{11} - a_{12}x(t) - a_{13}V(t, x(t - \tau), y(t - \tau))(x(t) - x_I), \qquad (6.1)$$

$$\dot{y}(t) = -a_{21} - a_{22}y(t) + a_{23}V(t, x(t - \tau), y(t - \tau))(y_I - y(t)), \qquad (6.2)$$

where $x(t)$ and $y(t)$ denote the arterial CO_2 and O_2 concentrations, respectively, $V(\cdot, \cdot, \cdot)$ is the ventilation function, τ is the transport delay, x_I and y_I are inspired CO_2 and O_2 concentrations. For detailed information on modeling considerations we refer the reader to [5] and [3]. We assume that the ventilation function has the form

$$V(t, x, y) = G_P(t)W(x, y), \qquad (6.3)$$

where the control gain, $G_P(t)$, is a function of time. For simplicity we assume that the time dependency of G_P is piecewise constant, and in particular,

$$G_P(t) = \begin{cases} G_{P1}, & 0 \le t < \theta_1, \\ G_{P2}, & \theta_1 \le t < \theta_2, \\ G_{P3}, & \theta_2 \le t, \end{cases} \qquad (6.4)$$

where $\theta_1, \theta_2 > 0$, $G_{P1} \ge 0$, $G_{P2} \ge 0$ and $G_{P3} \ge 0$ are constant parameters. W is given by

$$W(x, y) = e^{-0.05y}(x - I_P). \qquad (6.5)$$

Moreover, in (6.1)–(6.2) we have that

$$a_{11} = 863 \frac{\dot{Q} K_{CO_2} P_{VCO_2}}{M_{LCO2}},$$

$$a_{12} = 863 \frac{\dot{Q} K_{CO_2}}{M_{LCO2}},$$

Table 6.1 Normal parameter values

Quantity	Unit	Value
τ	min	0.1417
$\dot{Q}$	l/min	6.0
K_{CO_2}		0.0057
P_{VCO_2}	mmHg	46.0
P_{VO_2}	mmHg	41.0
M_{LCO_2}	1	3.2
M_{LO_2}	1	2.5
m_v		0.0021
m_a		0.00025
B_v		0.0662
B_a		0.1728
G_{P1}	l/min/mmHg	45.0
I_P	mmHg	35.0
x_I		0
y_I		146.0

$$a_{13} = \frac{E_F}{M_{LCO_2}},$$

$$a_{21} = 863\frac{\dot{Q}}{M_{LO2}}(-m_v P_{VO_2} + B_a - B_v),$$

$$a_{22} = 863\frac{\dot{Q}m_a}{M_{LO_2}},$$

$$a_{23} = \frac{E_F}{M_{LO_2}},$$

where the normal values of the parameters appearing on the right hand side of the above equations are listed in Table 6.1 (see also [3] for more detailed information on system parameter values).

Substitution of the normal values into Eqs. (6.1)–(6.2) yields

$$\dot{x}(t) = 422.4277 - 9.2233x(t)$$
$$- 0.21875V(t, x(t - 0.1417), y(t - 0.1417))x(t), \tag{6.6}$$

$$\dot{y}(t) = -42.8946 - 0.5178y(t)$$
$$+ 0.28V(t, x(t - 0.1417), y(t - 0.1417))(146 - y(t)) \tag{6.7}$$

with ventilation function

$$V(t, x, y) = G_P(t)e^{-0.05y}(x - 35), \tag{6.8}$$

where G_P is defined by (6.3).

6.3 Numerical Approximation

In this section we define a simple numerical scheme to approximate solutions of
(6.1)–(6.2). This method is based on equations with piecewise constant argument
(EPCA) and was introduced in [10] for linear scalar delay and neutral differential
equations. Later this scheme was extended for a large class of nonlinear delay
systems in [11, 12]. In our presentation we follow the framework described in
[11, 12]. Let h be a fixed positive constant, and define the notation

$$[t]_h = \left[\frac{t}{h}\right] h,$$

where $[\cdot]$ is the greatest integer function. Then $[t]_h$ as a function of t is piecewise
constant, since $[t]_h = nh$ for $t \in [nh, (n+1)h)$. For a fixed $h > 0$ we associate the
system

$$\dot{x}_h(t) = a_{11} - a_{12}x_h([t]_h)$$
$$- a_{13}V([t]_h, x_h([t]_h - [\tau]_h), y_h([t]_h - [\tau]_h))(x_h([t]_h) - x_I), \qquad (6.9)$$
$$\dot{y}_h(t) = -a_{21} - a_{22}y_h([t]_h)$$
$$+ a_{23}V([t]_h, x_h([t]_h - [\tau]_h), y_h([t]_h - [\tau]_h))(y_I - y_h([t]_h)) \qquad (6.10)$$

for $t \geq 0$. For negative t we associate the initial functions of (6.1) and (6.2) to (6.9)
and (6.10), respectively. System (6.9)–(6.10) is a system of equations with piecewise
constant argument (EPCA). Such equations were introduced and first studied by
Cooke and Wiener [6–8, 20]. The solutions, x_h and y_h, of (6.9)–(6.10) are defined
as continuous functions, which are differentiable and satisfy system (6.9)–(6.10) on
each interval $(nh, (n+1)h)$ $(n = 0, 1, 2, \ldots)$. Since the right-hand-side of both (6.9)
and (6.10) are constant on each interval $[nh, (n+1)h)$, we get that both x_h and y_h
are piecewise linear continuous functions (linear spline functions). Therefore, they
are determined by their values at the mesh points nh. Introduce the sequences

$$u_n = x_h(nh) \quad \text{and} \quad v_n = y_h(nh),$$

and let

$$k = \left[\frac{\tau}{h}\right].$$

Then integrating (6.9) and (6.10) from nh to t and taking the limit $t \to (n+1)h-$,
we get by simple calculation that u_n and v_n satisfy

$$u_{n+1} = u_n + h\left(a_{11} - a_{12}u_n - a_{13}V(nh, u_{n-k}, v_{n-k})(u_n - x_I)\right), \qquad (6.11)$$

$$v_{n+1} = v_n + h\left(-a_{21} - a_{22}v_n + a_{23}V(nh, u_{n-k}, v_{n-k})(y_I - v_n)\right), \qquad (6.12)$$

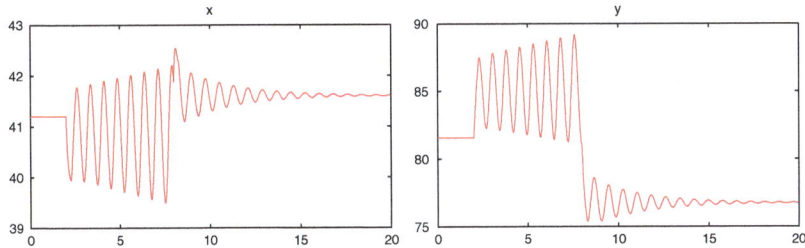

Fig. 6.1 Effect of changing the control gain with switching times $\tau_1 = 2$ and $\tau_2 = 8$ and gain constants $G_{p_1} = 45$, $G_{p_2} = 60$ and $G_{p_3} = 30$

for $n = 0, 1, 2, \ldots$, where for negative integer n the sequences u_n and v_n are defined by $u_n = x(nh)$ and $v_n = y(nh)$, i.e., the initial functions corresponding to the original system (6.1)–(6.2). Therefore the sequences u_n and v_n are well-defined and can be easily generated by the explicit delayed recurrence relations (6.11)–(6.12), so the solutions of (6.9)–(6.10) are uniquely determined. It is shown in [11] that

$$\lim_{h \to 0+} x_h(t) = x(t) \quad \text{and} \quad \lim_{h \to 0+} y_h(t) = y(t)$$

uniformly on each interval $[0, T]$ for any $T > 0$.

Example 6.1. In this example we study numerically the effect of changing the control gain for the stability of the solutions of the respiratory system (6.1)–(6.2). We assume normal table values except that we use $\tau = 0.25$, i.e., we consider (6.6)–(6.7) with ventilation (6.3)–(6.5). Furthermore, in (6.4) we select $\theta_1 = 2$, $\theta_2 = 8$ for the switching times, and $G_{P1} = 45$, $G_{P2} = 60$ and $G_{P3} = 30$ for the control gains. We start the system from its equilibrium corresponding to the $G_P(t) = G_{P1}$ constant gain, i.e., use constant initial functions

$$x(t) = 41.1906, \qquad t \le 0, \quad \text{and} \quad y(t) = 81.5645, \qquad t \le 0.$$

The numerical solution corresponding to the discretization constant $h = 0.001$ is shown in Fig. 6.1.

We can see from the figure that the equilibrium of the system with gain $G_P(t) = G_{P2}$ is unstable, (in fact, it is asymptotically periodic if we compute the solution for a long enough time interval), but after switching back to gain constant $G_P(t) = G_{P3}$, it is again asymptotically stable.

6.4 Parameter Estimation

We consider again system (6.1)–(6.2) with ventilation (6.3)–(6.4). We assume that some of the parameters in this system are not known, and we denote the unknown parameters by $\gamma_1, \ldots, \gamma_m$. We can consider, for example, the control gain constants

G_{P1}, G_{P2} and G_{P3} as the unknown parameters (in that case $m = 3$ and $\gamma_i = G_{Pi}$ for $i = 1, 2, 3$), or the transport delay τ can be the only unknown parameter ($m = 1$, $\gamma_1 = \tau$), but we can consider any other parameters in Eqs. (6.1) and (6.2), or in the ventilation function (6.3)–(6.4) to be unknown. The goal is to determine the values of these unknown parameters, assuming we know the measurements of the solutions at finitely many times, $t_1, t_2, \ldots, t_M$. One standard approach to this problem is to define a least-square cost function, and then find the parameter values with the least possible cost.

First we need to introduce the following notation. We shall assume that some system parameters are fixed and denote by $\gamma_1, \ldots, \gamma_m$ unknown parameters to be identified in (6.1)–(6.2), (6.3)–(6.4). The solution corresponding to a particular selection of $\gamma_1, \ldots, \gamma_m$ of this problem is denoted by

$$x(t; \gamma_1, \ldots, \gamma_m) \quad \text{and} \quad y(t; \gamma_1, \ldots, \gamma_m),$$

where for the remaining parameters the values from Table 6.1 are used. Suppose the measurements of x and y at the time t_i are denoted by X_i and Y_i, respectively, for $i = 0, \ldots, M$. We will use equally spaced measurements over a time interval $[T_0, T]$, i.e.,

$$t_i = T_0 + \frac{T - T_0}{M} i, \qquad i = 0, 1, \ldots, M. \tag{6.13}$$

Of course, any time values could be used. Then we define the cost function by

$$J(\gamma_1, \ldots, \gamma_m) = \sum_{i=1}^{M} (x(t_i; \gamma_1, \ldots, \gamma_m) - X_i)^2 + \sum_{i=1}^{M} (y(t_i; \gamma_1, \ldots, \gamma_m) - Y_i)^2.$$
$$\tag{6.14}$$

Then the mathematical problem is to find the parameter values $\gamma_1, \ldots, \gamma_m$ which minimize the cost function J.

One standard approach to solve this problem used, e.g., in [1, 2, 13, 14, 16] is the following: find finite dimensional approximate solutions x^N, y^N of (6.1)–(6.2), and define the corresponding cost J^N as

$$J^N(\gamma_1, \ldots, \gamma_m) = \sum_{i=1}^{M} (x^N(t_i; \gamma_1, \ldots, \gamma_m) - X_i)^2 + \sum_{i=1}^{M} (y^N(t_i; \gamma_1, \ldots, \gamma_m) - Y_i)^2,$$

and find the minimizer $(\gamma_1^N, \ldots, \gamma_m^N)$ of J^N. One can show (see, e.g., [14]) that, under minor assumptions, a subsequence of $(\gamma_1^N, \ldots, \gamma_m^N)$ approaches to the minimizer of J.

In this paper we consider a sequence of discretization constants, h_N, tending to 0, and use the approximation scheme defined in the previous section corresponding to h_N as the numerical scheme in the above process. Then if N is large enough, i.e., equivalently, h_N is small enough, we find the minimizer of the corresponding cost function J^N by a nonlinear least square minimization code, based on a secant method with Dennis–Gay–Welsch update, combined with a trust region technique. See Sect. 10.3 in [9] for detailed description of this method. Then we consider the result as the approximation of the minimizer of J. Here we know that for the "true parameters" the value of the cost function is 0, so if the numerical method stops at a parameter value where the cost function is not close to 0, then we can conclude that the method is terminated at a local minimum instead of a global minimum. Then we restart the method from a different initial parameter value. Of course, we know that the numerical method converges only locally, so we have to find initial guesses close enough to the true parameter values in order to observe convergence. Another important issue in the parameter estimation process is that whether two different parameter sets can generate the same measurements, i.e., the question of identifiability of the parameters. This is a difficult theoretical problem (see, e.g., [18, 19] or [14, Example 5.4]). The lack of identifiability can be another reason for getting non converging approximations.

In the remaining part of this section we give several numerical examples to demonstrate the applicability of the above parameter estimating process for the respiratory system (6.1)–(6.4). In all these examples we achieved good recovery of the original parameters, which also indicated that we numerically observed identifiability of the considered parameters.

Example 6.2. In this example we generated measurements of (6.1)–(6.4) corresponding to the normal parameter values listed in Table 6.1 and using a constant gain coefficient function $G_P(t)$, i.e.,

$$G_{P1} = G_{P2} = G_{P3} = 45.$$

We assume that the system is at the equilibrium, so we use initial conditions $x(t) = 41.1906$ and $y(t) = 81.5645$ which correspond to the equilibrium values. The measurements are taken over the interval $[T_0, T] = [0, 2]$ using formula (6.13) with $M = 11$. We consider the coefficients a_{12}, a_{13}, a_{22} and a_{23} to be unknown, and the goal in this example is to estimate these parameter values using the measurements. In this example we used the discretization step size $h = 0.01$ and the initial parameters

$$a_{12} = 8.5, \qquad a_{13} = 0.3, \qquad a_{22} = 0.6, \quad \text{and} \quad a_{23} = 0.4.$$

The first three steps of the numerical method can be seen in Fig. 6.2. The solid line is the solution x, y and the ventilation function W along the solutions corresponding to the true parameters, and the circles are the measurements of the respective functions at sample time points. The dotted curves are the solutions x, y and the ventilation V along the solutions corresponding to parameter values generated by the numerical scheme in the first two steps. We can see that the graphs approach to the graph corresponding to the true parameter values even in the first few steps. Table 6.2 contains the value of the cost function, the actual parameter value, and the error of the particular parameter when compared to the true parameter value at each

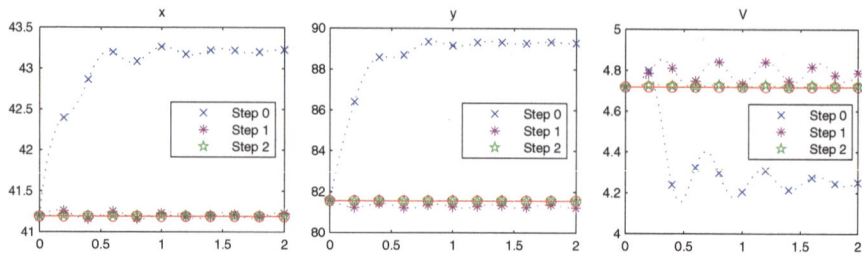

Fig. 6.2 Estimation of $a_{12}, a_{13}, a_{22}, a_{23}$, case $G_{P1} = G_{P2} = G_{P3} = 45$

Table 6.2 Estimation of $a_{12}, a_{13}, a_{22}, a_{23}$, case $G_{P1} = G_{P2} = G_{P3} = 45$

Step	Cost	a_{12}	a_{13}	a_{22}	a_{23}	$\Delta(a_{12})$	$\Delta(a_{13})$	$\Delta(a_{22})$	$\Delta(a_{23})$
0	287.84623772	8.50000	0.30000	0.60000	0.40000	0.72330	0.08125	0.08220	0.12000
1	0.38419648	8.80052	0.30321	0.86988	0.36664	0.42278	0.08446	0.35208	0.08664
2	0.00077480	8.81127	0.30535	0.86543	0.37275	0.41203	0.08660	0.34763	0.09275
3	0.00056159	8.81142	0.30538	0.86538	0.37282	0.41188	0.08663	0.34758	0.09282
4	0.00043748	8.81160	0.30541	0.86535	0.37287	0.41170	0.08666	0.34755	0.09287
5	0.00034371	8.81177	0.30544	0.86533	0.37290	0.41153	0.08669	0.34753	0.09290
6	0.00034371	8.81177	0.30544	0.86533	0.37290	0.41153	0.08669	0.34753	0.09290

step. (We denote the error in the parameter γ by $\Delta(\gamma)$). The method converges in five steps, but in each parameter value a small error can be observed. Our explanation for this error (which can be seen running the code from different initial values, as well) is that the constant solution is not "rich enough" for better estimation.

Example 6.3. In this example we change the gain constants in the ventilation to move the solutions away from the equilibrium. We use switching times $\theta_1 = 0.2$ and $\theta_2 = 0.4$ and gain constants

$$G_{P1} = 45, \quad G_{P2} = 0, \quad G_{P3} = 60.$$

This corresponds to the physical case when one takes normal breaths, then stops breathing for 12 s (between time 0.2 and 0.4 min), but then takes larger breaths for a while. We again try to estimate a_{12}, a_{13}, a_{22} and a_{23}. We used the same initial parameter values, measurements and $h = 0.01$ as in Example 6.2. The numerical results can be seen in Fig. 6.3 and in Table 6.3. In this case we achieved perfect recovery of the true parameter values up to five decimal digits accuracy in the fifth step.

Example 6.4. Now we use the same measurements and $h = 0.01$ as in Example 6.3, but this time we consider G_{P1}, G_{P2} and G_{P3} as the unknown parameters in the system. (The switching times are the same as in the previous example). Starting from the initial guess $G_{P1} = G_{P2} = G_{P3} = 40$, we again get good approximation of the true parameters, as can be seen in Fig. 6.4 and in Table 6.4.

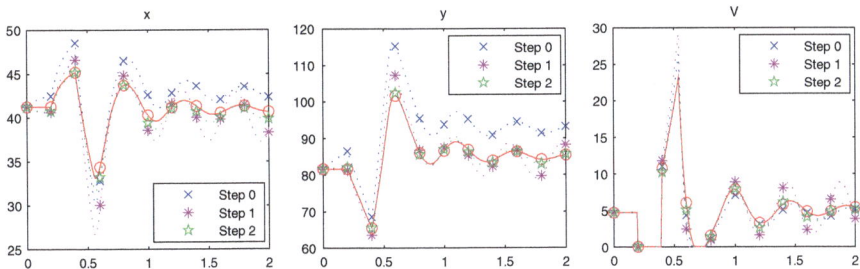

Fig. 6.3 Estimation of $a_{12}, a_{13}, a_{22}, a_{23}$, case $G_{P1} = 45$, $G_{P2} = 0$, $G_{P3} = 60$

Table 6.3 Estimation of $a_{12}, a_{13}, a_{22}, a_{23}$, case $G_{P1} = 45$, $G_{P2} = 0$, $G_{P3} = 60$

Step	Cost	a_{12}	a_{13}	a_{22}	a_{23}	$\Delta(a_{12})$	$\Delta(a_{13})$	$\Delta(a_{22})$	$\Delta(a_{23})$
0	350.31169443	8.50000	0.30000	0.60000	0.40000	0.72330	0.08125	0.08220	0.12000
1	51.72675038	8.86517	0.33654	0.63847	0.30804	0.35813	0.11779	0.12067	0.02804
2	3.54921161	9.21149	0.25766	0.53943	0.29705	0.01181	0.03891	0.02163	0.01705
3	0.00525638	9.21676	0.22070	0.51609	0.27984	0.00654	0.00195	0.00171	0.00016
4	0.00000009	9.22328	0.21876	0.51779	0.28000	0.00002	0.00001	0.00001	0.00000
5	0.00000000	9.22330	0.21875	0.51780	0.28000	0.00000	0.00000	0.00000	0.00000

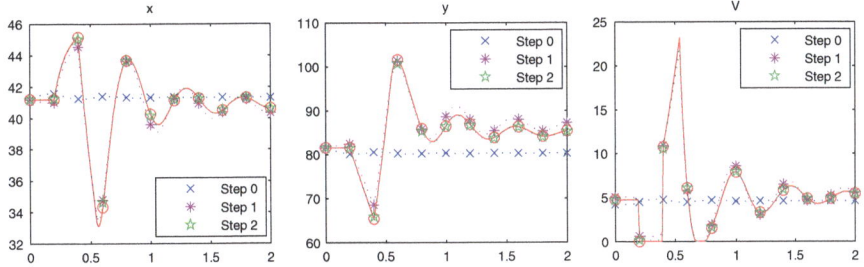

Fig. 6.4 Estimation of G_{P1}, G_{P2} and G_{P3}

Table 6.4 Estimation of G_{P1}, G_{P2} and G_{P3}.

Step	Cost	G_{P1}	G_{P2}	G_{P3}	$\Delta(G_{P1})$	$\Delta(G_{P2})$	$\Delta(G_{P3})$
0	483.75379438	40.00000	40.00000	40.00000	5.00000	40.00000	20.00000
1	14.01499742	47.97180	5.86922	68.70514	2.97180	5.86922	8.70514
2	0.90980645	44.53413	1.28130	59.25469	0.46587	1.28130	0.74531
3	0.04368879	45.07905	0.28985	59.98344	0.07905	0.28985	0.01656
4	0.03010176	44.99316	0.28580	60.00549	0.00684	0.28580	0.00549
5	0.01160245	44.80199	0.28644	60.08301	0.19801	0.28644	0.08301
6	0.01160203	44.80199	0.28644	60.08302	0.19801	0.28644	0.08302

Example 6.5. In this example we repeat the previous experiment with the only difference that in the measurements of x and y additive random errors with normal distributions of absolute values less than 0.3 are included. The corresponding numerical results can be seen in Fig. 6.5 and in Table 6.5. With these noisy

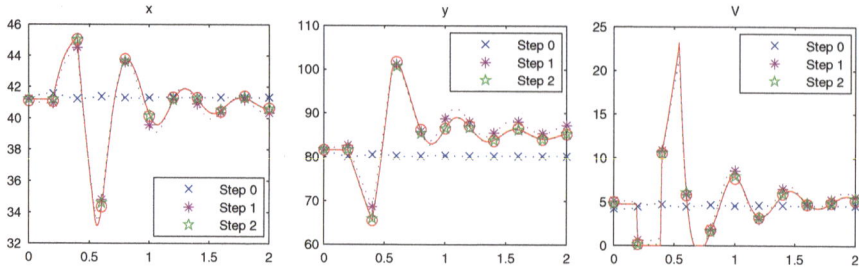

Fig. 6.5 Estimation of G_{P1}, G_{P2} and G_{P3} using noisy measurements of x and y

Table 6.5 Estimation of G_{P1}, G_{P2} and G_{P3} using noisy measurements of x and y

Step	Cost	G_{P1}	G_{P2}	G_{P3}	$\Delta(G_{P1})$	$\Delta(G_{P2})$	$\Delta(G_{P3})$
0	485.59595635	40.00000	40.00000	40.00000	5.00000	40.00000	20.00000
1	14.52482670	47.17405	5.54416	68.35639	2.17405	5.54416	8.35639
2	1.15085189	43.70021	1.16266	58.68728	1.29979	1.16266	1.31272
3	0.24101584	44.26992	0.20717	59.44590	0.73008	0.20717	0.55410
4	0.22474717	44.17809	0.20403	59.46826	0.82191	0.20403	0.53174
5	0.20156319	43.98286	0.20541	59.55063	1.01714	0.20541	0.44937
6	0.20156185	43.98286	0.20541	59.55066	1.01714	0.20541	0.44934

measurements the numerical results still converge, but we can observe larger errors, in G_{P1} and G_{P3}, than in the previous example. Repeating this experiment with different noisy measurements, even with larger random errors, always yield similar observations.

Example 6.6. In this example we assume that we do not have direct measurements of the solutions x and y, instead, we suppose we can measure the value of the ventilation function along the solution. Let $\bar{G}_{P1}, \bar{G}_{P3}, \bar{G}_{P3}$ denote the true parameters,

$$V_i = V(t_i, x(t_i; \bar{G}_{P1}, \bar{G}_{P3}, \bar{G}_{P3}), y(t_i; \bar{G}_{P1}, \bar{G}_{P3}, \bar{G}_{P3})), \qquad i = 0, 1, \ldots, M,$$

and now we use the following cost function

$$\tilde{J}(G_{P1}, G_{P3}, G_{P3}) = \sum_{i=0}^{M} (V(t_i, x(t_i; G_{P1}, G_{P3}, G_{P3}), y(t_i; G_{P1}, G_{P3}, G_{P3})) - V_i)^2$$

instead of the one defined by (6.14). Otherwise we used the same initial parameters and discretization constant as in the previous example. The corresponding results can be found in Fig. 6.6 and in Table 6.6. We can see that the measurements of the ventilation contained enough information on the parameters to guarantee the convergence of the method. In fact, in this case the last step was even better than that in the previous example.

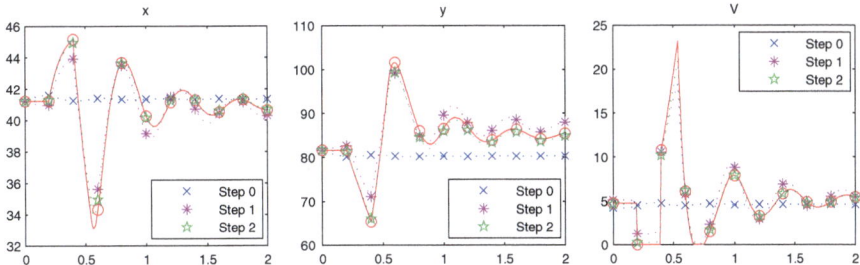

Fig. 6.6 Estimation of G_{P1}, G_{P2} and G_{P3} using measurements of V

Table 6.6 Estimation of G_{P1}, G_{P2} and G_{P3} using measurements of V

Step	Cost	G_{P1}	G_{P2}	G_{P3}	$\Delta(G_{P1})$	$\Delta(G_{P2})$	$\Delta(G_{P3})$
0	42.10899178	40.00000	40.00000	40.00000	5.00000	40.00000	20.00000
1	2.63937045	48.52067	12.50361	71.69544	3.52067	12.50361	11.69544
2	0.22630530	44.09782	2.03095	57.15278	0.90218	2.03095	2.84722
3	0.00867380	45.50737	0.13858	60.12323	0.50737	0.13858	0.12323
4	0.00443630	45.31892	0.13395	60.05083	0.31892	0.13395	0.05083
5	0.00044917	45.01624	0.13081	59.99976	0.01624	0.13081	0.00024
6	0.00025245	44.95685	0.12889	60.00478	0.04315	0.12889	0.00478

Table 6.7 Estimation of G_{P1}, G_{P2} and G_{P3} using noisy measurements of V

Step	Cost	G_{P1}	G_{P2}	G_{P3}	$\Delta(G_{P1})$	$\Delta(G_{P2})$	$\Delta(G_{P3})$
0	41.06549791	40.00000	40.00000	40.00000	5.00000	40.00000	20.00000
1	2.69930168	47.91054	12.67821	71.49751	2.91054	12.67821	11.49751
2	0.26392223	43.53271	2.32477	56.08888	1.46729	2.32477	3.91112
3	0.10162997	43.60376	0.87665	57.97591	1.39624	0.87665	2.02409
4	0.10161041	43.59960	0.87671	57.97828	1.40040	0.87671	2.02172
5	0.10148228	43.54956	0.88325	57.97853	1.45044	0.88325	2.02147
6	0.10142830	43.52050	0.89136	57.96415	1.47950	0.89136	2.03585
7	0.10135694	43.47035	0.90820	57.93390	1.52965	0.90820	2.06610

Example 6.7. We repeat the previous experiment but adding a random error of normal distribution with absolute value less than 0.3 to the measurements of V used in the previous example. With these noisy measurements the numerical approximations still converge, but the rate of convergence is very slow. We listed only the first seven steps of the numerical method in Table 6.7, and the first two iterates in Fig. 6.7. We can observe larger error than in the previous example.

Example 6.8. In this example we assume that the transport delay τ is the only unknown parameter. If we start the system from its equilibrium, then changing the time delay has no effect on the solution, therefore it is not possible to identify the delay from such measurement. Therefore it is necessary to move the system away

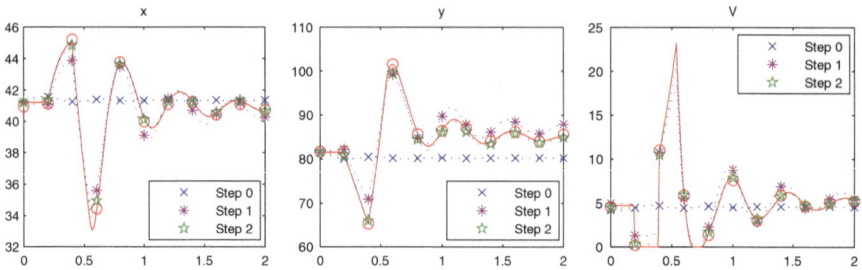

Fig. 6.7 Estimation of G_{P1}, G_{P2} and G_{P3} using noisy measurements of V

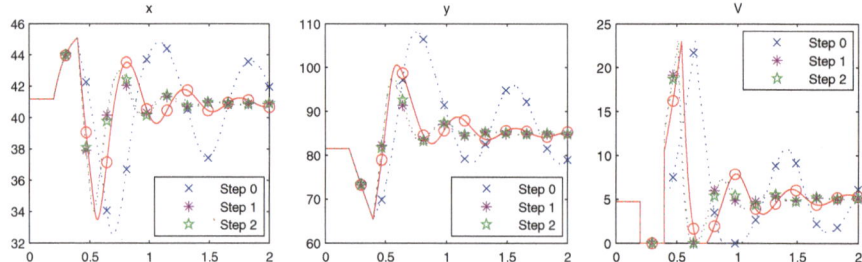

Fig. 6.8 Estimation of τ

Table 6.8 Estimation of τ

Step	Cost	τ	$\Delta(\tau)$
0	458.59350832	0.25000	0.10830
1	53.96078682	0.10704	0.03466
2	42.51290806	0.11226	0.02944
3	20.15586956	0.12306	0.01864
4	0.29816598	0.14364	0.00194
5	0.00000000	0.14178	0.00008

from the equilibrium. We apply the same procedure as before, i.e., we change the gain values at the switching times as follows

$$\theta_1 = 0.2, \quad \theta_2 = 0.4, \quad G_{P1} = 45, \quad G_{P2} = 0, \quad G_{P3} = 60.$$

We also observed that if we use measurements on the interval where the solution is still constant, i.e., on $[0, 0.2]$, then at these points the solution again does not depend on the delay, and numerical minimization method will not usually converge. Therefore now we used the interval $[T_0, T] = [0.3, 2]$ to make measurements using equidistant time points with $M = 11$. Starting from $\tau = 0.25$ and using $h = 0.0005$ we obtained a convergent sequence, what can be seen in Fig. 6.8 and in Table 6.8. We get again a very good approximation of the original delay value, $\tau = 0.1417$. In this experiment the convergence of the scheme is sensitive for the selection of the initial parameter value. The reason of it is that if at any step the numerical scheme produces a "large" τ, then using that τ the corresponding solution will be constant on $[0.3, 1]$, therefore the minimization will fail. Also, in identifying the delay the discretization constant has to be very small, since otherwise small change in the delay has no effect on the approximate solution, so the minimization will fail. For the same reason, in the minimization code the parameter which determines the time steps of computing approximate derivatives has to be relatively large (compared to the previous examples) otherwise again the change in the delay will not effect the solution, so the minimization will fail.

6.5 Conclusions

We have investigated parameter identification issues in a simplified model of the respiratory system. Case studies indicated identifiability of various system parameters, e.g., coefficients, gains, and transport delay. We obtained strong evidence that "information rich" input data significantly improves the accuracy of the determination of unknown parameters. Our numerical simulations also showed that identification of system parameters is more or less equally possible either by measuring O_2, CO_2 concentrations or ventilation data. The method presented here is applicable to models with multiple state-dependent delays [16, 17].

Acknowledgements This research was partially supported by the National Science Foundation under grant DMS-0705247 (FH and JT) and by Hungarian NFSR Grant No. K101217 (FH).

References

1. Banks, H.T., Burns, J.A., Cliff, E.M.: Parameter estimation and identification for systems with delays. SIAM J. Contr. Optim. **19**(6), 791–828 (1981)
2. Banks, H.T., Daniel Lamm, P.K.: Estimation of delays and other parameters in nonlinear functional differential equations. SIAM J. Contr. Optim. **21**(6), 895–915 (1983)
3. Batzel, J.J., Tran, H.T.: Stability of the human respiratory control system: I. Analysis of a two-dimensional delay state-space model. J. Math. Biol. **41**, 45–79 (2000)
4. Batzel, J.J., Tran, H.T.: Stability of the human respiratory control system: II. Analysis of a three-dimensional delay state-space model. J. Math. Biol. **41**, 80–102 (2000)
5. Cooke, K.L., Turi, J.: Stability, instability in delay equations modeling human respiration. J. Math. Biol. **32**, 535–543 (1994)
6. Cooke, K.L., Wiener, J.: Retarded differential equations with piecewise constant delays. J. Math. Anal. Appl. **99**, 265–297 (1984)
7. Cooke, K.L., Wiener, J.: Stability regions for linear equations with piecewise continuous delays. Comput. Math. Appl. **12A**, 695–701 (1986)
8. Cooke, K.L., Wiener, J.: A survey of differential equations with piecewise continuous arguments. In: Busenberg, S., Martelli, M. (eds.) Delay Differential Equations and Dynamical Systems. Lecture Notes in Mathematics, vol. 1475, pp. 1–15. Springer, Berlin (1991)
9. Dennis, J.E., Schnabel, R.B.: Numerical methods for unconstrained optimization and nonlinear equations. Prentice-Hall, Englewood Cliffs, NJ (1983)

10. Győri, I.: On approximation of the solutions of delay differential equations by using piecewise constant arguments. Int. J. Math. Math. Sci. **14**(1), 111–126 (1991)

11. Győri, I., Hartung, F., Turi, J.: Numerical approximations for a class of differential equations with time- and state-dependent delays. Appl. Math. Lett. **8**(6), 19–24 (1995)

12. Hartung, F., Herdman, T.L., Turi, J.: On existence, uniqueness and numerical approximation for neutral equations with state-dependent delays. Appl. Numer. Math. **24**, 393–409 (1997)

13. Hartung, F., Herdman, T.L., Turi, J.: Parameter identification in classes of hereditary systems of neutral type. Appl. Math. Comput. **89**(1-3), 147–160 (1998)

14. Hartung, F., Herdman, T.L., Turi, J.: Parameter identifications in classes of neutral differential equations with state-dependent delays. Nonlinear Anal. **39**(3), 305–325 (2000)

15. Hartung, F., Krisztin, T., Walther, H.O., Wu, J.: Functional differential equations with state-dependent delays: Theory and applications. In: Canada, A., Drabek, P., Fonda, A. (eds.) Handbook of Differential Equations, vol. 3, pp. 435–545. Elsevier, Amsterdam (2006)

16. Hartung, F., Turi, J.: Identification of parameters in delay equations with state-dependent delays. Nonlinear Anal. **29**(11), 1303–1318 (1997)

17. Hartung, F., Turi, J.: Linearized stability in functional differential equations with state-dependent delays. Discrete and Continuous Dynamical Systems (Added Volume), 416–425 (2001). Dynamical Systems and Delay Differential Equations (Kennesaw, GA, 2000)

18. Lunel, S.M.V.: Parameter identifiability of differential delay equations. Int. J. Adapt. Contr. Signal Process. **15**, 655–678 (2001)

19. Nakagiri, S., Yamamoto, M.: Identifiability of linear retarded systems in Banach spaces. Funkcial. Ekvac. **31**, 315–329 (1988)

20. Wiener, J.: Generalized Solutions of Functional Differential Equations. World Scientific, Singapore (1993)

Part II
Practice

Chapter 7
Experimental Studies of Respiration and Apnea

Eugene N. Bruce

Abstract The use of physiologically-based computational models of chemoreflex control of ventilation has provided general insights into the roles of specific mechanisms in the genesis of periodic breathing and apneas. Our early studies utilized formal mathematical approaches to simplify complex models of this type so that their behaviors could more easily be predicted from various combinations of physiological and environmental parameters. Because it is difficult to apply such models to individual patients, we subsequently pursued a "black-box" approach in which the objective was to characterize the dynamic properties of the system for individual subjects, then relate these properties to physiological and environmental parameters. By stimulating ventilation through pseudorandom variations in inspired CO_2 (or O_2) level, we estimated input–output models, both open-loop (i.e., from end-tidal P_{CO_2} to ventilation) and closed-loop (i.e., from inspired CO_2 to ventilation). We found that the dynamic properties of the resulting models differ between normal subjects and both sleep apnea patients and heart failure patients. We also demonstrated in normal subjects that the closed-loop model does not change between wakefulness and quiet sleep, even though the gain of the open-loop (or controller) model decreases. To explore the mechanistic basis for these findings using a detailed, physiologically-based, chemoreflex model, we enhanced the typical model of this type by improving the representation of O_2 transport and distribution beyond the usual, single lumped-compartment, approach. In our new model, brain and muscle tissue each comprise two subcompartments with intercompartmental diffusion and arterio-venous shunting, as well as O_2 binding to myoglobin in muscle. We use this model to predict changes in brain tissue P_{O_2} during sleep apnea. Chapter 8 provides another approach to respiratory control system modeling while Chap. 6 discusses the role of transport delay in respiratory control.

E.N. Bruce (✉)
Center for Biomedical Engineering, University of Kentucky, Lexington, KY, USA
e-mail: Eugene.Bruce@uky.edu

J.J. Batzel et al. (eds.), *Mathematical Modeling and Validation in Physiology*,
Lecture Notes in Mathematics 2064, DOI 10.1007/978-3-642-32882-4_7,
© Springer-Verlag Berlin Heidelberg 2013

7.1   Introduction

The use of physiologically-based computational models of chemoreflex control of breathing has provided general insights into the roles of specific mechanisms involved in the feedback control of ventilation in sleep apnea and periodic breathing (see also Chap. 8). These models are based on the principle of conservation of mass, usually applied to oxygen and carbon dioxide. Typically the models include lumped, uniform compartments representing the lungs, arterial blood, various large tissues (e.g., skeletal muscles, brain), and venous blood. In addition, these models usually incorporate equations to predict the chemoreflex control of ventilation ($\dot{V}$) and cardiac output (Q) as functions of partial pressures of CO_2 and O_2 in brain or arterial blood compartments . Periodic breathing often is viewed as a form of oscillation, and the models are examined for conditions which alter the stability of the system. On this basis, loop gain and time delays have received considerable emphasis [2, 8, 10, 20]. Because ventilatory control is subjected to frequent random disturbances, the potential contributions of transient "instabilities" (e.g., damped oscillations leading to stochastic resonances) have also been noted [14].

To model sleep apnea, it is necessary to distinguish between obstructive and central apneas and to include mechanisms for both. Thus, the usual approach is to represent the chemoreflex control of patency of the upper airway as a separate feedback pathway, with actual ventilation being dependent on both the chemoreflex-specified (or desired) ventilation and the degree of airway patency [13]. Obstructive apnea occurs when the airway patency falls to zero while desired ventilation is still nonzero.

In the cases of both periodic breathing and sleep apnea, these modeling studies have motivated general concepts which are difficult to validate when applied to individual patients. The major difficulty is the large number of parameters involved in such models. It is not possible to measure all of them in one patient, so the model might include a few which are measured and many which are typical of normal, healthy subjects. The quantitative differences between the model and data from individual patients are often too large to permit conclusions about specific mechanisms.

Other studies utilized formal mathematical approaches to simplify complex models of this type so that their behaviors could more easily be predicted for various combinations of physiological and environmental parameters [4, 9]. These approaches improved the ability to anticipate the general consequences of altering model parameters (such as changing the inspired gas) on the occurrence of periodic breathing or apnea; however, even when the model was simplified to one having, for example, six parameters, it was not possible to measure these parameter values in individual patients. Furthermore, these parameters of reduced models represent broad conceptual relationships between variables rather than basic physiological properties and the design of experimental protocols to estimate them in human subjects is challenging.

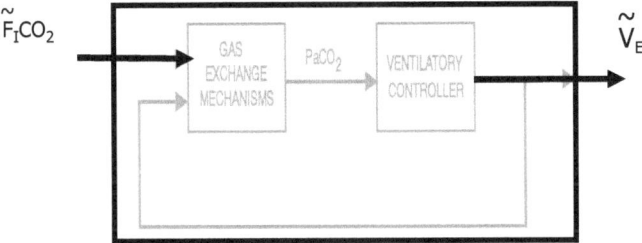

Fig. 7.1 General scheme for estimating a "black-box" model for control of ventilation: Apply variations in inspired CO_2 ($F_I\widetilde{C}O_2$) and measure resulting ventilatory fluctuations ($\widetilde{V}_E$). The internal mechanisms are not detected (such as the role of arterial carbon dioxide partial pressure P_{aCO_2}

7.2 "Black-box" Models of Chemoreflex Control of Ventilation

The black-box approach identifies properties of the ventilatory control system in individual subjects by applying a known stimulus and measuring the ventilatory response. This method had been utilized for many years by investigators who applied a stepwise change in the composition of the inspired gas. The method was most effective when used with a protocol for forcing a stepwise change in end-tidal gas pressure (rather than inspired gas pressure) developed by Swanson [19] and Slessarov [18] and refined by Poulin and Robbins [17]. However, given the well-known problems with estimating parameters based on persistently exciting stimuli [12, Chap. 3], we utilized a pseudorandom stimulus for the same purpose [15]. Figure 7.1 indicates the application of this method for determining the ventilatory response to CO_2 stimulation. The light gray lines represent feedback control of ventilation by CO_2 in its most abstract form, and the heavy black lines indicate that this entire system comprises the contents of the "black box". A computer monitors the subject's breathing in real time and varies the inspired CO_2 concentration according to a pseudorandom binary sequence (indicated by the tilde) [15]. The variations in ventilation (also indicated by a tilde) are related to the stimulus using a Box-Jenkins form of model [12], as shown in Fig. 7.2. The general input–output relation for this model is:

$$y(t) = \frac{B(q)}{F(q)}u(t-k) + \frac{C(q)}{D(q)}e(t), \qquad (7.1)$$

where q is the unit delay operator, t has units of discrete time, and $e(t)$ is uncorrelated random noise. $B(q), C(q), D(q)$ and $F(q)$ are polynomials in q whose coefficients are to be estimated. From continuous measurements of airflow and airway CO_2 level (Fig. 7.3) one derives ventilation ($\widetilde{V}_E$ (t)) and end-tidal P_{CO_2} ($\widetilde{P}_{CO_2}$) for each breath. (One may convert the discrete time interval "breath number"

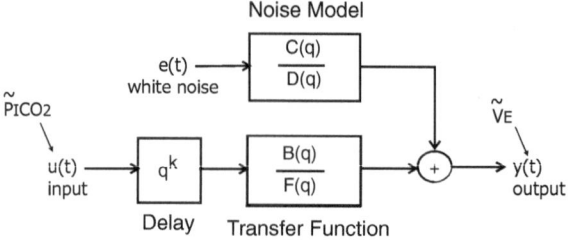

Fig. 7.2 Box-Jenkins model structure used to estimate parameters of the black-box model of ventilatory control. $P_I\tilde{C}O_2$ denotes applied variations of inspired CO_2 pressure; $\tilde{V}_E$ denotes variations of ventilation from its pre-stimulus mean level

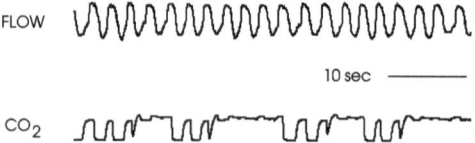

Fig. 7.3 Examples of data signals during pseudo-random CO_2 stimulation (awake human subject)

into sampled "continuous time" data, if desired.) The polynomial coefficients are estimated using an iterative, least squares method [12]. Polynomial orders are chosen on the basis of an AIC measure, and the same orders are used for data which are to be compared (e.g., before and after an intervention). In order to compare the models in different situations, we evaluate the unit-pulse response of the model (i.e., the response to a single pulse of $u(t)$).

We applied this approach to the question of why periodic breathing occurs more frequently during sleep than during wakefulness. The usual explanations have assumed that the loop gain of the ventilatory controller increases in sleep, leading to instability. Our analysis of closed-loop ventilatory response to a pulse of CO_2 in hyperoxia found no increase in oscillatory behavior during NREM sleep in young subjects, on average [16], although some individual subjects showed a slight increase (Fig. 7.4). Also, there may be a decrease in closed-loop stability when peripheral chemoreceptors are not depressed by hyperoxia (Chap. 1). Thus, the mechanisms by which ventilation becomes unstable near sleep onset probably involve fluctuations in state of arousal rather than state-dependent changes in controller properties [13]. To the contrary, elderly subjects typically exhibited a change from a non-oscillatory unit-pulse response awake to a damped oscillation in sleep (Fig. 7.4). Thus, we conclude that the ventilatory control system is less stable in NREM sleep in the elderly. In other comparisons we found that, compared to normal subjects, the ventilatory response is more unstable in COPD patients [11] and congestive heart failure patients [unpublished observations].

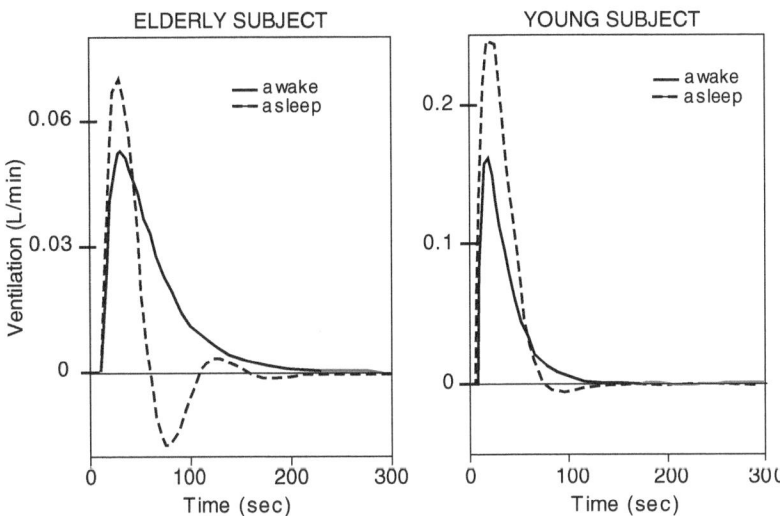

Fig. 7.4 Calculated unit-pulse responses of black-box models of ventilatory control for an elderly subject and a young subject awake (*solid*) and during NREM sleep (*dashed*)

7.3   A Model for Predicting Brain Tissue Oxygen Tensions

It is often assumed that the most deleterious aspect of sleep apnea is the impaired delivery of oxygen to the brain [1, 6]. The mechanisms which contribute to central and obstructive apneas, and the physiological responses which ensue and affect gas transport, are numerous and highly complex. Even models of ventilatory control based on simple principles can become very complicated [5]. Because of the difficulty of producing central and obstructive apneas in a model of ventilatory control in a manner that would mimic an actual patient, we proposed to develop a model of oxygen delivery to brain tissue that would be driven by a patient's actual ventilation (Fig. 7.5). Because the typical single, lumped, compartment representation of skeletal muscle tissue is inadequate for evaluating the range of P_{O_2} values that occur in a tissue, recently we developed a model of skeletal muscle tissue having two tissue and three vascular subcompartments [3]. Subsequently, we also developed a similar model for brain tissue; however, because gray matter and white matter have such different levels of metabolism and perfusion, they were studied separately. The basic structure of the brain tissue model is shown in Fig. 7.6 and the model equations are presented in the appendix.

To evaluate the model, its resting values were compared to values from the literature (Table 7.1) for white matter (WM), gray matter (GM), and whole brain (G+W). Of particular importance, the predicted oxygen tensions in the two brain tissue subcompartments lie in the lower half and upper half of the ranges of tissue P_{O_2} found in the literature. Therefore, it was concluded that the model should be

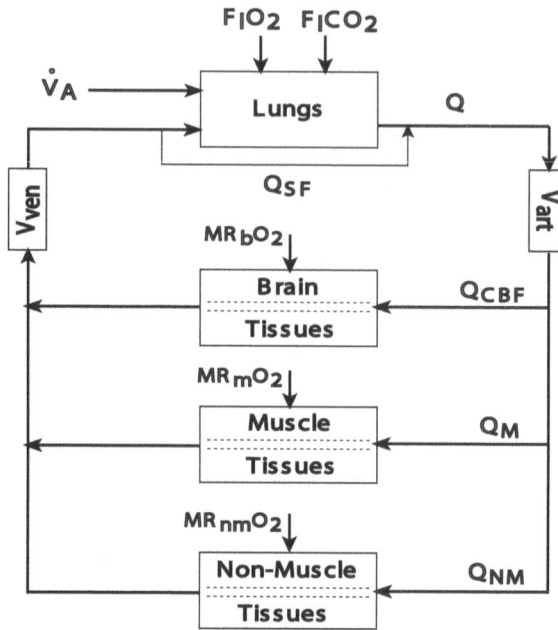

Fig. 7.5 Schematic of a model for predicting tissue oxygen levels (P_{O_2}) in the brain using a subject's measured alveolar ventilation ($\dot{V}_A$). Q, Q_{CBF}, Q_M, Q_{NM} denote cardiac output, and blood flows to brain (CBF), muscle (M), and nonmuscle (NM) compartments respectively. Q_{SF} denotes pulmonary shunt blood flow. V_{art}, V_{ren} represent volumes of arterial and mixed venous blood compartments respectively. $F_I\tilde{C}O_2$ and $F_I\tilde{O}_2$ denote fractions of CO_2 and O_2 in inspired gas respectively. MR_bO_2, MR_nO_2, and $MR_{nm}O_2$ denote metabolic rates in the brain tissue, muscle tissue and non-muscle tissue, respectively

useful for predicting brain tissue oxygen tensions. Steady states were simulated for various constant ventilation levels to determine the effect of reducing arterial P_{O_2} by hypoventilation (Fig. 7.7) during Waking and NREM sleep. The lower values of both brain and muscle P_{O_2} are relatively resistant to arterial hypoxia for $P_{aO_2} > 45$ Torr; however, in the model muscle P_{O_2} drop in NREM sleep whereas brain tissue P_{O_2} are preserved. Therefore, the decrease in ventilation accompanying NREM sleep alters brain P_{O_2} through its effect on P_{aO_2}, but there is no additional effect from the other factors that differ between waking and sleep. Subsequently we examined the effect on brain P_{O_2} of an apnea lasting one minute and of a 2 min hypoventilation followed by transient hyperventilation (Fig. 7.8). Apnea produces a rapid fall in brain P_{O_2} in both tissue subcompartments. P_{O_2} during apnea falls below 10 Torr, a level which may reflect even lower tensions and impairment to aerobic metabolism at some locations in the tissue. There is a tendency for the oxygen tensions of the two subcompartments to equalize.

Finally, we extracted the rib cage and abdominal signals from a polysomnogram of an elderly female subject (Fig. 7.9). Relative gain of these signals was determined

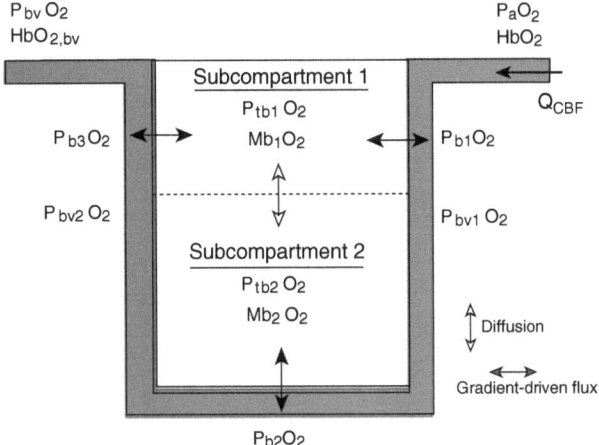

Fig. 7.6 Structure of model of brain tissue, comprising two tissue sub-compartments and three vascular sub-compartments. $P_{bk}O_2$ denotes oxygen partial pressure in vascular sub-compartment k; $P_{bvk}O_2$ denotes oxygen partial pressure at the distal (venous) end of vascular sub-compartment k. $P_{bv}O_2$ and $HbO_{2,bv}$ denote oxygen partial pressure and oxyhemoglobin content in venous outflow from the brain tissue vascular bed, respectively. $P_{tbk}O_2$ and Mb_kO_2 denote oxygen partial pressure and metabolic rate in tissue sub-compartment k, respectively. Q_{CBF} denotes brain blood flow

Table 7.1 Measured (from various studies in the literature) and model predicted values of oxygen related variables in gray (GM) and white (WM) matter and for the whole brain (G + M). CBF denotes cerebral blood flow and MRO_2 denotes metabolic rates. $P_{sag}O_2$ and $HbO_{2,sag}$ denote oxygen partial pressure and oxyhemoglobin saturation measured in sagittal sinus respectively. P_tO_2 denotes various brain tissue partial pressures of oxygen. $P_{cap}O_2$ and $P_{ven}O_2$ denote oxygen partial pressure measured near brain capillaries and venules, respectively. P_aO_2 represents arterial partial pressure of oxygen. For other symbols, see text in appendix

Parameter	Units	Measured	Model
MRO_2 (GM)	ml O_2/min/100 mg	5.7, 5.9	5.0
CBF (GM)	ml/min/100 mg	61, 65.3, 69, 62.0, 66.5	75
MRO_2 (WM)	ml O_2/min/100 mg	1.8, 1.43	
CBF (WM)	ml/min/100 mg	19.0, 21.4, 22.2	
MRO_2 (G + W)	ml O_2/min/100 mg	3.65, 3.2	
CBF (G + W)	ml/min/100 mg	55, 53.5	
$P_t O_2$ (GM)	Torr	5–15, 42, 7–42 (mean 23)	18, 32
$P_{cap}O_2$ (GM)	Torr		37.0
$P_t O_2$ (G + W)	Torr	10–40, 27–47, 12–48	
$P_{ven}O_2$ (G + W)	Torr	37.9–40.9	
$P_{ven}O_2$ (GM)	Torr		31.5
$P_{sag}O_2$ (G + W)	Torr	43.5, 44–60	
$P_{sag}O_2$ (GM)	Torr	y	31.6
% $HbO_{2,sag}$ (G + W)	Torr	68, 71.5	
% $HbO_{2,sag}$ (GM)	Torr		63.4

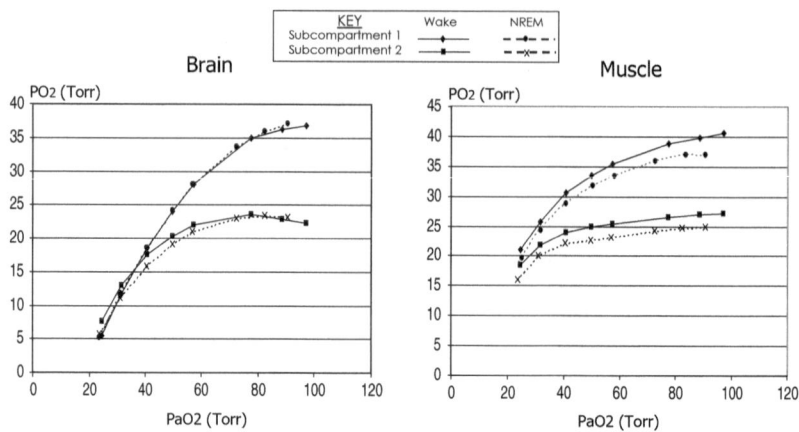

Fig. 7.7 Model predicted steady-state P_{O_2} values in brain and skeletal muscle tissues awake and in NREM sleep at various arterial oxygen partial pressures (P_{aO_2}) levels resulting from hypoventilation

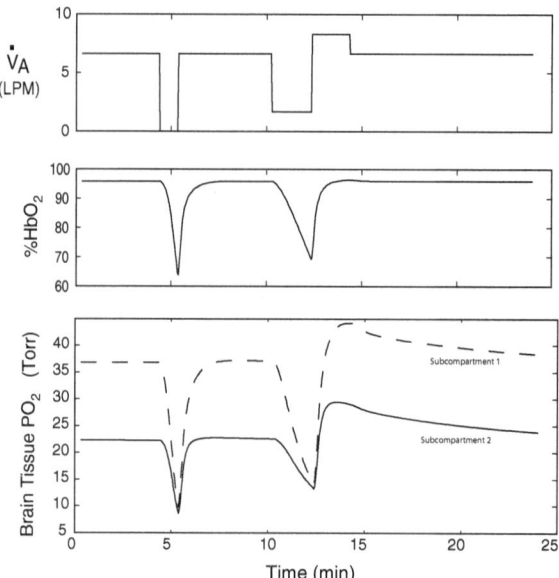

Fig. 7.8 Model predicted responses to an apnea lasting one minute and a subsequent 2 min hypopnea followed by a transient hyperventilation

by their excursions during an obstructive apnea, and amplitude scaling was evaluated by matching the HbO_2 predicted by the model when it was driven by this subject's ventilation signal during a period of uniform breathing. The predicted P_{O_2} in the two brain tissue compartments are also shown in Fig. 7.9. During the first

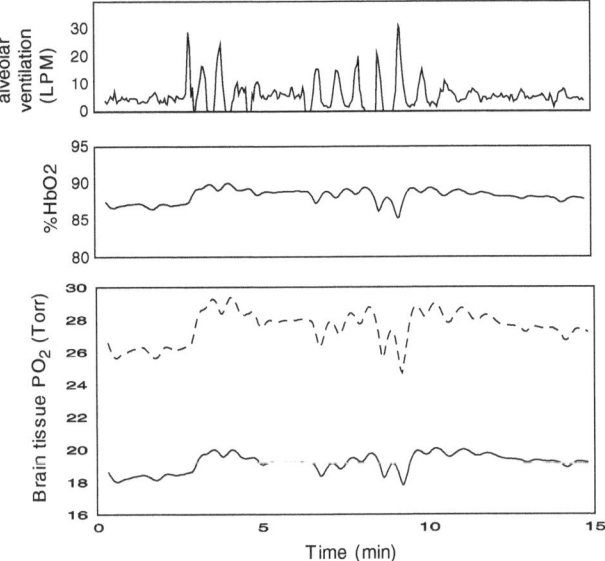

Fig. 7.9 Model predicted brain tissue P_{O_2} in response to two episodes of periodic breathing in an elderly female subject. Ventilation (Vdota) $\dot{V}_A$ was calculated from an actual polysomnogram as explained in the text

episode of periodic breathing brain P_{O_2} actually rise because of the hyperventilatory periods, whereas during the second episode these P_{O_2} fall progressively but still remain well above 10 Torr. The model predicts that brain tissue P_{O_2} will be highly correlated with % HbO_2.

It would be unwise to extrapolate from one subject but this simulation raises the question whether profound swings in ventilation during sleep will affect brain tissue P_{O_2} as profoundly. On the other hand, the predicted fluctuations in % HbO_2 are more attenuated than the measured changes in oxygen saturation in the polysomnogram, suggesting that predictions from the model about brain P_{O_2} might be underestimating the actual changes.

7.4 Conclusions

Although integrative models of chemoreflex control of ventilation predict that periodic breathing is related to factors which alter stability of feedback loops that control ventilatory demand and upper airway patency, it is not possible to test these concepts directly in individual patients. Indirect approaches using black-box models applied to individual subjects generally support these concepts; however, these latter approaches do not support the idea that ventilatory control is less stable in sleep than during wakefulness. Thus, the mechanisms by which ventilation becomes unstable near sleep onset probably involve fluctuations in state rather than state-dependent changes in physiological properties [7, 13]. The deleterious effects of longstanding sleep-disordered breathing include cognitive dysfunction [1, 6] which may be due to episodic severe brain hypoxia. We propose to utilize a model of oxygen delivery to the brain which is driven by the patient's own ventilation in order to estimate the degree of brain hypoxia experienced by the patient. Such a model will need to be "tuned" to each patient so that its dynamic behaviors represent the patient rather than merely mimicking average population behaviors.

Appendix

Conservation of mass is applied in each subcompartment for oxygen in the same manner as described in previous Chaps. 5 and 6, the difference being that oxygen can diffuse between the two tissue subcompartments via a flux driven by the concentration gradient. Thus, the typical mass balance equation for a tissue subcompartment has the form

$$\frac{dC_{vmk,g}(t)}{dt} = \frac{Flux_{mk,g}(t)}{V_{mk}} + \frac{D_{mk,g}(C_{mk,g}(t) - C_{mk',g}(t))}{Dx},$$

where

$$Flux_{mk,g}(t) = [D_{bmk,g}(P_{tck,g}(t) - P_{mk,g}(t))] - MR_{mk,g}$$

and g refers to either O_2 or CO_2, k represents subcompartment k (1 or 2), V_{mk} is the volume of that subcompartment, $C_{vmk,g}$ and $P_{mk,g}$ are the concentration and partial pressure of gas g in subcompartment k, $D_{mk,g}$ is the intratissue diffusion coefficient for gas g. $D_{bmk,g}$ is the blood-tissue conductance for gas g. $C_{mk,g}$ is the dissolved gas concentration, and $MR_{mk,g}$ is the metabolic rate of consumption ($MR > 0$) or production ($MR < 0$) of gas g in subcompartment k.

$P_{tck,g}$ is the effective partial pressure of gas g in the vascular compartment of subcompartment k driving the flux of g from the blood to the tissue. For O_2, this effective pressure is found by first determining the average HbO_2 concentration in the vascular compartment, then evaluating the corresponding pressure from the oxygen dissociation curve using an iterative method. The within-tissue diffusion coefficient for O_2 is set to 10 times the standard diffusivity [3]. D_{bmk,O_2} is the product of PSm, solubility, and tissue mass for both compartments. PSm is the permeability-surface area for muscle, which was set to 37.5. For exchange of O_2 between subcompartment 1 and the venular blood compartment, all blood-tissue conductances were set to 7.5 % of their values for subcompartment 1.

In vascular compartments O_2 exists in dissolved and combined forms so that in compartment i, the total O_2 concentration is $C_i O_2 = HbO_2 + S_{O_2} P_i O_2$. Other physiological factors are taken into account: (i) control of Q by changes in arterial P_{O_2}; (ii) control of brain blood flow (Q_{CBF}) by changes in arterial P_{O_2} and P_{CO_2} (the latter estimated from changes in ventilation by assuming that metabolic rate is unchanged); (iii) changes in muscle blood flow due to changes in arterial P_{O_2}; (iv)

effect of sleep state on brain and muscle tissue metabolic rates and perfusion. The first two factors are discussed by Topor [20]; the third is discussed in [3], and the effects of sleep state are represented as follows: whole body metabolic rate decreases by 10 % in NREM sleep; both Q and Q_{CBF} are determined as functions of arterial partial pressure of oxygen and carbon dioxide P_{aO_2} and P_{aCO_2} respectively, and both decrease by 10 % in NREM (relative to awake values). Changes in either blood flow occur with a time constant of 2 min.

Acknowledgements This work was supported by grants OH008651 from NIOSH and NS050289 from NIH.

References

1. Ayalon, L., Ancoli-Israel, S., Aka, A.A., McKenna, B.S., Drummond, S.P.: Relationship between obstructive sleep apnea severity and brain activation during a sustained attention task. Sleep **3**(32), 373–381 (2009)
2. Batzel, J.J., Tran, H.T.: Stability of the human respiratory control system. II. Analysis of a three-dimensional delay state-space model. J. Math. Biol. **1**(41), 80–102 (2000)
3. Bruce, E.N., Bruce, M.C., Erupaka, K.: Prediction of the rate of uptake of carbon monoxide from blood by extravascular tissues. Respir. Physiol. Neurobiol. **161**, 142–159 (2008)
4. Carley, D.W., Shannon, D.C.: A minimal mathematical model of human periodic breathing. J. Appl. Physiol. **3**(65), 1400–1409 (1988)
5. Cheng, L., Ivanova, O., Fan, H.H., Khoo, M.C.: An integrative model of respiratory and cardiovascular control in sleep-disordered breathing. Respir. Physiol. Neurobiol. **174**, 4–28 (2010)
6. Cohen-Gogo, S., Do, N.T., Levy, D., Métreau, J., Mornand, P., Parisot, P., Fauroux, B.: Sleep-disordered breathing in children. Arch. Pediatr. **2**(16), 123–131 (2009)
7. Deegan, P.C., McNicholas, W.T.: Pathophysiology of obstructive sleep apnoea. Eur. Respir. J. **8**(7), 1161–1178 (1995)
8. ElHefnawy, A., Saidel, G.M., Bruce, E.N.: CO2 control of the respiratory system: Plant dynamics and stability analysis. Ann. Biomed. Eng. **16**(5), 445–461 (1988)
9. ElHefnawy, A., Saidel, G.M., Bruce, E.N., Cherniack, N.S.: Stability analysis of CO2 control of ventilation. J. Appl. Physiol. **69**(2), 498–503 (1990)
10. Grodins, F.S., Buell, J., Bart, A.J.: Mathematical analysis and digital simulation of the respiratory control system. J. Appl. Physiol. **22**(2), 260–276 (1967)
11. Hudgel, D.W., Gordon, E.A., Thanakitcharu, S., Bruce, E.N.: Instability of ventilatory control in patients with obstructive sleep apnea. Am. J. Respir. Crit. Care Med. **158**(4), 1142–1149 (1998)
12. Ljung, L.: System Identification: Theory for the User. Prentice Hall, Englewood Clifffs, NJ (1987)
13. Longobardo, G.S., Evangelisti, C.J., Cherniack, N.S.: Analysis of the interplay between neurochemical control of respiration and upper airway mechanics producing upper airway obstruction during sleep in humans. Exp. Physiol. **93**(2), 271–287 (2008)
14. Modarreszadeh, M., Bruce, E.N.: Ventilatory variability induced by spontaneous variations of PaCO2 in humans. J. Appl. Physiol. **76**(6), 2765–2775 (1994)
15. Modarreszadeh, M., Bruce, E.N.: Long-lasting ventilatory response of humans to a single breath of hypercapnia in hyperoxia. J. Appl. Physiol. **72**(1), 242–250 (1992)
16. Modarreszadeh, M., Bruce, E., Hamilton, H., Hudgel, D.: Ventilatory stability to CO2 disturbances in wakefulness and quiet sleep. J. Appl. Physiol. **79**(4), 1071–1081 (1995)

17. Poulin, M., Liang, P., Robbins, P.: Dynamics of the cerebral blood flow response to step changes in end-tidal PCO2 and PO2 in humans. J. Appl. Physiol. **81**(3), 1084–1095 (1996)
18. Slessarev, M., Han, J., Mardimae, A., Prisman, E., Presii, D., Volgyesi, G., Ansel, C., Duffin, J., Fisher, J.A.: Prospective targeting and control of end-tidal CO2 and O2 concentrations. J. Physiol. **581**, 1207–1219 (2007)
19. Swanson, G.D.: Evaluation of the Grodins respiratory model via dynamic end-tidal forcing. Am. J. Physiol. **233**(1), R66–72 (1977)
20. Topor, Z.L., Pawlicki, M., Remmers, J.E.: A computational model of the human respiratory control system: Responses to hypoxia and hypercapnia. Ann. Biomed. Eng. **32**(11), 1530–1545 (2004)

Chapter 8
Model Validation and Control Issues in the Respiratory System

James Duffin

Abstract This chapter develops static and dynamic models of the chemoreflex control of breathing based on experimental measurements. A graphical concept model of the steady state based on current physiology is built up first, which demonstrates key concepts in the control of breathing such as loop gain and its clinical partner CO_2 reserve. The Stewart approach to modeling acid-base is used to convert this static model to handle the effects of acid-base changes on respiratory control. Finally, this static model of the chemoreflex control system is incorporated into a dynamic simulation of the control of breathing and acid-base balance using a graphical programming language. The dynamic model demonstrates the instabilities observed during sleep at altitude and the effects of changes in cerebrovascular reactivity on loop gain and stability that are a part of the sleep apnoea syndrome. Hence this chapter will also draw connections to the chapter by Bruce (Chap. 7).

8.1 Introduction

8.1.1 The Respiratory Control System

Breathing is responsible for supplying sufficient oxygen (O_2) for metabolism and eliminating the carbon dioxide (CO_2) produced by metabolism. The respiratory control system accomplishes this aim by altering pulmonary ventilation so that at equilibrium, i.e., steady state, O_2 uptake at the lungs equals O_2 consumption by the tissues, and CO_2 elimination at the lungs equals CO_2 production by the tissues.

As Fig. 8.1 illustrates, when pulmonary gas exchange matches metabolism, tissue partial pressures of oxygen (P_{O_2}) and carbon dioxide (P_{CO_2}) remain constant, and so

J. Duffin (✉)

Thornhill Research Inc., 210 Dundas St. W. Suite 200, Toronto, ON, Canada, M5G 2E8

e-mail: j.duffin@utoronto.ca

J.J. Batzel et al. (eds.), *Mathematical Modeling and Validation in Physiology*,
Lecture Notes in Mathematics 2064, DOI 10.1007/978-3-642-32882-4_8,
© Springer-Verlag Berlin Heidelberg 2013

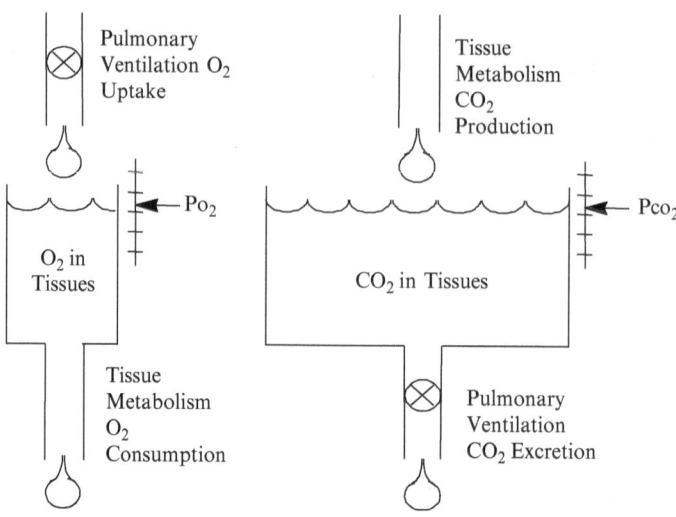

Fig. 8.1 A conceptual model showing the balance between metabolic requirements and pulmonary gas exchange of O_2 and CO_2. Notice the difference in storage compartment sizes for O_2 and CO_2

the control system is set up to accomplish its main goal by constraining these partial pressures within limits. The diagram also displays the difference in the storage capacities for O_2 and CO_2; as a consequence changes in P_{O_2} are fast but those for P_{CO_2} are slow.

What are the constraints for the partial pressures? The requirement for P_{O_2} is relatively simple; it should be kept at a partial pressure that saturates arterial haemoglobin and provides a gradient that is sufficient to supply tissue metabolic requirements for O_2. Since the carriage of oxygen in blood is such that saturation can be achieved over a wide range of P_{O_2}, it need not be closely regulated unless saturation falls (hypoxia). By contrast the requirement for P_{CO_2} is not based on providing a gradient sufficient for CO_2 elimination, because CO_2 diffuses through an aqueous environment much more easily that oxygen. Instead, P_{CO_2} must be at a level that ensures hydrogen ion concentrations ($[H^+]$) remain within the limits necessary for protein function (about 40 nM/l). This requirement is a major goal of the respiratory control system, and is so important that it also involves another control system; renal adjustment of ionic concentrations to control $[H^+]$.

$[H^+]$ depends directly on P_{CO_2}. So by controlling P_{CO_2} (normally 40 mmHg) pulmonary ventilation controls $[H^+]$. These changes may be produced rapidly, but if they fail or are inappropriate then renal and other control mechanisms operate the long term to control the ionic content of blood and tissues, thereby altering the balance of concentrations between negative and positive strongly dissociated ions known as the strong ion difference [SID] [12,38,40]. Since the renal control system is slow relative to the respiratory system its effects can be modeled as parameter changes for different conditions in a dynamic model with a time base in minutes.

Forward Part of Loop

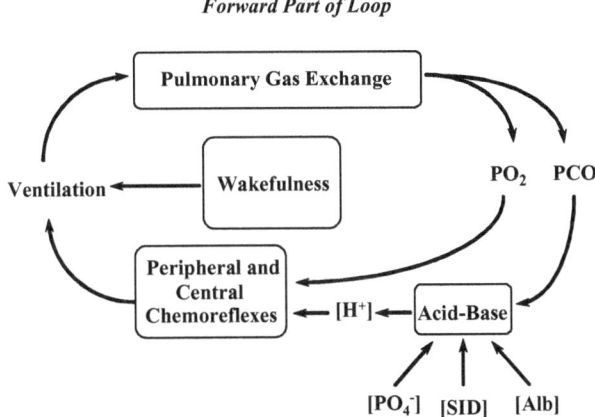

Fig. 8.2 The respiratory chemoreflex regulator; a negative feedback system. Note that the stimulus to the chemoreceptors is $[H^+]$ while pulmonary gas exchange controls P_{CO_2}; acid-base changes will alter this relationship

The respiratory control system responsible for maintaining P_{CO_2} is a negative feedback regulator with two chemoreceptors that sense $[H^+]$, one in the brain (central chemoreceptors) and one in the carotid bodies (peripheral chemoreceptors) [8] as pictured in the block diagram of Fig. 8.2. Increases in $[H^+]$ at the chemoreceptors stimulate breathing via a chemoreflex arc that includes the chemoreceptors, central nervous system integration of these signals, transmission to the respiratory muscles and consequent production of pulmonary ventilation. This chemoreflex control system also guards against asphyxia by increasing the sensitivity of the peripheral chemoreceptors to $[H^+]$ during hypoxia.

During exercise metabolism increases and therefore ventilation must also increase quickly if pulmonary gas exchange is to match metabolism. The chemoreflex control system does not provide a sufficiently fast regulation of P_{CO_2} and P_{O_2}, and so another control system intervenes. Fast neural response mechanisms including afferent feedback from the exercising muscles and a parallel activation of respiration with the exercising muscles termed central command provide the major drives to increase breathing during exercise. Modeling these responses to exercise is beyond the scope of the model developed here.

The chemoreflex control system is not the only user of the respiratory muscles. Other uses such as speech [31] may interrupt. These non-chemoreflex inputs or drives to the respiratory muscles are often termed behavioral [15, 28, 35]. Since these drives disappear during sleep, they can be considered as a wakefulness drive to breathe, and in any particular state can be modeled as a constant. In between these systems are other control systems, such as that responsible for the choice of efficient patterns of breathing in terms of tidal volume and respiratory rate, and the protection of the airways by pulmonary reflexes such as cough.

8.1.2   The Central Chemoreceptors

The central chemoreceptors are located in the medulla (near the ventrolateral surface and scattered within the brain tissue) and respond to the $[H^+]$ of their local environment [14, 27]. They are often thought of as CO_2 receptors because central $[H^+]$ is directly dependent on the P_{CO_2} of chemoreceptor tissue. Two factors complicate the physiology of the central chemoreceptors. First, the blood-brain barrier to polar solutes prevents changes in arterial $[H^+]$ from reaching the central environment easily, whereas CO_2 passes freely across the barrier. As a result, central $[H^+]$ may differ from arterial $[H^+]$. The central chemoreceptors are therefore somewhat isolated from arterial acid-base disturbances, except as they involve changes in arterial P_{CO_2}. Indeed, control of central acid-base to keep central $[H^+]$ at normal values may result in differences in the relation between P_{CO_2} and central $[H^+]$ that produce shifts in the central chemoreflex ventilatory response to CO_2 [8].

Second, because the central chemoreceptors are in brain tissue, they are relatively slow in responding even to changes in P_{CO_2} in arterial blood so that the time constant for the central chemoreceptor response to a change in alveolar P_{CO_2} is about 100 s; hence, it takes 5 min (three time constants) for the system to respond fully to changes in inspired CO_2. Furthermore, increases in cerebral blood flow in response to increases in arterial P_{CO_2} will affect the relation between arterial P_{CO_2} and central chemoreceptor P_{CO_2}. If cerebral blood flow increases markedly with P_{CO_2} the difference between central chemoreceptor P_{CO_2} and arterial P_{CO_2} declines as arterial P_{CO_2} increases.

8.1.3   The Peripheral Chemoreceptors

The peripheral chemoreceptors, located in the carotid bodies at the bifurcation of the carotid arteries, "taste" the blood approaching the brain. Signals from the receptors are sent to the respiratory controller in the medulla via the carotid sinus nerve (a branch of the glossopharyngeal or IX-th cranial nerve) [18,22]. In many textbooks the peripheral chemoreceptors are described as hypoxia sensors with their sensitivity to $[H^+]$ largely ignored. This view is misleading as it implies an independent drive to breathe associated with hypoxia. Such a drive only becomes significant at extreme hypoxia and is not evident in moderate hypoxia [25, 32].

Hypoxia's primary role is to increase the sensitivity of the peripheral chemoreceptors to arterial $[H^+]$. This feature of the peripheral chemoreceptors has two important implications: if the P_{O_2} is high, there is little (if any) peripherally-mediated ventilatory response to P_{CO_2}; and if the P_{CO_2} is low (and therefore below a ventilatory recruitment threshold), there is little (if any) response to hypoxia [20]. These receptors are therefore maximally stimulated by a simultaneous increase in $[H^+]$ and decrease in P_{O_2}, i.e., by asphyxia [39].

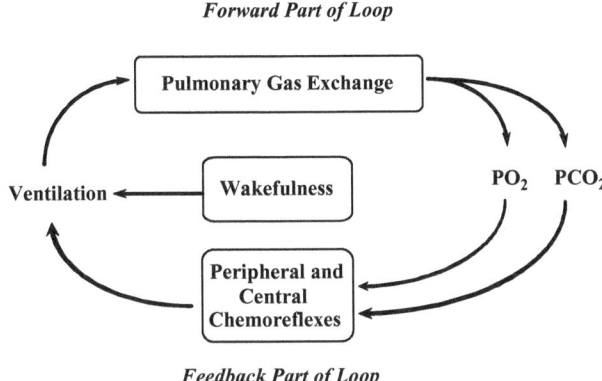

Fig. 8.3 The respiratory chemoreflex regulator in terms of inputs P_{CO_2} and P_{O_2}

8.2   Modeling the Chemoreflex Control System

8.2.1   Measurement Techniques

Ideally the input–output relationships for a model of the chemoreflexes would be determined by measuring the chemoreceptor inputs of $[H^+]$ and P_{aO_2} and the resulting ventilation as shown in Fig. 8.2. Such measurements cannot be made non-invasively and so indirect measurements are substituted. Instead of $[H^+]$, P_{CO_2} is measured, with end-tidal P_{CO_2} and P_{O_2} as reasonably good estimates of arterial P_{CO_2} and P_{O_2} in healthy young individuals. The respiratory feedback control system becomes that shown in Fig. 8.3.

While end-tidal P_{CO_2} and P_{O_2} will measure the inputs to the peripheral chemoreceptors, they do not measure the central P_{CO_2}, the input to the central chemoreceptors. Read [33] described a rebreathing method to overcome this difficulty. Individuals rebreathe from a small bag, which at the start of rebreathing contains enough CO_2 to equilibrate the P_{CO_2} of alveolar gas, and arterial blood to the P_{CO_2} of mixed venous blood before recirculation occurs. After recirculation, CO_2 produced by metabolism in the tissues slowly accumulates, and with cardiac output, cerebral blood flow and ventilation acting as mixers, the P_{CO_2} of the alveolar gas, arterial blood and venous blood all rise together. Since venous P_{CO_2} can be taken as a good estimate of central P_{CO_2} during rebreathing, end-tidal P_{CO_2} during rebreathing is a measure of the input to the central chemoreceptors. Thus, rebreathing as described by Read provides a means of measuring the inputs to the central and peripheral chemoreceptors while the slowly increasing P_{CO_2} stimulates the output ventilation, and so the input–output relationship can be determined.

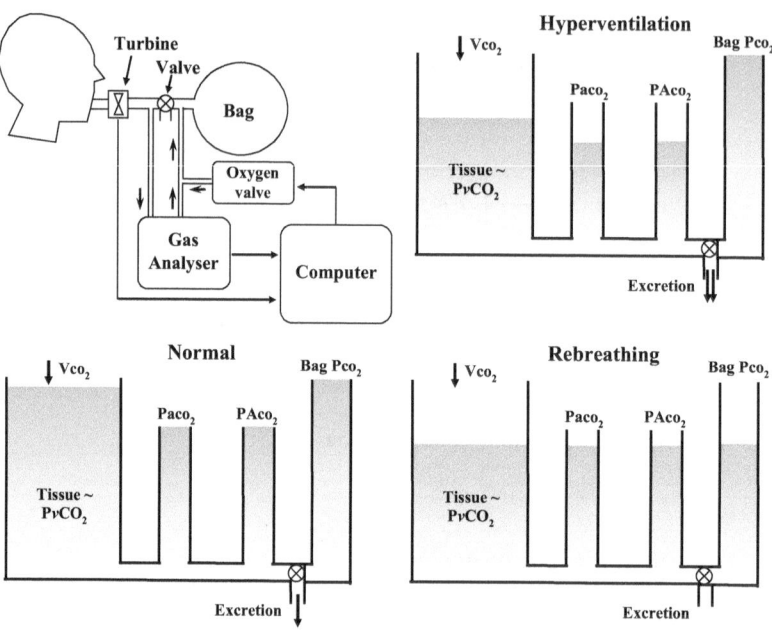

Fig. 8.4 Rebreathing: The apparatus is a shown at the *upper left*. The compartment diagrams illustrate the changes in P_{CO_2} during a modified (Duffin) rebreathing test. During normal breathing, metabolism produces CO_2 that enters the tissue compartment which flows via the venous blood to the lungs where it is eliminated so that arterial (P_{a,CO_2}) and alveolar (P_{A,CO_2}) partial pressures are less than those in venous and tissue compartments. Hyperventilation increases pulmonary excretion to quickly lower P_{a,CO_2} and P_{A,CO_2}, and over the course of 5 min, tissue and venous P_{CO_2} tensions decline. When the valve is operated so that rebreathing begins, the CO_2 in the bag quickly increases P_{a,CO_2} and P_{A,CO_2} to venous and tissue tensions while decreasing bag P_{CO_2} to the same tension. As rebreathing continues the metabolic production of CO_2 is low enough at rest so that circulation and breathing keep all P_{CO_2} levels equilibrated. In this way measurement of end-tidal P_{CO_2} at the mouth is a good estimate of both central and peripheral chemoreceptor inputs

Read's method is a hyperoxic method and so the ventilatory contribution of the peripheral chemoreceptors is minimized. In 1988, I introduced a rebreathing method [10] that was capable of measuring the input–output relations of the chemoreceptors during isoxia. In this method (Fig. 8.4) the rebreathing bag initially holds the target isoxia, and oxygen is delivered to the bag to supply the oxygen consumption and maintain isoxia. Another modification was to start the rebreathing after a 5 min period of hyperventilation so that the initial equilibration was hypocapnic. As P_{CO_2} slowly increases during rebreathing, ventilation remains unchanged until P_{CO_2} reaches a ventilatory recruitment threshold, above which it increases with P_{CO_2}. This method thus controls the peripheral and central respiratory chemoreceptor stimuli, in terms of P_{CO_2} and P_{O_2}, independent of both pulmonary gas exchange and cerebral blood flow. Figure 8.5 illustrates the input–output relations measured by this method.

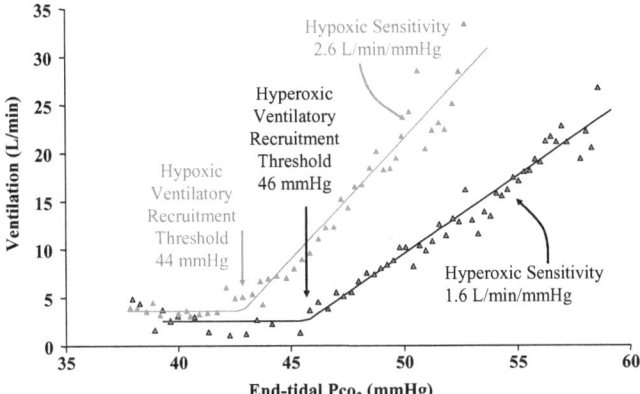

Fig. 8.5 Duffin rebreathing test results for an isoxic hypoxic ($P_{O_2} = 50$ mmHg) test (*grey*) and a hyperoxic ($P_{O_2} = 150$ mmHg) test (*black*). The points are breath-by-breath values and the lines are the fitted input–output relations that define the ventilatory recruitment threshold and the slope/sensitivity parameters. The sub-threshold or basal ventilation measures the wakefulness drive to breathe

8.2.2 Chemoreflex Model Parameter Estimation

The input–output relations are interpreted as follows. The constant ventilation below the ventilatory recruitment threshold is termed the basal ventilation and taken as a measure of the wakefulness or non-chemoreflex drive to breathing [13, 30, 35]. The slope of the linear increase in ventilation with P_{CO_2} above the ventilatory recruitment threshold is taken as the sensitivity of the chemoreflex response. Accumulated data from many rebreathing tests was used to determine an average input–output relation for several levels of isoxia [11]. While the ventilatory recruitment thresholds at various isoxic P_{O_2} levels varied little between individuals, the sensitivity varied considerably between individuals [25], but nevertheless followed a similar trend with the isoxic level as shown in Fig. 8.6. Both measures varied rectangular hyperbolically with isoxic P_{O_2}.

Using these fits an average chemoreflex input–output relationship was constructed as shown in Fig. 8.7. The various isoxic lines were constrained to converge to a point that determined the thresholds for the central and peripheral chemoreceptors; defined as that P_{CO_2} above which the chemoreceptor neural activity output increases with P_{CO_2}. However, ventilation does not increase until a chemoreceptor neural activity threshold is exceeded, and that neural activity threshold sets the ventilatory recruitment thresholds. The hyperoxic isoxic lines were considered to reflect the response of the central chemoreceptors alone since hyperoxia can be considered to minimize the peripheral chemoreceptors contribution [6]. With that assumption and the assumption that the central and peripheral contributions to the ventilation response are additive, the difference between the hyperoxic response and the normoxic and hypoxic responses can be considered as the peripheral neural

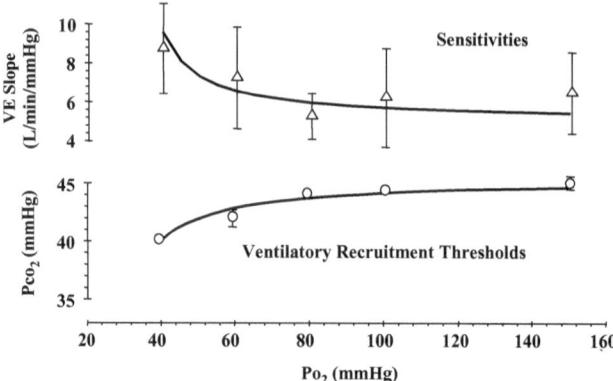

Fig. 8.6 Average (SEM error bars) sensitivities and ventilatory recruitment thresholds from many Duffin rebreathing tests at different isoxic P_{O_2}

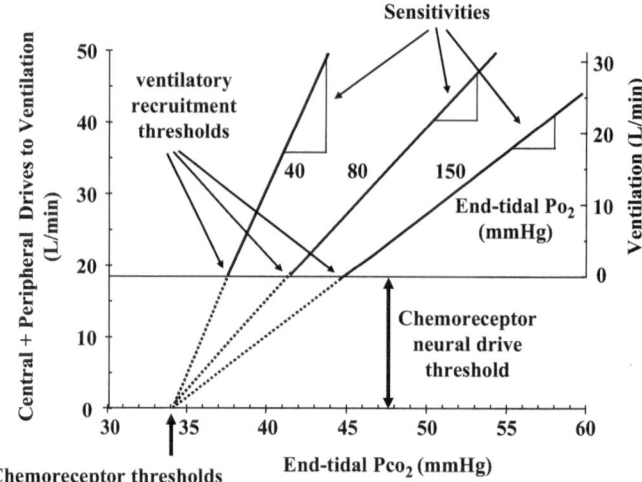

Fig. 8.7 Chemoreflex input–output relationships for an average individual. The linear relationships (*solid lines*) between ventilation output and P_{CO_2} input were constructed using the fitted data of Fig. 8.6 for ventilatory recruitment thresholds and slopes/sensitivities for the isoxic P_{O_2} tensions (three shown here) with the constraint that when extended (*dotted lines*) they met at a point defined as the chemoreceptors' thresholds. Conceptually, the chemoreceptors were considered to provide a neural activity drive to breathe when the chemoreceptor threshold was exceeded but ventilation remained unaffected until a chemoreceptor neural drive threshold was exceeded

drive to breathe. Although the assumption of additive responses has been challenged recently in animal experiments [4], those from humans [37] support it.

This estimate of the parameters for the chemoreflex model of input–output in terms of inputs P_{CO_2} and P_{O_2} and output ventilation for an average individual served to set the overall characteristics; linear ventilatory drive responses above a chemoreceptor threshold for each of the central and peripheral chemoreceptors, which add to produce a neural drive to breathe that acts to increase pulmonary ventilation after a neural drive threshold is exceeded. The point at which ventilation begins to increase as neural drive exceeds the neural drive threshold determines a ventilatory recruitment threshold. It now remains to convert the model inputs of P_{O_2} and P_{CO_2} to the actual chemoreceptor inputs of P_{O_2} and $[H^+]$.

8.3 Modeling Acid-Base Relations

The traditional model of acid-base is based on the equilibrium reaction of CO_2 and water expressed by the Henderson equation and later its logarithmic version, the Henderson–Hasselbach equation, which defines pH. This approach is easy to understand; $[H^+]$ can be considered to be controlled by the respiratory control of P_{CO_2} and the renal control of $[HCO_3^-]$. Because $[HCO_3^-]$ is in such excess, changes in P_{CO_2} produce changes in $[H^+]$ but hardly any in $[HCO_3^-]$ so that there is a virtually linear relationship between P_{CO_2} and $[H^+]$. However, this approach considers $[HCO_3^-]$ to be an independently controlled variable whereas it is actually a dependent variable; it is the movement of the strongly-dissociated ions like sodium and potassium that are transported across cell membranes rather than bicarbonate and hydrogen.

The model of acid-base relations used here is therefore based on the physico-chemical approach introduced by Peter Stewart [38]. In Stewart's approach, P_{CO_2} and two other variables, SID (strong ion difference; the concentration difference of strongly dissociated positive and negative ions in solution) and [Atot] (the total concentration of weakly dissociated anions in solution) are the independent variables that determine the dependent variables $[H^+]$ and $[HCO_3^-]$.

This model has been updated to consider the concentration of weakly dissociated anions in plasma in more detail, with phosphate and albumin concentrations substituted for [Atot] [40]. The resulting modified Stewart model equations are listed in Appendix I. The chemoreflex model can now be developed using the modified Stewart model to convert the P_{CO_2} resulting from pulmonary ventilation into the $[H^+]$ inputs to the central and peripheral chemoreceptors as shown in Fig. 8.2 with the relevant equations given in Appendix II.

8.4 The Chemoreflex Control System Model

8.4.1 The Central Chemoreflex

The drive to breathe from the central chemoreceptors can be modeled as a linear increase with $[H^+]$ above a chemoreceptor threshold and the slope of this response is the central sensitivity (Eq. (8.9); Appendix II). The central chemoreflex response to any particular $[H^+]$ is therefore determined by both its sensitivity and chemoreceptor threshold (see Fig. 8.8).

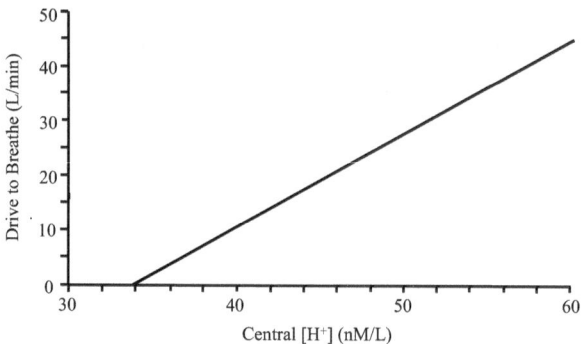

Fig. 8.8 The response of the central chemoreceptors in terms of drive to breathe expressed as l/min of pulmonary ventilation resulting from a central $[H^+]$ input stimulus

8.4.2   The Peripheral Chemoreflex

The peripheral chemoreceptors respond to both the $[H^+]$ and the P_{O_2} of arterial blood, such that, in response to hypoxia, the sensitivity of the chemoreceptor response to $[H^+]$ is increased. Like the central chemoreflex, the drive to breathe from peripheral chemoreceptors increases linearly with $[H^+]$ above a chemoreceptor threshold, with the slope of the response representing its sensitivity, so the response to any particular $[H^+]$ is determined by both sensitivity and threshold. However, for the peripheral chemoreceptors, the sensitivity of the response also depends on the P_{O_2}, in a rectangular hyperbolic relation such that hypoxia increases the P_{CO_2} sensitivity markedly, and hyperoxia decreases it almost to zero; as a result the response is represented by a series of isoxic lines (see Fig. 8.9). In addition to these responses a tonic peripheral chemoreceptor drive to breathe may be present ([7] and can be included in the model (Eq. (8.12); Appendix II)). It is in the presence of prolonged continuous or intermitted hypoxia, that the tonic peripheral chemoreflex drive may undergo modifications [9]; these changes are beyond the scope of the model developed here.

8.4.3   The Complete Chemoreflex Control System Model

The central and peripheral chemoreceptor drives to breathe summate to provide the total chemoreflex drive to breathe [37], but do not affect pulmonary ventilation until the total ventilatory drive exceeds a drive threshold; only then is breathing affected. Thus, an increase in $[H^+]$ does not increase ventilation until a threshold $[H^+]$ has been exceeded; referred to as the ventilatory recruitment threshold. This ventilatory recruitment threshold is equivalent to the apnoeic threshold when the wakefulness drive is absent during sleep [24] (Eq. (8.12); Appendix II). Above the

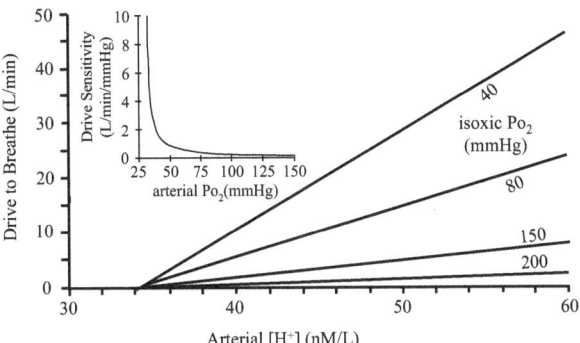

Fig. 8.9 The response of the peripheral chemoreceptors in terms of drive to breathe expressed as l/min of pulmonary ventilation resulting from an arterial $[H^+]$ stimulus. The *inset* shows the relation between P_{O_2} and the sensitivity to $[H^+]$

ventilatory recruitment threshold, the ventilatory response is usually linear, with a slope (sensitivity) varying with P_{O_2} [25]. When P_{O_2} is high, the ventilatory to response CO_2 is due almost entirely to the central chemoreflex [5].

A complete graphical picture of the control of breathing in a resting individual may be developed by assuming a difference between central and arterial $[H^+]$ of 6 nM/l. This value was chosen partly as a compromise between the measured differences between CSF and arterial P_{CO_2} of about 10 mmHg, and suggestions that the actual medullary P_{CO_2} in the region of the central chemoreceptors may be only a few mmHg above arterial due to its differential perfusion [17]. This assumption has the effect of displacing the central chemoreceptor threshold downwards, because central $[H^+]$ is greater than arterial. Figure 8.10 shows these combined responses and the effect of the drive threshold in determining the ventilatory recruitment thresholds.

When $[H^+]$ is below the ventilatory recruitment threshold, ventilation is maintained by the waking neural drive [13, 29, 30, 35]. However, when this drive is lost during sleep, a $[H^+]$ below the ventilatory recruitment threshold results in apnea (see Chap. 7). The basal ventilations measured in the Duffin rebreathing tests were averaged to produce the wakefulness drive parameter for the model. The chemoreflex and wakefulness drives to breathe are added to produce a graph of the dependence of ventilation on $[H^+]$. The resulting equations are detailed in Appendix II.

8.4.4   The Graphical Model of the Chemoreflex Control of Breathing

Finally, to estimate the resting ventilation and P_{CO_2} in an individual at rest the model may be used to calculate the dependence of ventilation on P_{CO_2} for any constant

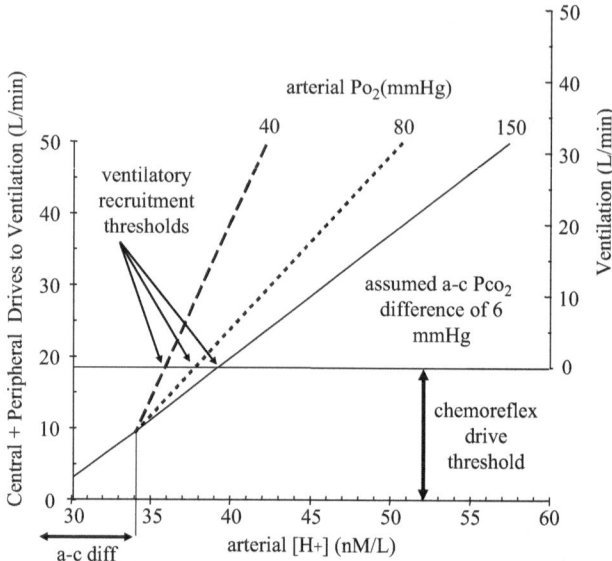

Fig. 8.10 The response of the combined peripheral and central chemoreceptors in terms of drive to breathe expressed as l/min of pulmonary ventilation resulting from an arterial [H$^+$] stimulus at constant P_{O_2}. The central and peripheral chemoreceptor thresholds are 34 nM/l but differ in this graph because it is plotted vs. arterial [H$^+$] and central P_{CO_2} is higher than arterial P_{CO_2} by an amount dependent on cerebral blood flow and central metabolism (here assumed as 6 mmHg)

acid-base state and P_{O_2} as shown in Fig. 8.11. This graph can also display the dependence of P_{CO_2} on ventilation, i.e., the forward part of the control loop shown in Fig. 8.2, which is termed the iso-metabolic hyperbola; a relationship resulting from the dependence of the amount of CO_2 excreted on the lung ventilation and P_{CO_2}. The intersection of the two relations, forward, and chemoreflex feedback (plus the independent or wakefulness drive), determines the control system equilibrium point; i.e., resting P_{CO_2} and ventilation as shown in Fig. 8.11.

This steady state model of the chemoreflex control system can be used to predict the ventilation and P_{CO_2} for many situations and is therefore useful for understanding the changes in respiratory control physiology that may occur under different conditions. As an example consider the changes that occur on ascent to altitude. The peripheral chemoreflex sensitivity to [H$^+$] increases on ascent because the inspired P_{O_2} is decreased and this in turn stimulates ventilation, which lowers P_{CO_2} withdrawing the [H$^+$] stimulus somewhat. After some days of acclimatization with increasing peripheral chemoreceptor drive, acid-base changes act to restore normal central [H$^+$], but now the relation between central P_{CO_2} and central [H$^+$] has changed so that the ventilatory recruitment threshold P_{CO_2} is lower than at sea level. Figure 8.12 illustrates these changes.

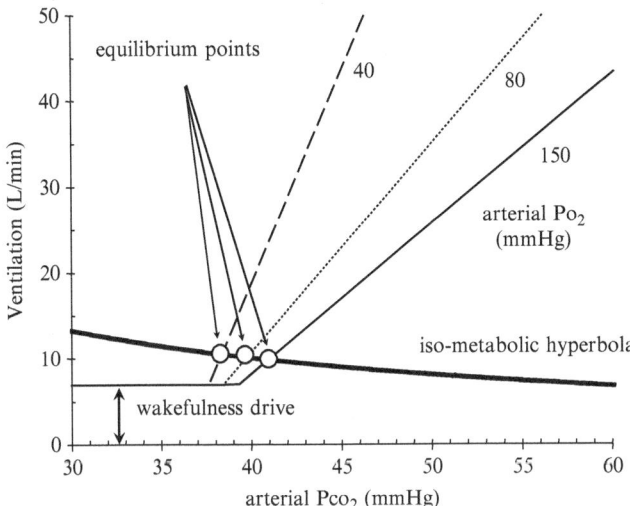

Fig. 8.11 The graphical model of the steady state chemoreflex control system expressed as l/min of pulmonary ventilation resulting from an arterial P_{CO_2} stimulus for constant P_{O_2}. The forward part of the loop is shown by the metabolic hyperbola and the intersection with the feedback shows the equilibrium point (*circles*) of the system in terms of resting ventilation and P_{CO_2}. The model assumes an arterial to central difference in P_{CO_2} of 6 mmHg, which is partly determined by cerebral blood flow

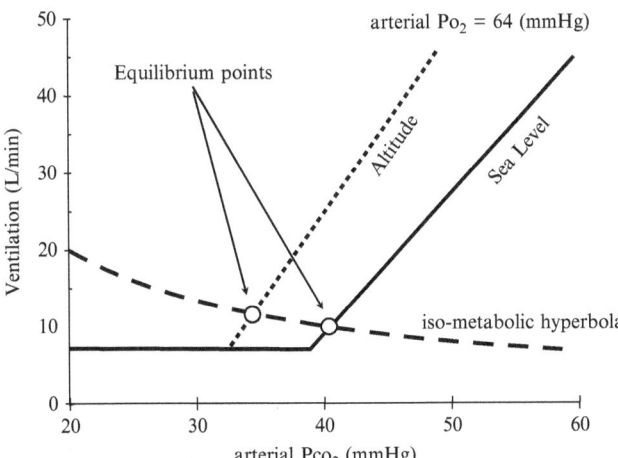

Fig. 8.12 The chemoreflex control of breathing model incorporating the modified Stewart model of acid-base predicts the decrease in the ventilatory recruitment threshold P_{CO_2}, as well as the increase in resting ventilation and decrease of resting arterial P_{CO_2} resulting from a 5 day sojourn at altitude (*solid line* to *dashed line*) that increased plasma albumin and phosphate. It was assumed that central [H$^+$] was defended by altering central SID; there were no changes in the chemoreflex parameters or the metabolic hyperbola or cerebral blood flow. Arterial [Alb] and [PO$_4^-$] were significantly elevated by an average of 1.12 g/dL and 0.38 mM/l respectively as measured experimentally [36]

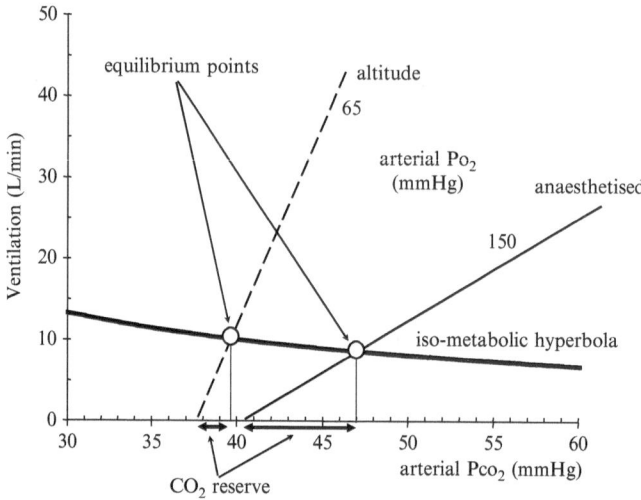

Fig. 8.13 The graphical model of the steady state chemoreflex control system illustrating two scenarios; that of an anesthetized patient breathing oxygen (*solid line*) and a mountaineer at altitude sleeping (*dashed line*). The CO_2 reserve is a graphical measure of system stability

8.5   Chemoreflex Control System Stability

This graphic representation of the chemoreflex control of breathing is also useful for predicting ventilation and P_{CO_2} under conditions that affect the stability of the control system (Fig. 8.13). For example, a sleeping patient recovering from anesthesia and breathing supplemental oxygen lacks the wakefulness drive, the peripheral chemoreflex drive is minimal because the P_{O_2} is high, and the central chemoreflex sensitivity is depressed by the anaesthesia. As a result, the ventilatory recruitment threshold becomes the apnoea threshold and the equilibrium point is at a higher P_{CO_2}. By contrast, a sleeping mountaineer, newly-arrived at altitude, with enhanced peripheral chemoreflex sensitivity has a lower resting P_{CO_2}. The stability of the chemoreflex control system in these two conditions is markedly different. The CO_2 reserve, the difference between resting P_{CO_2} and the apnoeic threshold, is a measure of stability; the lower the CO_2 reserve the less stability. It shows the amount by which P_{CO_2} must be decreased to produce apnoea.

The stability of the chemoreflex control system is governed by the sensitivity or gain of the entire control loop, both forward and feedback portions (see Fig. 8.2) as well as by the delay between the pulmonary control of arterial P_{CO_2} and P_{O_2} and the sensing of the resulting [H$^+$] changes by the chemoreceptors. Increases in loop gain are associated with instability [21, 43]. One useful measure dependent on loop gain is the CO_2 reserve [41], which shows graphically how far the P_{CO_2} must be depressed from resting to produce apnoea in a sleeping individual. As the anesthetized patient considered above illustrates, moving the equilibrium point to increasing P_{CO_2} slightly increases the ability of ventilation to change P_{CO_2}, i.e., the sensitivity or gain of the forward part of the loop is increased a little. However at the same time the chemoreflex sensitivity (the feedback portion of the loop) has decreased markedly in the sleeping patient, so that overall the loop gain, which is the combination of the forward and feedback gains, remains low; stability and the CO_2 reserve are therefore high [23].

By contrast the mountaineer on ascent to high altitude has an enhanced peripheral sensitivity to P_{CO_2} and therefore a higher feedback gain (Fig. 8.13). As a result the equilibrium point is at a lower P_{CO_2} on the metabolic hyperbola so that the forward gain is reduced slightly. However the overall loop gain is increased, and so when the wakefulness drive is removed during sleep the CO_2 reserve is low and the system stability is decreased. As the dynamic model developed below demonstrates, a decreased stability can produce oscillations in ventilation (Cheyne–Stokes breathing pattern) with alternating periods of apnoea and increased ventilation.

8.6   The Cerebral Blood Flow Control System

8.6.1   Interaction with the Chemoreflex Control System

The effect of cerebral blood flow must also be considered in this model of the chemoreflex control system; as cerebral blood flow increases, the arterial to central P_{CO_2} difference decreases and CO_2 is washed out from brain tissue reducing the stimulus to the central chemoreceptors. In this way cerebral blood flow control participates in mitigating the effects of changing arterial P_{CO_2} on central P_{CO_2}, thereby defending central $[H^+]$. Thus, while the input–output relation of the central chemoreceptors for central $[H^+]$ inputs remains unchanged by cerebral blood flow changes, the input–output relation of the central chemoreceptors for arterial P_{CO_2} inputs does change.

It is for this reason that measuring the ventilatory response of the central chemoreflex using Duffin's modified rebreathing technique [19], which measures the central chemoreflex ventilatory response to central P_{CO_2}, may differ from that using a steady state technique, which measures the central chemoreflex ventilatory response to arterial P_{CO_2} and therefore includes the effect of the cerebral blood flow response. Some studies have found the steady state chemoreflex sensitivity less than the rebreathing sensitivity [3] but not others [26] likely due to inter-subject variability in the sensitivity of the cerebral blood flow to arterial P_{CO_2}. Indeed, a reduction of the sensitivity of cerebral blood flow to arterial P_{CO_2} has been shown to increase the sensitivity of the central chemoreflex response to changes in arterial P_{CO_2} [42].

8.6.2   The Cerebral Blood Flow Control System Model

The regulation of cerebral blood flow is extremely complex and not fully understood at present [2]. Nevertheless, because arterial P_{CO_2} is a major influence [16],

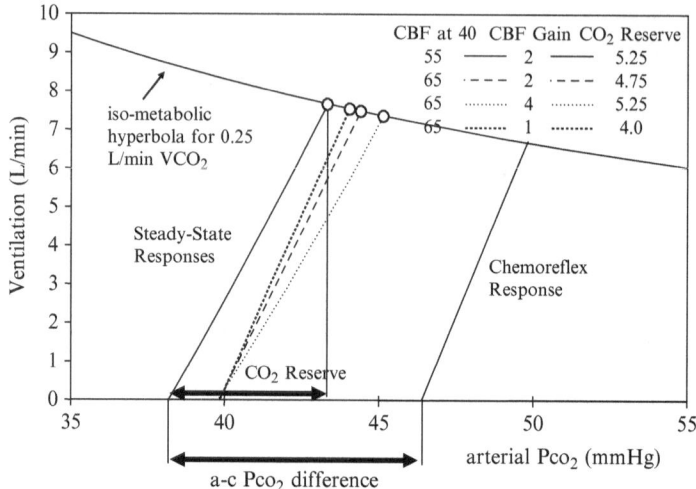

Fig. 8.14 The graphical model of the steady state chemoreflex control system illustrating the effect of different cerebral blood control parameters in a sleeping individual. Increasing the resting cerebral blood flow decreases the arterial to central (a-c) P_{CO_2} difference and decreases the CO_2 reserve. At any resting cerebral blood flow, increasing the cerebral blood flow sensitivity to arterial P_{CO_2} decreases the respiratory chemoreflex sensitivity to arterial P_{CO_2} and increases the CO_2 reserve

a simplified model can be attempted; the model assumed here is a linear relationship between cerebral blood flow and P_{CO_2} [34] with a gain parameter and a resting cerebral blood flow at a P_{CO_2} of 40 mmHg. This cerebral blood flow model was used to calculate the arterial to central difference in P_{CO_2} and applied to the chemoreflex model of a sleeping individual, and estimates of CO_2 reserve calculated. These are shown in Fig. 8.14, which demonstrates that increasing the resting cerebral blood flow with no change in cerebral blood flow sensitivity to arterial P_{CO_2} reduces the CO_2 reserve and stability, while increasing the cerebral blood flow sensitivity to arterial P_{CO_2} with no change in resting cerebral blood flow increases the CO_2 reserve and stability.

8.7 Dynamic Simulation of the Chemoreflex and Cerebral Blood Flow Control Systems

The chemoreflex model including cerebral blood flow control was combined with a simple 3-compartment model of lungs, brain and other tissues; the equations are detailed in Appendix III (see also Fig. 8.17). This dynamic model was simulated using a graphical programming language (LabVIEW, National Instruments) to solve the resulting mass balance equations with Euler integration and display the dynamic

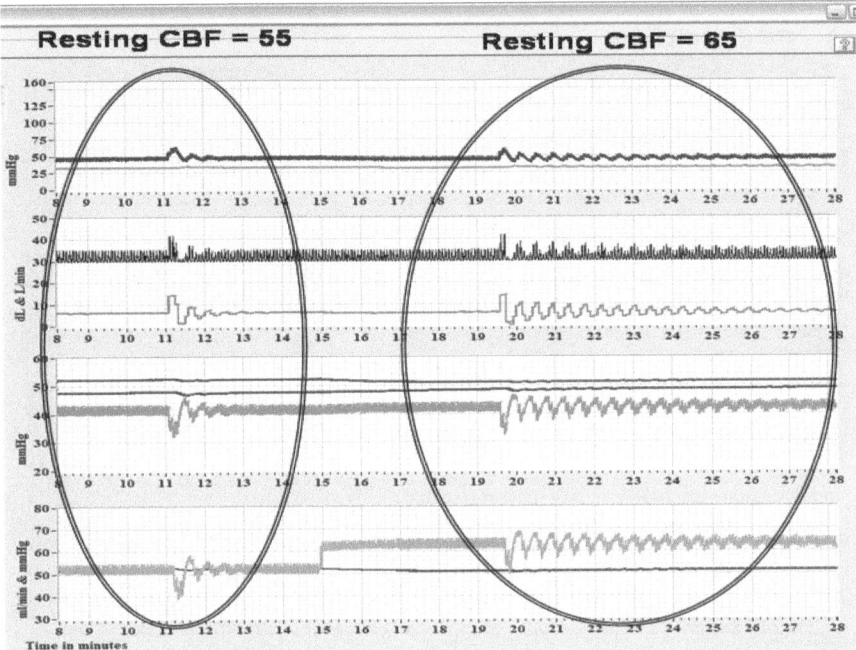

Fig. 8.15 The dynamic simulation model panel for a sleeping individual at altitude (inspired air oxygen fraction $F_{I,O_2} = 14\%$). The *top graph* show the venous and arterial P_{O_2} (mmHg) vs. time (min), the *second graph* shows the lung volume (dL) and ventilation (l/min) vs. time (min), the third graph shows the venous, arterial and brain (central) P_{CO_2} (mmHg) vs. time (min), and the lowest graph shows the cerebral (brain) blood flow (CBF ml/min) and brain (central) P_{CO_2} (mmHg) vs. time (min). The first transient hyperventilation is followed by a brief oscillation when the resting CBF is normal at 55 (ml/min) and with a sensitivity of 2 (ml/min/mmHg). After the resting CBF was increased to 65 (ml/min) the brief hyperventilation produced a longer period of oscillation recovery demonstrating that the increase in resting CBF had reduced the stability of the ventilatory control system

changes in several respiratory variables in a manner similar to an experimental data acquisition system. The model subject characteristics such as oxygen consumption can be set, as can the chemoreflex and cerebral blood flow control parameters. While the simulation is running it can be paused to allow the application of challenges such as changes in inspired gases or changes in the parameter settings. Switches to remove the wakefulness drive (sleep) or hyperventilate are also provided.

This dynamic model was used to simulate a sleeping individual at altitude. To disturb the system and observe its stability a brief hyperventilation was applied such as might occur during sleep and the resulting transient disturbance in breathing observed. Examples demonstrating the effects of changes in the regulation of cerebral blood flow in a sleeping individual are shown in Figs. 8.15 and 8.16, which are screenshots of the dynamic simulation program. They demonstrate the same

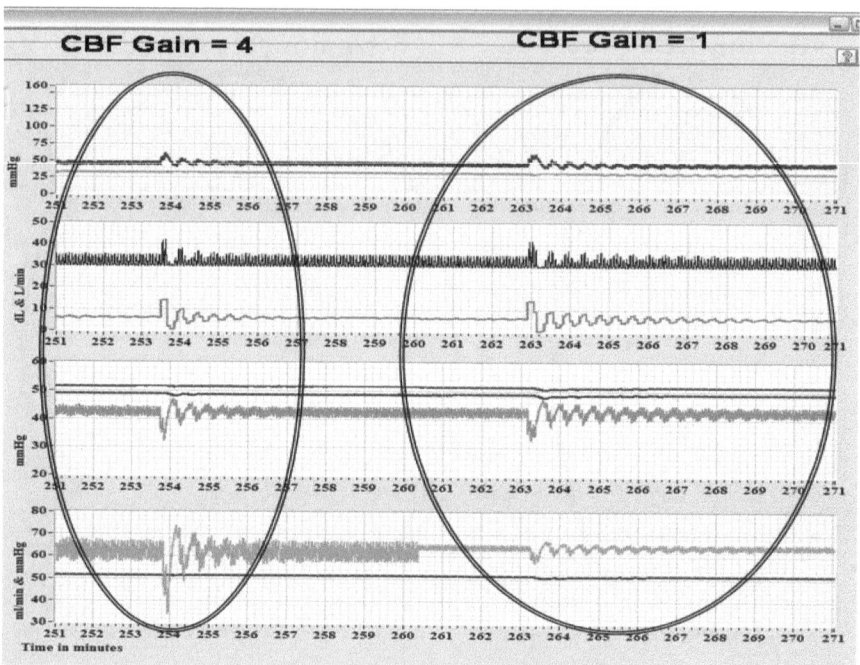

Fig. 8.16 The dynamic simulation model panel for a sleeping individual at altitude (inspired air oxygen fraction $F_{I,O_2} = 14\%$). The *graphs* are as described in Fig. 8.15. The first transient hyperventilation is followed by a brief oscillation when the resting CBF is at 65 (ml/min) and with a sensitivity of 4 (ml/min/mmHg). After the CBF sensitivity was decreased to 1 (ml/min/mmHg) the brief hyperventilation produced a longer period of oscillation recovery demonstrating that the decrease in CBF sensitivity reduced the stability of the ventilatory control system

conclusions as the steady state graphical model and are qualitatively similar to those observed in sleeping individuals at altitude whose cerebral blood flow control has been altered by indomethacin [1].

8.7.1 Concluding Remarks

These modeling experiments demonstrate some of the validation and control issues in modeling the respiratory chemoreflex control system. While the chemoreflex control system parameters are reasonably well estimated for a sea level population from experiments, it remains to incorporate experimental findings characterizing the chemoreflexes in other populations such as high altitude residents and patients with respiratory disease. Such findings have yet to be definitively obtained at least partly because of the use of inadequate measurement techniques. The complexities involved in the control of cerebral blood remain poorly understood so that a definitive model lies in the future. Nevertheless the model experiments did demonstrate the interaction between cerebral blood flow control and the respiratory chemoreflexes and the implications for system stability that reflect current findings from experiments altering cerebral blood flow.

Appendix

In the following appendices, equations are written in a style adaptable to computer implementation such as in Matlab. Hence equation symbols are written without subscripts as would be the case in a program. For example, P_{CO_2} appears as PCO2. The one exception are the chemical symbols such as H^+ which appears here as [H+] in the equations. Such symbols would need to be written, e.g., as Hplus. The symbols are written to maintain ease of identification while not emphasizing the superscript and subscripts appearing in the main text.

Appendix I: The Modified Stewart Equations for Acid-Base Balance

Water dissociation: [H+] [OH−] = K'W   (8.1)
Bicarbonate formation: [H+] [HCO3−] = KC PCO2   (8.2)
Carbonate formation: [H+] [CO32−] = K3 [HCO3−]   (8.3)
Phosphoric acid dissociation: [Pi] = Pi,Tot {2 − [H+] / (K2 + [H+])}   (8.4)
Serum albumin dissociation: [A−] = [A-Fixed] − [H+] [AH,Tot] / (KH + [H+]) (8.5)
[A-Fixed] = 21 [Alb] 10 / 66500   (8.6)
[AH,Tot] = 16 [Alb] 10 / 66500   (8.7)
Electrical neutrality: [SID]+[H+]−[HCO3−]−[A−]−2[CO32−]−[OH−]−[Pi−] = 0   (8.8)

Here:

[H+] = hydrogen ion concentration (M/l)
[OH−] = hydroxyl ion concentration (M/l)
K'W is the ion product for water (2.39E−14)
KC combines equilibrium & solubility constants (2.45E−11)
PCO2 = partial pressure of carbon dioxide (mmHg)
[CO32−] = concentration of carbonate form of CO2 (M/l)
K3 is the dissociation constant for carbonate (1.16E−10)
[Pi,Tot] is the concentration of Phosphate (M/l)
K2 is the phosphoric acid dissociation constant (2.19E−7)
SID = [Na+]+[K+]+[Ca2+]−[Cl−]−[La−]−[other anions] (M/l)

[Alb] is albumin concentration (g/dl)

66,500 is the molecular weight of albumin

[A-Fixed] is the fixed negative charge concentration with 21 fixed negative charges per mole of albumin (M/l)

[AH,Tot] is the concentration of histidine residues with 16 residues per mole of albumin (M/l)

$[A-]$ is the concentration of net charges on albumin (M/l)

KH is the histidine dissociation constant (1.77E−7)

These equations are solved iteratively until equation 8 < 1.0E−5.

Note: when calculating blood values of $[H+]$ using the modified Stewart equations above, it is necessary to include the effect of a change in SID as Chloride shifts from plasma to red cells. The following empirically determined relation may be used:

$$SID = SIDn + (PCO2 - 24)/7$$

where SIDn = the SID at PCO2 = 24 (37.5 mM/l)

Important references for these equations can be found in:

- Stewart, P.A. (1983). Modern quantitative acid-base chemistry. Canadian Journal of Physiology & Pharmacology 61, 1444–1461.
- Watson, P. D. (1999). Modeling the effects of proteins on pH in plasma. J. Appl. Physiol. 86, 1421–1427.

Appendix II: The Chemoreflex Equations

Central chemoreceptor drive:

If $([H+]c - Tc) < 0$

   then Dc = 0

   else Dc = Sc $([H+]c - Tc)$ (8.9)

where:

Dc = the drive to breathe from the central chemoreceptors (l/min)

$[H+]c$ = Hydrogen ion concentration at the central chemoreceptors (nM/l)

Tc = central chemoreceptor threshold (default 34 nM/l)

Sc = central chemoreceptor sensitivity (default 1–3 l/min/nM/l)

Peripheral chemoreceptor drive:

Sp = Sp0+ A / (PO2−PO20)   (8.10)

If $([H+]p - Tp) < 0$

   then Dp = 0

   else Dp = Dp0 + Sp $([H+]p - Tp)$ (8.11)

where:

Sp = peripheral chemoreceptor sensitivity to [H+]p (l/min/nM/l)

Sp0 = peripheral chemoreceptor sensitivity to [H+]p in hyperoxia (default 0 l/min/nM/l)

A = Area constant for the rectangular hyperbolic relation between Sp and PO2 (default 17.8 l/min/(mmHg) (nM/l))

P0 = PO2 for maximum sensitivity before failure (default 30 mmHg)

Dp = the drive to breathe from the peripheral chemoreceptors (l/min)

Dp0 = tonic drive to breathe from the peripheral chemoreceptors (default 0 l/min)

[H+]p = Hydrogen ion concentration at the peripheral chemoreceptors (nM/l)

Tp = peripheral chemoreceptor threshold (default 34 nM/l)

Total drives to breathe:

If (Dc + Dp) < Td
    then Ventilation = Dw
    else Ventilation = Dc + Dp + Dw (8.12)

where:

Td = Chemoreflex drive threshold (default 18.5 l/min)

Dw = Wakefulness drive to breathe (default 7 l/min)

Appendix III: The Dynamic Model

The model consists of a pulmonary gas exchange compartment and two tissue compartments for CO2 and O2; a brain or central chemoreceptor compartment and an all other tissues compartment as illustrated in Fig. 8.17. Ventilation is controlled by brain compartment [H+] and arterial [H+] and PO2 via the central and peripheral chemoreflexes respectively. Cerebral blood flow is controlled by arterial PCO2, and cardiac output is controlled by arterial PCO2 and PO2. Lung volume changes are modeled as a sine wave inspiration and exponential expiration.

Blood Flows

Variables:

Q = cardiac output ml/min

Qi = initial cardiac output; default value of 4,500 ml/min

QCO2 = Q sensitivity to PaCO2; default 40

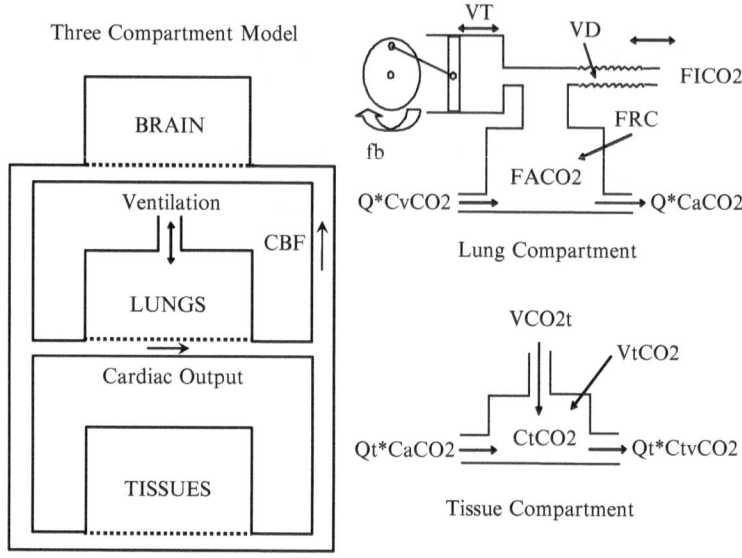

Fig. 8.17 A block diagram of the three compartment model for simulating the exchange of CO_2 and O_2 between body tissues and the environment via pulmonary ventilation. CO_2 variables are shown to illustrate the nomenclature

QO2 = Q inverse sensitivity to PaO2; default 56,490
Qb = brain blood flow ml/min/100g
Qbi = initial brain blood flow ml/min/100g at 40 mmHg PCO2; default 55
QbCO2 = Qb sensitivity to PaCO2 ml/min/100g/mmHg; default 2
DELb = delay to brain s; default 10
DELt = delay to tissue s; default 10
DELc = delay to carotid bodies s; default 6

Equations:

Cardiac Output [Rothe (1990)]:

$$Q = Qi + QCO2 * PaCO2 + QO2/PaO2;$$

Brain Blood Flow [Ainslie (2009)]

$$Qb = Qbi + QbCO2(PaCO2 - 40)ml/min$$

Ventilation

Variables:

t = time in s
dt = time increment s
Ti = inspiratory time s
TI = inspiratory phase duration s
Te = expiratory time s
TE = expiratory phase duration s
I/E = Boolean inspiratory or expiratory flag
VE = ventilation in l/min from chemoreflex equations
VL = Lung Volume
eiVL = end-inspiratory Lung Volume
fb = breathing frequency breaths/min; default 12
VT = tidal volume ml = 1,000*VE/fb set at end expiration
VA = alveolar volume ml
etVA = end-expiratory VA ml
eiVA = end-inspiratory VA ml
dVA = change in lung volume in time dt
VD = dead space volume; default 150 ml
FRC = functional residual capacity; default 3,000 ml
ExpTau = Time constant for expiration s; default 0.5 s

Equations:

TI + TE = 60/fb s
TI/(TI + TE) = 0.5 default
TI = 30/fb s
TE = TI
During inspiration: VL = etVA + 0.5*VT* sin(PI*Ti/TI − PI/2) + VT/2
During expiration: VL = (eiVA-FRC)*exp(-Te/0.5) +FRC.

Pulmonary Gas Exchange of CO$_2$

Variables:

FICO2 = fraction of inspired CO2; default 0 %
FACO2 = fraction of CO2 in alveolar volume
etFACO2 = end-expiratory FACO2

CaCO2 = arterial concentration of CO2 ml/ml
CvCO2 = mixed venous concentration of CO2 ml/ml
PACO2 = FRC partial pressure of CO2 mmHg
PaCO2 = arterial blood partial pressure of CO2 mmHg
PmCO2 = mouth partial pressure of CO2 mmHg
PmiCO2 = mouth partial pressure of CO2 mmHg during inspiration
PmeCO2 = mouth partial pressure of CO2 mmHg during expiration
PetCO2 = end-expiratory partial pressure of CO2 mmHg
QCO2 = CO2 entering alveolar volume from blood each dt
dVA = increment of lung volume change in time increment dt
VCO2i = CO2 entering the alveolar volume during inspiration for each dVA
VCO2e = CO2 leaving the alveolar volume during expiration for each dVA

Equations:

$$QCO2 = Q * dt * (CvCO2 - oldCaCO2);$$

During inspiration, dead space with end-tidal FACO2 enters first, then fresh gas with
FICO2. The amount of CO2 entering the alveolar volume for each dVA is:
if ((VA-etVA) <= VD)
   then VCO2i = dVA*etFACO2;
   else VCO2i = dVA*FICO2;
During expiration, the amount of CO2 leaving the alveolar volume each dVA is:
   VCO2e = dVA*oldFACO2;
   IE flag = TRUE for inspiration, FALSE for expiration
if IE TRUE
   then VCO2 = VCO2i;
   else VCO2 = VCO2e;
  End
   newFACO2 = (oldFACO2*(VA−dVA) + QCO2 + VCO2)/VA;
   PACO2 = newFACO2*713;
   PaCO2= PACO2;

During inspiration, gas entering the mouth is at FICO2:
PmiCO2 = FICO2*713;

During expiration, gas leaving the mouth is dead space with FICO2, then gas with
FACO2.
if ((eiVL-VL) <= VD)
   then PmeCO2 = FICO2*713;
   else PmeCO2 = newFACO2*713;
if (IE > 0)
   then PmCO2 = PmiCO2;
   else PmCO2 = PmeCO2;

Pulmonary Gas Exchange of O_2

Variables:

FIO2 = fraction of inspired O2; default 20.9 %
FAO2 = fraction of O2 in FRC
etFAO2 = end-tidal FAO2
CaO2 = arterial concentration of O2 ml/ml
CvO2 = mixed venous concentration of O2 ml/ml
PAO2 = FRC partial pressure of O2 mmHg
PaO2 = arterial blood partial pressure of O2 mmHg
PmO2 = mouth partial pressure of O2 mmHg
PmiO2 = mouth partial pressure of O2 mmHg during inspiration
PmeO2 = mouth partial pressure of O2 mmHg during expiration
PetO2 = end-expiratory partial pressure of O2 mmHg
QO2 = O2 entering alveolar volume from blood each dt
dVA = increment of lung volume change in time increment dt
VO2i = O2 entering the alveolar volume during inspiration for each dVA
VO2e = O2 leaving the alveolar volume during expiration for each dVA

Equations:

$$QO2 = Q * dt * (CvO2 - oldCaO2);$$

During inspiration, dead space with end-tidal FAO2 enters first, then fresh gas with FIO2. The amount of O2 entering the alveolar volume for each dVA is:
  if ((VA-etVA) <= VD)
   then VO2i = dVA*etFAO2;
   else VO2i = dVA*FIO2;
During expiration, the amount of O2 leaving the alveolar volume each dVA is:

$$VO2e = dVA * oldFAO2;$$

IE flag = TRUE for inspiration, FALSE for expiration
  if IE TRUE
   then VO2 = VO2i;
   else VO2 = VO2e;
  End
  newFAO2 = (oldFAO2*(VA-dVA) + QO2 + VO2)/VA;
  PAO2 = newFAO2*713;
  PaO2 = PAO2;
During inspiration, gas entering the mouth is at FIO2:
  PmiO2 = FIO2*713;
During expiration, gas leaving the mouth is dead space with FIO2, then gas with FAO2.

```
    if ((eiVL-VL) <= VD)
       then PmeO2 = FIO2*713;
       else PmeO2 = newFAO2*713;
 if (IE > 0)
    then PmO2 = PmiO2;
    else PmO2 = PmeO2;
```

O_2 and CO_2 Carriage in Blood

These equations are applied to both arterial and venous blood.

Variables:

Hb = concentration of hemoglobin g/l; default 150
[O2] = blood oxygen concentration
PO2 = blood partial pressure of O2 mmHg
PCO2 = blood partial pressure of CO2 mmHg
SO2 = blood O2 saturation
n = 2.6
k1 = 1.312E−3
k2 = 3.03E−5
k3 = 26.6
SID = plasma strong ion difference mM/l; default 37.5
PHOS = plasma [Phosphate] mM/l; default 4.2
ALB = plasma [Albumin] g/dL; default = 1.4
H = plasma [H+] nM/l
pH = plasma pH
[CO2] = blood CO2 concentration ml/ml
CO2p = plasma concentration of CO2 mM/l = Sum of all forms of CO2
[HCO3−] = blood bicarbonate concentration mM/l
[CO32−] = blood carbonate concentration mM/l
d = 0.0307 aqueous solubility constant for CO2
c = 0.00011 carbonic acid solubility constant

Equations:

The O2 dissociation curve [Chiari (1997)]:

PO2vir = PO2 (40/PCO2)^0.3
SO2 = (PO2vir^n) / (k3^n + PO2vir^n)
[O2] = k1 Hb SO2 + k2 PO2

The CO_2 dissociation curve [Douglas (1988)]:

pH is calculated from the modified Stewart equations detailed in Appendix I
$CO2p = [HCO3-] + [CO32-] + d*PCO2 + c*PCO2$
$[CO2] = 0.022260\ CO2p*(1-0.0289*Hb/((3.352-0.456*SO2)*(8.142-pH)));$

Compartments CO_2 Exchange

One difference from other compartment models was the treatment of compartment gas exchange. Using mass balance equations for compartment gas exchange requires knowledge of tissue dissociation curves; difficult to ascertain. Instead the venous concentrations were assumed to change exponentially with time because in the steady state the arterial-venous concentration difference equals the ratio of metabolism/blood flow. When this steady state equilibrium is disturbed the inequality between the arterial-venous concentration difference and the metabolism/blood flow ratio is restored exponentially with a time constant equal to the ratio of compartment volume to blood flow.

Variables:

VCO2 = tissue compartment CO2 production ml/min; default 0.8*VO2
CtvCO2 = tissue venous concentration of CO2
Qt = Q-Qb
VtCO2 = volume constant for CO2 in the tissue compartment ml; default 10,000
VCO2b = brain compartment CO2 production ml/min; default 3 ml/min
VbCO2 = volume constant for CO2 in the brain (central chemoreceptors) compartment ml; default 100

Equations:

dCtvCO2/dt = {VCO2 – Qt*(CtvCO2–CaCO2)}/VtCO2;
dCbvCO2/dt = {VCO2b – Qc*(oldCbvCO2–CaCO2)}/VbCO2;
CvCO2 = CtvCO2(Qt/Q) + CbvCO2(Qb/Q); delayed by DELt for pulmonary compartment

Compartments O_2 Exchange

Variables:

VO2 = tissue compartment O2 consumption ml/min; default 300
CtvO2 = tissue venous concentration of O2

Qt = Q − Qb
VtO2 = volume constant for O2 in the tissue compartment ml; default 3,528
VO2b = brain compartment O2 consumption ml/min; default 3 ml/min
CbvO2 = brain (central chemoreceptors) venous concentration of O2
VbO2 = volume constant for O2 in the brain (central chemoreceptors) compartment ml; default 14

Equations:

dCtvO2/dt = {VO2 − Qt*(CaO2−CtvO2)}/VtCO2;
dCbvO2/dt = Qb*(CbvO2 − CaO2) − VO2b/VbO2;
CvO2 = CtvO2(Qt/Q) + CbvO2(Qb/Q); delayed by DELt for pulmonary compartment

Chemoreflex Inputs

Peripheral chemoreceptor inputs

PO2 = arterial PO2 after delay DELc
[H+] = arterial [H+] after delay DELc
SID = central strong ion difference mM/l; default 31
PHOS = central [Phosphate] mM/l; default 1.17
ALB = central [Albumin] g/dL; default = 0.61
PbCO2 = Function of (CbvCO2); Inverse CO2 dissociation curve
[H+] is calculated from the modified Stewart equations using the above values

References

1. Ainslie, P., Cotter, J.D., Dawson. A., Fan, J.L., Lucas, R.A.I., Lucas, S.J.E., Peebles, K.N., Bilson, K., Swart, M., Thomas, K.N., Burgess, K.R.: Influence of cerebral blood flow on central sleep apnea at high altitude. In: International Hypoxia Conference, Lake Louise, Canada 2009
2. Ainslie, P.N., Duffin, J.: Integration of cerebrovascular CO2 reactivity and chemoreflex control of breathing: mechanisms of regulation, measurement, and interpretation. Am J Physiol Regul Integr Comp Physiol. **296**, R1473–R1495 (2009)
3. Berkenbosch, A., Bovill, J.G., Dahan, A., DeGoede, J., Olievier, I.C.: The ventilatory CO2 sensitivities from Read's rebreathing method and the steady-state method are not equal in man. J. Physiol. **411**, 367–377 (1989)
4. Blain, G.M., Smith, C.A., Henderson, K.S., Dempsey, J.A.: Contribution of the carotid body chemoreceptors to eupneic ventilation in the intact, unanesthetized dog. J Appl Physiol. **106**, 1564–1573 (2009)
5. Cunningham, D.J.C.: Review lecture: Studies on arterial chemoreceptors in man. J. Physiol. **384**, 1–26 (1987)
6. Cunningham, D.J.C.: Studies on arterial chemoreceptors in man. J. Physiol. **384**, 1–26 (1987)

7. Dahan, A., Nieuwenhuijs, D., Teppema, L.: Plasticity of central chemoreceptors: Effect of bilateral carotid body resection on central CO_2 sensitivity. PLoS Med. **4**(7), e239 (2007)

8. Duffin, J.: Role of acid-base balance in the chemoreflex control of breathing. J. Appl. Physiol. **99**(6), 2255–2265 (2005)

9. Duffin, J., Mahamed, S.: Adaptation in the respiratory control system. Can. J. Physiol. Pharmacol. **81**(8), 765–773 (2003)

10. Duffin, J., McAvoy, G.V.: The peripheral-chemoreceptor threshold to carbon dioxide in man. J. Physiol. **406**, 15–26 (1988)

11. Duffin, J., Mohan, R.M., Vasiliou, P., Stephenson, R., Mahamed, S.: A Model of the chemoreflex control of breathing in humans: Model parameters measurement. Respir. Physiol. **120**, 13–26 (2000)

12. Fencl, V., Leith, D.E.: Stewart's quantitative acid-base chemistry: Applications in biology and medicine. Respir. Physiol. **91**, 1–16 (1993)

13. Fink, B.R.: Influence of cerebral activity in wakefulness on regulation of breathing. J. Appl. Physiol. **16**, 15–20 (1961)

14. Guyenet, P.G., Bayliss, D.A., Mulkey, D.K., Stornetta, R.L., Moreira, T.S., Takakura, A.T.: The retrotrapezoid nucleus and central chemoreception. Adv. Exp. Med. Biol, **605**, 327–332 (2008)

15. Horn, E.M., Waldrop, T.G.: Suprapontine control of respiration. Respir. Physiol. **114**(3), 201–211 (1998)

16. Ide, K., Eliasziw, M., Poulin, M.J.: Relationship between middle cerebral artery blood velocity and end-tidal PCO2 in the hypocapnic-hypercapnic range in humans. J Appl Physiol. **95**, 129–137 (2003)

17. Irsigler, G.B., Stafford, M.J., Severinghaus, J.W.: Relationship of $CSF\ pH$, O_2, and CO_2 responses in metabolic acidosis and alkalosis in humans. J. Appl. Physiol. **48**(2), 355–361 (1980)

18. Iturriaga, R., Varas, R., Alcayaga, J.: Electrical and pharmacological properties of petrosal ganglion neurons that innervate the carotid body. Respir. Physiol. Neurobiol. **157**(1), 130–139 (2007)

19. Jensen, D., Duffin, J., Lam, Y.-M., Webb, K.A., Simpson, J.A., Davies, G.A.L., Wolfe, L.A., O'Donnell, D.E.: Physiological mechanisms of hyperventilation during human pregnancy. Respir. Physiol. Neurobiol. **161**(1), 76–86 (2008)

20. Jounieaux, V., Parreira, V.F., Aubert, G., Dury, M., Delguste, P., Rodenstein, D.O.: Effects of hypocapnic hyperventilation on the response to hypoxia in normal subjects receiving intermittent positive-pressure ventilation. Chest **121**(4), 1141–1148 (2002)

21. Khoo, M.C.K.: Determinants of ventilatory instability and variability. Respir. Physiol. **122** (2-3), 167–182 (2000)

22. Kumar, P.: Sensing hypoxia in the carotid body: from stimulus to response. Essays Biochem. **43**, 43–60 (2007)

23. Mahamed, S., Hanly, P.J., Gabor, J., Beecroft, J., Duffin, J.: Overnight changes of chemoreflex control in obstructive sleep apnoea patients. Respir. Physiol. Neurobiol. **146**(2-3), 279–290 (2005)

24. Mateika, J.H., Omran, Q., Rowley, J.A., Zhou, X.S., Diamond, M.P., Badr, M.S.: Treatment with leuprolide acetate decreases the threshold of the ventilatory response to carbon dioxide in healthy males. J. Physiol. **561**(Pt 2), 637–646 (2004)

25. Mohan, R., Duffin, J.: The effect of hypoxia on the ventilatory response to carbon dioxide in man. Respir. Physiol. **108**(2), 101–115 (1997)

26. Mohan, R.M., Amara, C.E., Cunningham, D.A., Duffin, J.: Measuring the central-chemoreflex sensitivity in man: rebreathing and steady-state methods compared. Respir. Physiol. **115**(1), 23–33 (1999)

27. Nattie, E.E., Li, A.: Central chemoreception is a complex system function that involves multiple brainstem sites. J. Appl. Physiol. (2008)

28. Orem, J.: The nature of the wakefulness stimulus for breathing. Prog. Clin. Biol. Res. **345**, 23–30; discussion 31 (1990)

29. Orem, J., Lovering, A.T., Dunin-Barkowski, W., Vidruk, E.H.: Tonic activity in the respiratory system in wakefulness, NREM and REM sleep. Sleep **25**(5), 488–496 (2002)

30. Orem, J., Trotter, R.H.: Behavioral control of breathing. News Physiol. Sci. **9**, 228–232 (1994)

31. Phillipson, E.A., McClean, P.A., Sullivan, C.E., Zamel, N.: Interaction of metabolic and behavioral respiratory control during hypercapnia and speech. Am. Rev. Respir. Dis. **117**, 903–909 (1978)

32. Rapanos, T., Duffin, J.: The ventilatory response to hypoxia below the carbon dioxide threshold. Can. J. Appl. Physiol. **22**(1), 23–36 (1997)

33. Read, D.J.C.: A clinical method for assessing the ventilatory response to CO_2. Australas. Ann. Med. **16**, 20–32 (1967)

34. Reivich, M.: Arterial Pco2 and Cerebral Hemodynamics. Am. J. Physiol. **206**, 25–35 (1964)

35. Shea, S.A.: Behavioural and arousal-related influences on breathing in humans. Exp. Physiol. **81**(1), 1–26 (1996)

36. Somogyi, R.B., Preiss, D., Vesely, A., Fisher, J.A., Duffin, J.: Changes in respiratory control after 5 days at altitude. Respir. Physiol. Neurobiol. **145**(1), 41–52 (2005)

37. St. Croix, C.M., Cunningham, D.A., Paterson, D.H.: Nature of the interaction between central and peripheral chemoreceptor drives in human subjects. Can. J. Physiol. Pharmacol. **74**(6), 640–646 (1996)

38. Stewart, P.A.: Modern quantitative acid-base chemistry. Can. J. Physiol. Pharmacol. **61**, 1444–1461 (1983)

39. Torrance, R.W.: Prolegomena. Chemoreception upstream of transmitters. In: Zapata (ed.) Frontiers in Arterial Chemoreception, vol. 410, pp. 13–38. Plenum, New York (1996)

40. Watson, P.D.: Modeling the effects of proteins on pH in plasma. J. Appl. Physiol. **86**(4), 1421–1427 (1999)

41. Xie, A., Skatrud, J.B., Barczi, S.R., Reichmuth, K., Morgan, B.J., Mont, S., Dempsey, J.A.: Influence of cerebral blood flow on breathing stability. J. Appl. Physiol. **106**(3), 850–856 (2009)

42. Xie, A., Skatrud, J.B., Morgan, B., Chenuel, B., Khayat, R., Reichmuth, K., Lin, J., Dempsey, J.A.: Influence of cerebrovascular function on the hypercapnic ventilatory response in healthy humans. J. Physiol. (Lond.) **577**(1), 319–329 (2006)

43. Younes, M., Ostrowski, M., Atkar, R., Laprairie, J., Siemens, A., Hanly, P.: Mechanisms of breathing instability in patients with obstructive sleep apnea. J. Appl. Physiol. **103**(6), 1929–1941 (2007). doi: 10.1152/japplphysiol.00561.2007

Chapter 9
Experimental Studies of the Baroreflex

Clive M. Brown

Abstract In this chapter we seek to provide the reader with some insight into the physiology and physiological methods for examining the baroreflex control system which is the system most studied in the modeling chapters of this volume. In particular the reader is directed to the models presented in Chaps. 10 and 11.

9.1 Introduction

The baroreceptors are sited in the deep layers of adventitia in the carotid sinus and aortic arch and are the main sensory regions in the reflex control of blood pressure. There is also experimental evidence in favor of the existence of baroreceptors in the coronary circulation [1–4]. The baroreceptors are mechanoreceptors, meaning that they respond to changes in arterial stretch (rather than to pressure per se). Changes in blood pressure deform the arterial walls, altering the discharge frequency in the afferent nerve and resulting in the appropriate changes in autonomic activity to help maintain cardiovascular stability. The early work of Eberhard Koch demonstrated that there is a sigmoid relationship between carotid sinus pressure and electrocardiographic RR interval output that can be quantified in terms of gains, threshold and operating points [5]. In 1932, Bronk & Stella recorded the nervous discharge from carotid sinus baroreceptors in rabbits [6]. They found that with each arterial pulse there was a burst of afferent impulses, the frequency of which was dependent on the mean blood pressure and the waveform characteristics of the arterial pulse. Over the course of a single arterial pulse, the rate of impulse firing was higher while pressure was increasing than when it was falling [6]. Thus, the carotid baroreceptors are sensitive to both the mean level of arterial pressure and the rate of pressure

C.M. Brown (✉)
Department of Medicine, Division of Physiology, University of Fribourg, Switzerland
e-mail: clivembrown@yahoo.co.uk

J.J. Batzel et al. (eds.), *Mathematical Modeling and Validation in Physiology*,
Lecture Notes in Mathematics 2064, DOI 10.1007/978-3-642-32882-4_9,
© Springer-Verlag Berlin Heidelberg 2013

change. Since then, numerous animal studies have investigated the characteristics of the stimulus-response curves of the carotid baroreflex, often by means of isolating the carotid sinus and measuring responses to changes in carotid sinus pressure [7–10]. The vagal response to baroreceptor stimulation is very rapid, allowing the heart rate to be adjusted on a beat-to-beat basis [11]. Because the resting level of arterial pressure is on the central part of the sigmoid baroreflex curve [12], the baroreflex can respond rapidly to a blood pressure change in either direction. However, if a change in arterial pressure is maintained, the baroreceptors may adapt their set point to the new resting level within minutes. Baroreceptors also modulate vascular resistance through changes in sympathetic activity [13], but his response occurs more slowly, with blood pressure not changing at all until at least two seconds after the onset of the stimulus [14].

9.2   Physiological Relevance of the Baroreflex

In everyday life the body is subjected to a multitude of environmental situations that have the potential to challenge the control of blood pressure and organ perfusion. One of the most important challenges to the maintenance of blood pressure concerns the effects of gravitational forces on the circulation. Postural change from supine to upright results in a decrease in hydrostatic pressure in the upper part of the body and an increase in the lower body. A consequence of the hydrostatic pressure gradient is the translocation of circulating blood to the dependent regions. Since veins are about 30 times more compliant than arteries [15], they are much more susceptible to changes in transmural pressure. Furthermore, as veins contain approximately 70 % of the total blood volume [16] and are therefore considered as capacitance vessels, the distribution of blood volume is dependent on variations in their compliance. On standing, up to 500 ml of blood is displaced to the veins of the lower limbs [17]. Additionally, the increase in capillary hydrostatic pressure in the lower body causes an increase in the capillary filtration rate and therefore an overall decrease in plasma volume [18,19]. Taken together, the venous pooling and increased capillary filtration rate during upright standing result in a reduction in the effective circulating volume and a decrease in cardiac output of about 25 % [20]. Without the intervention of compensatory reflex mechanisms, the fall in cardiac output would lead to a drop in mean arterial pressure, eventually falling to levels that are insufficient for continued cerebral perfusion. The most important mechanism responsible for regulating blood pressure during orthostatic stress on a beat-to-beat basis is the arterial baroreflex.

9.3   Baroreceptor Denervation Studies

The relative importance of the baroreflex in blood pressure regulation can be demonstrated by experiments in which the baroreceptors are denervated. Cowley et al. [21] studied blood pressure in dogs that had undergone sino-aortic denervation.

They reported that average 24-h blood pressure was not different from control dogs. However, variability of blood pressure was much higher in the baroreceptor denervated dogs than in the baroreceptor intact dogs [21]. These experiments formed the basis for the viewpoint that the baroreceptors maintain blood pressure stability but do not regulate the long term blood pressure level. In studies where the carotid but not the aortic baroreceptors were denervated, blood pressure and blood pressure variability initially increased but then returned to normal levels within two weeks [22], suggesting that the aortic baroreceptors are able to take over the function of the carotid baroreceptors. The role of the baroreceptors in modulating the responses to orthostatic stress was emphasized by Cornish et al. [23], who studied responses to lower body negative pressure in primates after sino-aortic denervation. It was shown that in the baroreceptor denervated animals, but not in the baroreceptor intact animals, blood pressure fell markedly during lower body negative pressure [23]. Thus, the arterial baroreceptors are necessary in mediating the reflex responses to orthostatic stress. In humans, carotid sinus denervation following bilateral carotid body tumor resection resulted in postural drops in blood pressure and symptoms of orthostatic hypotension [24]. There were also longer-term elevations in both the mean level of blood pressure and in blood pressure variability [24], suggesting that the baroreflex may have a greater role in regulating the long-term blood pressure than previously thought.

9.4 Do Baroreceptors Have a Role in the Long Term Control of Blood Pressure?

As discussed above, denervation studies in animals [21] established the paradigm that baroreceptors modulate lability of blood pressure but not the mean level of blood pressure over time. However, an increasing number of studies suggest that the arterial baroreceptors may actually have a role in setting the long term level of blood pressure. Lohmeier et al. [25] performed elegant studies in dogs with one denervated kidney and used sodium excretion from the intact and denervated kidneys as an index of renal sympathetic activity. During a sustained rise in blood pressure, induced by 5 days of angiotensin II infusion, sodium excretion (and therefore renal sympathetic activity) was greater in the innervated compared with the denervated kidney. However, the difference in sodium excretion between the innervated and denervated kidney was reversed after cardiopulmonary and sino-aortic denervation. This, postulated the authors, was evidence for reflex inhibition of renal sympathetic nerve activity in response to a prolonged change in blood pressure [25]. Further studies performed by Thrasher [26] also support the concept that arterial baroreceptors may have a role in long term blood pressure control. Seven days of sustained baroreceptor unloading (ligation of the common carotid artery proximal to the carotid sinus) in dogs resulted in significant elevations in blood pressure [26]. The possibility that baroreceptors may contribute to the setting the long term level of blood pressure raises the question as to whether baroreflex dysfunction may be a contributing factor in the pathogenesis of hypertension. There are several reports of attenuated cardiovagal baroreflex function in patients with hypertension [27–30]. One study suggested that the baroreflex impairment was more substantial in hypertensive females than in males [28], though the reason for this gender effect was not established. Another study found that the cardiovagal baroreflex was impaired in hypertension but that the baroreflex control of sympathetic activity was intact [27]. Taken together, these studies suggest alterations of cardiovascular autonomic regulation in hypertension. It is, however, not yet determined whether alterations in baroreceptor function can actually contribute to the development of hypertension.

9.5   Testing Baroreflex Sensitivity in Humans

There are several methodologies available to assess baroreflex function in humans that may be applied in both research and clinical settings. The most commonly used techniques are outlined below.

9.5.1   Mechanical Methods of Assessing Baroreflex Function

A mechanical test of carotid sinus reflexes was described by Mandelstamm and Lifschitz, who in 1932 reported that compression of the carotid sinus against the vertebrae decreases heart rate and blood pressure [31]. Measuring the heart rate response to carotid sinus massage has become a fairly established clinical test, particularly for carotid sinus hypersensitivity, a syndrome characterized by inappropriate bradycardia in response to carotid baroreceptor stimulation [32, 33]. However, although manual mechanical compressions of the carotid sinus region can produce changes in heart rate and blood pressure, they cannot provide the quantitative responses necessary for the precise clinical or scientific evaluation of the baroreflex.

9.5.1.1   Neck Suction

An alternative approach is therefore to use an airtight chamber to apply positive or negative transmural pressures to the region of the neck overlying the carotid sinuses. The first such device was described by Ernsting and Parry in 1957 [34] and comprised a plastic chamber designed to fit around the entire neck of the test subject. Manipulating the pressure within the chamber resulted in changes in heart rate due to loading or unloading of the carotid baroreceptors, depending on whether positive or negative pressures were applied. The usefulness of this chamber was, however, somewhat limited by it being rather large, cumbersome and difficult to fit to a range of different subjects. Therefore, Eckberg et al. [35] devised a simplified neck suction device that was constructed from malleable lead and could be molded to fit a range of different subjects. Subatmospheric pressures are typically applied to the neck for 5s during held expiration (to avoid respiratory influences on responses) and the subsequent prolongation of the cardiac interval measured. Because the stimulus is applied for a short time period and the cardiac slowing occurs rapidly, the measured responses are unlikely to be modified by closed-loop responses. By measuring responses to various carotid distending pressures, it is possible to plot the linear range of the classical baroreflex curve. However, in order to characterize the whole sigmoid baroreflex curve (i.e., threshold, linear and saturation ranges) it is necessary to also apply positive pressures to the carotid sinus—a procedure that is more technically challenging than the application of negative pressures [36]. By applying brief repeated baroreceptor pulses comprising negative and positive pressures to a neck chamber, Eckberg could define the entire sigmoidal baroreceptor response curve [14]. A refinement of this method was subsequently developed to allow study of baroreflex responses in astronauts during spaceflight [37].

9.5.1.2 Oscillatory Neck Suction

Although the cardiac responses to neck suction are usually measured, vascular resistance responses have also been determined by measuring changes in brachial blood flow in response to baroreceptor loading or unloading [38, 39]. However, because the vascular responses to baroreceptor loading or unloading take around 10–15 s to develop, longer applications of positive or negative pressures to the neck are required. During this time, the reflex change in blood pressure might causes a modification in the stimulus to all baroreceptor areas including those in the aortic arch and also the carotid baroreceptors. The subsequent buffering of the responses could potentially result in an underestimation of the true effects of baroreceptor stimulation on vascular resistance. This buffering of baroreflex responses during neck suction is only an issue when the stimulus to the baroreceptors is static rather than dynamic. Bath et al. measured vasoconstrictor sympathetic activity in response to static and dynamic neck suction [40].

Application of neck suction in a sinusoidal manner induced rhythmical changes in sympathetic activity, whereas static neck suction application did not significantly change sympathetic activity [40]. Several clinical studies have subsequently used oscillatory neck suction to study both the vascular and cardiac components of the carotid arterial baroreflex. For example, Bernardi et al. [41] introduced a protocol whereby sinusoidal neck suction was applied at frequencies of 0.1 Hz and 0.2 Hz. Neck suction at 0.1 Hz causes fluctuations in blood pressure (sympathetic response) and in heart rate (sympathetic and sympathetic responses) at the same frequency. Neck suction at 0.2 Hz induces 0.2 Hz fluctuations in heart rate, mediated by parasympathetic modulation, but is too rapid to cause blood pressure fluctuations. Thus, separate application of both frequencies, with the breathing rate controlled at a separate frequency (usually 0.15 Hz), permits differentiation of the cardiovagal and vasomotor limbs of the baroreflex [41]. Because the stimulus to the baroreceptors is constantly varying, this technique minimizes the problem of baroreflex resetting or buffering of the responses from other baroreceptor regions. Application of oscillatory neck suction at 0.1 and 0.2 Hz has been used to study baroreflex function in ageing [42] and has demonstrated baroreflex impairments in conditions as diverse as glaucoma [43], diabetes [44], migraine [45] and Familal Dysautonomia [46]. It has also been used in more fundamental physiological studies, for example, to investigate the origin of respiration sinus arrhythmia [47, 48]. Although neck suction is a well-established technique that has contributed to numerous basic and clinical research studies it has never really been accepted as a routine test for clinical assessment of baroreflex function. The reasons for this are unclear, but may include a lack of standard methodology as well as possible technical difficulties in manufacturing the neck chamber, since the devices are usually constructed in-house. Other possible disadvantages of using the neck chamber technique include uncertainty over how much of the stimulus is actually transmitted through the neck to the baroreceptors [40, 49, 50] and the fact that only the carotid baroreceptors and not other baroreflex regions are stimulated.

9.5.2 Pharmacological Methods of Assessing Baroreflex Function

Another method of measuring baroreflex sensitivity is to induce blood pressure elevations or reductions by injecting vasoactive drugs and then to measure the subsequent effect on cardiac interval. This technique was originally developed by Smyth et al. [51] to study changes in baroreflex sensitivity in sleeping subjects without disturbing them. To induce a rise in blood pressure, angiotensin II was intravenously infused into the antecubital vein. Changes in systolic blood pressure were computed and related to the cardiac interval recorded during the subsequent heart beat. The slope of the line relating cardiac interval and blood pressure was calculated by linear regression and is considered the measure of baroreflex sensitivity [51]. Although angiotensin II was the original choice of pressor agent, it may have direct effects on the heart and subsequent studies used the $\alpha1$-adrenergic receptor agonist phenylephrine [52]. A major advantage of the pharmacological approach is that the stimulus for baroreflex activation is an actual increase in blood pressure, rather than the mechanical deformation of the carotid vessels as occurs with the neck suction technique. As such, the pharmacological method is widely considered as the gold-standard technique for the clinical assessment of baroreflex sensitivity. A variation of the pharmacological technique is to measure baroreflex sensitivity in response to falling, as well as rising blood pressure. Ebert and Cowley [53] introduced a method that involved sequentially infusing nitroprusside to lower and phenylephrine to raise blood pressure. The advantage of studying baroreflex during falling and rising blood pressure is that a much larger portion of the entire baroreflex curve can be evaluated. Further, in patients in whom inducing blood pressure rises may be considered risky, the cardiac interval responses to a fall in blood pressure induced by sodium nitroprusside may instead be studied. An important clinical application for baroreflex sensitivity assessment using the pharmacological method has been the stratification of cardiovascular risk in patients with ischemic heart disease. The hypothesis that measurement of baroreceptor sensitivity could have some prognostic value was originally tested in dogs in whom a myocardial infarction was induced [54]. It was found that not only did the dogs have reduced baroreceptor sensitivity after myocardial infarction but also that the dogs who were most susceptible to ventricular fibrillation during acute myocardial ischemia induced by a subsequent exercise test had the lowest baroreceptor reflex sensitivity [54]. A subsequent clinical study prospectively followed a series of 78 male patients after suffering a first myocardial infarction and showed a strong correlation between cardiovascular mortality and reduced baroreceptor sensitivity [55]. Further clinical studies have confirmed the usefulness of baroreflex sensitivity as a marker of prognosis in patients who have suffered myocardial infarction [56, 57].

9.5.3 Spontaneous Techniques

Baroreflex sensitivity in humans has classically been quantified either by the vasoactive drug infusion or by mechanical stimulation of the carotid baroreceptors. Although both techniques are well-established, their usefulness may be limited in certain situations. For example, administration of vasoactive drugs is invasive, while the neck suction technique requires specialized equipment that may not be readily available, particularly in the clinical setting. The last 25 years, however, have seen the introduction of analysis techniques that permit the non-invasive assessment baroreflex sensitivity from spontaneously-occurring fluctuations in blood pressure and heart rate.

9.5.3.1 Sequence Method

In 1988, Bertinieri et al. [58] described a technique that involved scanning blood pressure and heart rate recordings for sequences at least three heart beats where blood pressure is rising or falling, with changes of cardiac interval in the same direction occurring on the subsequent heart beat. For each of these so-called "baroreflex sequences," linear regression is applied to the changes in systolic blood pressure and cardiac interval and the average slope of all the sequences within a given time period is taken as the baroreflex sensitivity [58]. The technique is entirely non-invasive (providing a noninvasive blood pressure recording device is used) and is therefore a very attractive method for use in clinical and research studies. Because no specific baroreceptor stimulations are required, it is possible to analyze pre-recorded cardiovascular data to retrospectively assess baroreflex sensitivity. There is, however, conflicting information about whether baroreflex sensitivity calculated from the sequence technique is in agreement with results derived from the more established methods. Some studies have reported strong agreement between baroreflex sensitivity as measured with the pharmacological and sequence techniques [59, 60]. On the other hand, Pitzalis et al. [61] showed that the limits of agreement were quite wide, even though there was a strong correlation between pharmacological and sequence techniques. The authors therefore proposed that the sequence technique may not be suitable as a technique for the stratification of cardiovascular risk in post-myocardial infarction patients [61]. Nevertheless, because it is non-invasive and the possibility exists to apply the analysis to differing experimental conditions, numerous research studies have used the sequence technique to determine baroreflex sensitivity in various settings [62–65].

9.5.3.2   Frequency Domain Analysis

The underlying frequency components within a biosignal can be characterized using power spectral analysis. There are at least two physiologically relevant fluctuations. One is in the high-frequency range (0.15–0.40 Hz) and is synchronized with respiratory movements. High frequency oscillations in blood pressure derive from the mechanical effects of breathing-induced fluctuations in stroke volume. In heart rate, high-frequency oscillations are the result of parasympathetically-mediated respiratory sinus arrhythmia [66]. The second important cardiovascular fluctuation is in the low-frequency range (0.03–0.14 Hz). Low-frequency oscillations in blood pressure result from fluctuations in peripheral sympathetic vasoconstrictor tone [67]. In heart rate, low-frequency fluctuations are considered to derive from baroreflex-mediated adjustments to the sinus-node [68]. Therefore, one method of assessing baroreflex sensitivity is to analyze the relationship between blood pressure and cardiac interval fluctuations in the low frequency range. The transfer function gain (expressed as ms/mmHg) describes the ratio between changes in cardiac interval and systolic blood pressure within a specific frequency range and is analogous to the regression coefficient for time domain analysis. Robbe et al. [69] demonstrated that baroreflex sensitivity obtained from the transfer function gain correlated strongly with results of the phenylephrine technique. However, as with the sequence technique, there are some doubts about whether the indices derived from spectral analysis are adequate markers of baroreflex sensitivity [70].

9.6   Baroreflex Responses in Posturally-Related Syncope

Denervation studies in humans and animals have demonstrated the importance of the arterial baroreceptors in the maintenance of blood pressure during orthostatic stress [23, 24]. There are, however, conflicting reports on baroreflex sensitivity in patients with posturally-related syncope. It has been reported that individuals with a history of vasovagal syncope have augmented heart rate responses to neck suction compared to normal subjects [71, 72]. In contrast, Thomson et al. [73] found that baroreflex responses in syncope patients and controls were not different, while Morillo et al. [74] reported impaired baroreflex sensitivity in patients. Other studies have suggested that individuals with poor orthostatic tolerance have more latent baroreflex responses [75, 76]. Most of these studies investigated only the cardiovagal component of the baroreflex. During orthostatic stress, however, it is the vascular resistance responses to baroreceptor unloading that are much more relevant, since posturally-related syncope is associated with impaired vasoconstriction [77]. Cooper and Hainsworth [78] demonstrated that on moving from supine to a head-up position, the vascular responses to baroreceptor loading and unloading were augmented in normal subjects but not in the patients with orthostatic intolerance. Thus, the increase in sensitivity of the vascular limb of the baroreflex during orthostatic stress may have a role in maintaining blood pressure during orthostatic stress.

9.7 Role of Cardiopulmonary Receptors

Several studies have reported that baroreflex responsiveness increases during orthostatic stress [38, 78–81]. The exact reason for this phenomenon remains undetermined but one possibility is that unloading of so-called "cardiopulmonary receptors" by orthostatic stress causes an alteration of arterial baroreceptor sensitivity. The concept of cardiopulmonary receptors was proposed by McDowell in 1924 [82]. They are understood to exist in the low-pressure areas of the circulation and are responsive to changes in central venous pressure. They do not, however, comprise a single receptor type and different receptor areas elicit different responses [83]. In humans, cardiopulmonary receptors have been suggested to have a role in the reflex responses to changes in venous filling. For example, Roddie et al. [84] demonstrated that when the legs of a supine subject are raised passively, there is a reflex forearm vasodilatation. Since there was little effect on arterial pressure, they postulated that the vasodilatory response was due to stimulation of baroreceptors in the low pressure regions rather than arterial baroreceptors [84]. Similarly, low levels of lower body negative pressure (below −20 mmHg) elicit reflex responses with minimal change in arterial pressure—a phenomenon that has been attributed to unloading of cardiopulmonary receptors [85]. There is, however, also evidence against cardiopulmonary receptors having an important role in modulating responses to orthostatic stress. Experiments in dogs have shown that unloading of atrial mechanoreceptors decreases heart rate, which is an inappropriate response to orthostasis, and there are no consistent vascular resistance changes in the hindlimb [86]. Also, stimulation of right ventricular receptors has no effect on heart rate or systemic blood pressure [87]. Results from a study in which conscious monkeys were exposed to low levels of lower body negative pressure indicated that the reflex responses were attenuated after denervation of the arterial baroreceptors [23], thus emphasizing the importance of the arterial baroreflex in blood pressure regulation. Taylor et al. [88] measured the dimensions of the ascending aorta during lower body negative pressure application using magnetic resonance imaging. They showed that during very low levels of lower body negative pressure (−5 mmHg) there was a reduction of aortic cross-sectional area even when mean arterial pressure was unchanged. Thus, responses that have been attributed to cardiopulmonary receptor unloading might actually be due to unloading of the carotid baroreceptors, even without changes in mean arterial pressure. The question of the role of the cardiopulmonary receptors and possible interaction with arterial baroreceptor reflexes remains open.

9.8 Conclusions

Baroreflex physiology has been widely studied using various techniques; yet testing of baroreflex sensitivity has not yet been established as part of routine clinical procedure. This is despite the potential usefulness of baroreflex sensitivity as an indicator of prognosis in cardiovascular diseases. Some aspects of baroreflex physiology remain to be understood, particularly the potential role of the baroreflex in the long term regulation of blood pressure, as well as the mechanisms of reduced baroreflex sensitivity in hypertension.

References

1. Drinkhill, M.J., Moore, J., Hainsworth, R.: Afferent discharges from coronary arterial and ventricular receptors in anaesthetized dogs. J. Physiol. **472**, 785–799 (1993)
2. Drinkhill, M.J., McMahon, N.C., Hainsworth, R.: Delayed sympathetic efferent responses to coronary baroreceptor unloading in anaesthetized dogs. J. Physiol. **497**, 261–269 (1996)
3. McMahon, N.C., Drinkhill, M.J., Hainsworth, R.: Vascular responses to stimulation of carotid, aortic and coronary artery baroreceptors with pulsatile and non-pulsatile pressures in anaesthetized dogs. Exp. Physiol. **81**, 969–981 (1996)
4. McMahon, N.C., Drinkhill, M.J., Myers, D.S., Hainsworth, R.: Reflex responses from the main pulmonary artery and bifurcation in anaesthetised dogs. Exp. Physiol. **85**, 411–420 (2000)
5. Koch, E.: Die Reflektorische Selbssteuring des Kreislaufes. Steinkopf, Leipzig (1931)
6. Bronk, D.W., Stelle, G.: Afferent impulses in the carotid sinus nerve. I. The reaction of the discharge from single end organs to arterial blood pressure. J. Cell Comp. Physiol. **1**, 113–130 (1932)
7. Donald, D.E., Edis, A.J.: Comparison of aortic and carotid baroreflexes in the dog. J. Physiol. **215**, 521–538 (1971)
8. Hainsworth, R., Karim, F.: Left ventricular inotropic and peripheral vasomotor responses from independent changes in pressure in the carotid sinuses and cerebral arteries in anaesthetized dogs. J. Physiol. **228**, 139–155 (1973)
9. James, J.E., Daly, M.B.: Comparison of the reflex vasomotor responses to separate and combined stimulation of the carotid sinus and aortic arch baroreceptors by pulsatile and non-pulsatile pressures in the dog. J. Physiol. **209**, 257–293 (1970)

10. Karim, F., Poucher, S.M., Summerill, R.A.: The reflex effects of changes in carotid sinus pressure upon renal function in dogs. J. Physiol. **355**, 557–566 (1984)
11. Eckberg, D.L.: Baroreflex inhibition of the human sinus node: importance of stimulus intensity, duration, and rate of pressure change. J. Physiol. **269**, 561–577 (1977)
12. Mancia, G., Ferrari, A., Gregorini, L., Valentini, R., Ludbrook, J., Zanchetti, A.: Circulatory reflexes from carotid and extracarotid baroreceptor areas in man. Circ. Res. **41**, 309–315 (1977)
13. Bjurstedt, H., Rosenhamer, G., Tyden, G.: Cardiovascular responses to changes in carotid sinus transmural pressure in man. Acta Physiol. Scand. **94**, 497–505 (1975)
14. Eckberg, D.L.: Nonlinearities of the human carotid baroreceptor-cardiac reflex. Circ. Res. **47**, 208–216 (1980)
15. Hainsworth, R.: The importance of vascular capacitance in cardiovascular control. NIPS **5**, 250–254 (1990)
16. Rowell, L.B.: Human Circulation Regulation During Physical Stress. Oxford University Press, Oxford (1986)
17. Gauer, O.H., Thron, H.L.: Postural changes in the circulation. In: Hamilton, W.F., Dow, P. (eds.) Handbook of Physiology, pp. 2409–2439. American Physiological Society, Baltimore (1965)
18. Brown, C.M., Hainsworth, R.: Assessment of capillary fluid shifts during orthostatic stress in normal subjects and subjects with orthostatic intolerance. Clin. Auton. Res. **9**, 69–73 (1999)
19. Fawcett, J.K., Wynn, V.: Effects of posture on plasma volume and some blood constituents. J. Clin. Pathol. **13**, 304–310 (1960)
20. Hainsworth, R., Al-Shamma, Y.M.: Cardiovascular responses to upright tilting in healthy subjects. Clin. Sci. (Lond.) **74**, 17–22 (1988)
21. Cowley, A.W. Jr., Liard, J.F., Guyton, A.C.: Role of baroreceptor reflex in daily control of arterial blood pressure and other variables in dogs. Circ. Res. **32**, 564–576 (1973)
22. Ito, C.S., Scher, A.M.: Regulation of arterial blood pressure by aortic baroreceptors in the unanesthetized dog. Circ. Res. **42**, 230–236 (1978)
23. Cornish, K.G., Gilmore, J.P., McCulloch, T.: Central blood volume and blood pressure in conscious primates. Am. J. Physiol. **254**, H693–H701 (1988)
24. Smit, A.A., Timmers, H.J., Wieling, W., Wagenaar, M., Marres, H.A., Lenders, J.W., van Montfrans, G.A., Karemaker, J.M.: Long-term effects of carotid sinus denervation on arterial blood pressure in humans. Circulation **105**, 1329–1335 (2002)
25. Lohmeier, T.E., Lohmeier, J.R., Haque, A., Hildebrandt, D.A.: Baroreflexes prevent neurally induced sodium retention in angiotensin hypertension. Am. J. Physiol. Regul. Integr. Comp. Physiol. **279**, R1437–R1448 (2000)
26. Thrasher, T.N.: Unloading arterial baroreceptors causes neurogenic hypertension. Am. J. Physiol. Regul. Integr. Comp. Physiol. **282**, R1044–R1053 (2002)
27. Mancia, G., Ludbrook, J., Ferrari, A., Gregorini, L., Zanchetti, A.: Baroreceptor reflexes in human hypertension. Circ. Res. **43**, 170–177 (1978)
28. Sevre, K., Lefrandt, J.D., Nordby, G., Os, I., Mulder, M., Gans, R.O., Rostrup, M., Smit, A.J.: Autonomic function in hypertensive and normotensive subjects: The importance of gender. Hypertension **37**, 1351–1356 (2001)
29. McGarry, K., Laher, M., Fitzgerald, D., Horgan, J., O'Brien, E., O'Malley, K.: Baroreflex function in elderly hypertensives. Hypertension **5**, 763–766 (1983)
30. Bristow, J.D., Honour, A.J., Pickering, G.W., Sleight, P., Smyth, H.S.: Diminished baroreflex sensitivity in high blood pressure. Circulation **39**, 48–54 (1969)
31. Mandelstamm, M., Lifschitz, S.: Die Wirkung der Karotissinusreflexe auf den Blutdruck beim Menschen. Wien Arch. f. Inn. Med. **22**, 397 (1932)
32. Kumar, N.P., Thomas, A., Mudd, P., Morris, R.O., Masud, T.: The usefulness of carotid sinus massage in different patient groups. Age Ageing **32**, 666–669 (2003)
33. Tan, M.P., Kenny, R.A.: Cardiovascular assessment of falls in older people. Clin. Interv. Aging. **1**, 57–66 (2006)
34. Ernsting, J., Parry, D.J.: Some observations on the effects of stimulating the carotid arterial stretch receptors in the carotid artery of man. J. Physiol. **137**, 145 (1957)

35. Eckberg, D.L., Cavanaugh, M.S., Mark, A.L., Abboud, F.M.: A simplified neck suction device for activation of carotid baroreceptors. J. Lab. Clin. Med. **85**, 167–173 (1975)
36. Cooper, V.L., Hainsworth, R.: Carotid baroreflex testing using the neck collar device. Clin. Auton. Res. **19**, 102–112 (2009)
37. Sprenkle, J.M., Eckberg, D.L., Goble, R.L., Schelhorn, J.J., Halliday, H.C.: Device for rapid quantification of human carotid baroreceptor-cardiac reflex responses. J. Appl. Physiol. **60**, 727–732 (1986)
38. Cooper, V.L., Hainsworth, R.: Carotid baroreceptor reflexes in humans during orthostatic stress. Exp. Physiol. **86**, 677–681 (2001)
39. Cooper, V.L., Pearson, S.B., Bowker, C.M., Elliott, M.W., Hainsworth, R.: Interaction of chemoreceptor and baroreceptor reflexes by hypoxia and hypercapnia – a mechanism for promoting hypertension in obstructive sleep apnoea. J. Physiol. **568**, 677–687 (2005)
40. Bath, E., Lindblad, L.E., Wallin, B.G.: Effects of dynamic and static neck suction on muscle nerve sympathetic activity, heart rate and blood pressure in man. J. Physiol. **311**, 551–564 (1981)
41. Bernardi, L., Bianchini, B., Spadacini, G., Leuzzi, S., Valle, F., Marchesi, E., Passino, C., Calciati, A., Vigano, M., Rinaldi, M.: Demonstrable cardiac reinnervation after human heart transplantation by carotid baroreflex modulation of RR interval. Circulation **92**, 2895–2903 (1995)
42. Brown, C.M., Hecht, M.J., Weih, A., Neundorfer, B., Hilz, M.J.: Effects of age on the cardiac and vascular limbs of the arterial baroreflex. Eur. J. Clin. Invest. **33**, 10–16 (2003)
43. Brown, C.M., Dutsch, M., Michelson, G., Neundorfer, B., Hilz, M.J.: Impaired cardiovascular responses to baroreflex stimulation in open-angle and normal-pressure glaucoma. Clin. Sci. (Lond.) **102**, 623–630 (2002)
44. Sanya, E.O., Brown, C.M., Dutsch, M., Zikeli, U., Neundorfer, B., Hilz, M.J.: Impaired cardiovagal and vasomotor responses to baroreceptor stimulation in type II diabetes mellitus. Eur. J. Clin. Invest. **33**, 582–588 (2003)
45. Sanya, E.O., Brown, C.M., von Wilmowsky, C., Neundorfer, B., Hilz, M.J.: Impairment of parasympathetic baroreflex responses in migraine patients. Acta Neurol. Scand. **111**, 102–107 (2005)
46. Stemper, B., Bernardi, L., Axelrod, F.B., Welsch, G., Passino, C., Hilz, M.J.: Sympathetic and parasympathetic baroreflex dysfunction in familial dysautonomia. Neurology **63**, 1427–1431 (2004)
47. Piepoli, M., Sleight, P., Leuzzi, S., Valle, F., Spadacini, G., Passino, C., Johnston, J., Bernardi, L.: Origin of respiratory sinus arrhythmia in conscious humans. An important role for arterial carotid baroreceptors. Circulation **95**, 1813–1821 (1997)
48. Keyl, C., Dambacher, M., Schneider, A., Passino, C., Wegenhorst, U., Bernardi, L.: Cardiocirculatory coupling during sinusoidal baroreceptor stimulation and fixed-frequency breathing. Clin. Sci. (Lond.) **99**, 113–124 (2000)
49. Eckberg, D.L.: Temporal response patterns of the human sinus node to brief carotid baroreceptor stimuli. J. Physiol. **258**, 769–782 (1976)
50. Ludbrook, J., Mancia, G., Ferrari, A., Zanchetti, A.: The variable-pressure neck-chamber method for studying the carotid baroreflex in man. Clin. Sci. Mol. Med. **53**, 165–171 (1977)
51. Smyth, H.S., Sleight, P., Pickering, G.W.: Reflex regulation of arterial pressure during sleep in man. A quantitative method of assessing baroreflex sensitivity. Circ. Res. **24**, 109–121 (1969)
52. Gribbin, B., Pickering, T.G., Sleight, P., Peto, R.: Effect of age and high blood pressure on baroreflex sensitivity in man. Circ. Res. **29**, 424–431 (1971)
53. Ebert, T.J., Cowley, A.W. Jr.: Baroreflex modulation of sympathetic outflow during physiological increases of vasopressin in humans. Am. J. Physiol. **262**, H1372–H1378 (1992)
54. Billman, G.E., Schwartz, P.J., Stone, H.L.: Baroreceptor reflex control of heart rate: A predictor of sudden cardiac death. Circulation **66**: 874–880 (1982)
55. La Rovere, M.T., Specchia, G., Mortara, A., Schwartz, P.J.: Baroreflex sensitivity, clinical correlates, and cardiovascular mortality among patients with a first myocardial infarction. A prospective study. Circulation **78**, 816–824 (1988)

56. De Ferrari, G.M., Landolina, M., Mantica, M., Manfredini, R., Schwartz, P.J., Lotto, A.: Baroreflex sensitivity, but not heart rate variability, is reduced in patients with life-threatening ventricular arrhythmias long after myocardial infarction. Am. Heart J. **130**, 473–480 (1995)

57. Farrell, T.G., Odemuyiwa, O., Bashir, Y., Cripps, T.R., Malik, M., Ward, D.E., Camm, A.J.: Prognostic value of baroreflex sensitivity testing after acute myocardial infarction. Br. Heart J. **67**, 129–137 (1992)

58. Bertinieri, G., Di Rienzo, M., Cavallazzi, A., Ferrari, A.U., Pedotti, A., Mancia, G.: Evaluation of baroreceptor reflex by blood pressure monitoring in unanesthetized cats. Am. J. Physiol. **254**, H377–H383 (1988)

59. Parlow, J., Viale, J.P., Annat, G., Hughson, R., Quintin, L.: Spontaneous cardiac baroreflex in humans. Comparison with drug-induced responses. Hypertension **25**, 1058–1068 (1995)

60. Watkins, L.L., Grossman, P., Sherwood, A.: Noninvasive assessment of baroreflex control in borderline hypertension. Comparison with the phenylephrine method. Hypertension **28**, 238–243 (1996)

61. Pitzalis, M.V., Mastropasqua, F., Passantino, A., Massari, F., Ligurgo, L., Forleo, C., Balducci, C., Lombardi, F., Rizzon, P.: Comparison between noninvasive indices of baroreceptor sensitivity and the phenylephrine method in post-myocardial infarction patients. Circulation **97**, 1362 1367 (1998)

62. Butler, G.C., Senn, B.L., Floras, J.S.: Influence of atrial natriuretic factor on spontaneous baroreflex sensitivity for heart rate in humans. Hypertension **25**, 1167–1171 (1995)

63. Steptoe, A., Vogele, C.: Cardiac baroreflex function during postural change assessed using non-invasive spontaneous sequence analysis in young men. Cardiovasc. Res. **24**, 627–632 (1990)

64. Mancia, G., Groppelli, A., Di Rienzo, M., Castiglioni, P., Parati, G.: Smoking impairs baroreflex sensitivity in humans. Am. J. Physiol. **273**, H1555–H1560 (1997)

65. Parati, G., Di Rienzo, M., Bonsignore, M.R., Insalaco, G., Marrone, O., Castiglioni, P., Bonsignore, G., Mancia, G.: Autonomic cardiac regulation in obstructive sleep apnea syndrome: Evidence from spontaneous baroreflex analysis during sleep. J. Hypertens. **15**, 1621–1626 (1997)

66. Saul, J.P., Berger, R.D., Chen, M.H., Cohen, R.J.: Transfer function analysis of autonomic regulation. II. Respiratory sinus arrhythmia. Am. J. Physiol. **256**, H153–H161 (1989)

67. Malliani, A., Pagani, M., Lombardi, F., Cerutti, S.: Cardiovascular neural regulation explored in the frequency domain. Circulation **84**, 482–492 (1991)

68. Bernardi, L., Leuzzi, S., Radaelli, A., Passino, C., Johnston, J. A., Sleight, P.: Low-frequency spontaneous fluctuations of R-R interval and blood pressure in conscious humans: A baroreceptor or central phenomenon? Clin. Sci. (Lond.) **87**, 649–654 (1994)

69. Robbe, H.W., Mulder, L.J., Ruddel, H., Langewitz, W.A., Veldman, J.B., Mulder, G.: Assessment of baroreceptor reflex sensitivity by means of spectral analysis. Hypertension **10**, 538–543 (1987)

70. Lipman, R.D., Salisbury, J.K., Taylor, J.A.: Spontaneous indices are inconsistent with arterial baroreflex gain. Hypertension **42**, 481–487 (2003)

71. El Sayed, H., Hainsworth, R.: Relationship between plasma volume, carotid baroreceptor sensitivity and orthostatic tolerance. Clin. Sci. (Lond.) **88**, 463–470 (1995)

72. Wahbha, M.M., Morley, C.A, Al Shamma, Y.M., Hainsworth, R.: Cardiovascular reflex responses in patients with unexplained syncope. Clin. Sci. (Lond.) **77**, 547–553 (1989)

73. Thomson, H.L., Wright, K., Frenneaux, M.: Baroreflex sensitivity in patients with vasovagal syncope. Circulation **95**, 395–400 (1997)

74. Morillo, C.A., Eckberg, D.L., Ellenbogen, K.A., Beightol, L.A., Hoag, J.B., Tahvanainen, K.U., Kuusela, T.A., Diedrich, A.M.: Vagal and sympathetic mechanisms in patients with orthostatic vasovagal syncope. Circulation **96**, 2509–2513 (1997)

75. Gulli, G., Cooper, V.L., Claydon, V., Hainsworth, R.: Cross-spectral analysis of cardiovascular parameters whilst supine may identify subjects with poor orthostatic tolerance. Clin. Sci. (Lond.) **105**, 119–126 (2003)

76. Gulli, G., Claydon, V.E., Cooper, V.L., Hainsworth, R.: R-R interval-blood pressure interaction in subjects with different tolerances to orthostatic stress. Exp. Physiol. **90**, 367–375 (2005)

77. Brown, C.M., Hainsworth, R.: Forearm vascular responses during orthostatic stress in control subjects and patients with posturally related syncope. Clin. Auton. Res. **10**, 57–61 (2000)
78. Cooper, V.L., Hainsworth, R.: Effects of head-up tilting on baroreceptor control in subjects with different tolerances to orthostatic stress. Clin. Sci. (Lond.) **103**, 221–226 (2002)
79. Bevegard, S., Castenfors, J., Lindblad, L.E., Tranesjo, J.: Blood pressure and heart rate regulating capacity of the carotid sinus during changes in blood volume distribution in man. Acta Physiol. Scand. **99**, 300–312 (1977)
80. Ebert, T.J.: Carotid baroreceptor reflex regulation of forearm vascular resistance in man. J. Physiol. **337**, 655–664 (1983)
81. Victor, R.G., Mark, A.L.: Interaction of cardiopulmonary and carotid baroreflex control of vascular resistance in humans. J. Clin. Invest. **76**, 1592–1598 (1985)
82. McDowell, R.J.S.: A vago-pressor reflex. J. Physiol. **59**, 41–47 (1924)
83. Hainsworth, R.: Reflexes from the heart. Physiol. Rev. **71**, 617–658 (1991)
84. Roddie, I.C., Shepherd, J.T., Whelan, R.F.: Reflex changes in vasoconstrictor tone in human skeletal muscle in response to stimulation of receptors in a low-pressure area of the intrathoracic vascular bed. J. Physiol. **139**, 369–376 (1957)
85. Zoller, R.P., Mark, A.L., Abboud, F.M., Schmid, P.G., Heistad, D.D.: The role of low pressure baroreceptors in reflex vasoconstrictor responses in man. J. Clin. Invest. **51**, 2967–2972 (1972)
86. Carswell, F., Hainsworth, R., Ledsome, J.R.: The effects of distension of the pulmonary vein-atrial junctions upon peripheral vascular resistance. J. Physiol. **207**, 1–14 (1970)
87. Crisp, A.J., Hainsworth, R., Tutt, S.M.: The absence of cardiovascular and respiratory responses to changes in right ventricular pressure in anaesthetized dogs. J. Physiol. **407**, 1–13 (1988)
88. Taylor, J.A., Halliwill, J.R., Brown, T.E., Hayano, J., Eckberg, D.L.: "Non-hypotensive" hypovolaemia reduces ascending aortic dimensions in humans. J. Physiol. **483**, 289–298 (1995)

Chapter 10
Development of Patient Specific Cardiovascular Models Predicting Dynamics in Response to Orthostatic Stress Challenges

Johnny T. Ottesen, Vera Novak, and Mette S. Olufsen

Abstract Physiological realistic models of the controlled cardiovascular system are constructed and validated against clinical data. Special attention is paid to the control of blood pressure, cerebral blood flow velocity, and heart rate during postural challenges, including sit-to-stand and head-up tilt. This study describes development of patient specific models, and how sensitivity analysis and nonlinear optimization methods can be used to predict patient specific characteristics when analyzed using experimental data. Finally, we discuss how a given model can be used to understand physiological changes between groups of individuals and how to use modeling to identify biomarkers.

10.1 Introduction

This manuscript summarizes a series of cardiovascular control models predicting baroreflex and cerebral autonomic regulation of blood flow and pressure during orthostatic postural challenges including sitting to standing (STS) and head-up tilt (HUT) [2, 5, 18–22, 24–30]. During these challenges gravity pools blood from the upper to the lower body. As a result venous return is reduced, leading to a decrease

J.T. Ottesen (✉)
Department of Science, Systems, and Models, Roskilde University, 4000 Roskilde, Denmark
e-mail: Johnny@ruc.dk

V. Novak
Division of Gerontology, Beth Israel Deaconess Medical Center and Harvard University, Boston, MA 02215
e-mail: vnovak@bidmc.harvard.edu

M.S. Olufsen
Center for Research in Scientific Computing and Department of Mathematics, North Carolina State University, Raleigh, NC 27695
e-mail: msolufse@ncsu.edu

J.J. Batzel et al. (eds.), *Mathematical Modeling and Validation in Physiology*,
Lecture Notes in Mathematics 2064, DOI 10.1007/978-3-642-32882-4_10,
© Springer-Verlag Berlin Heidelberg 2013

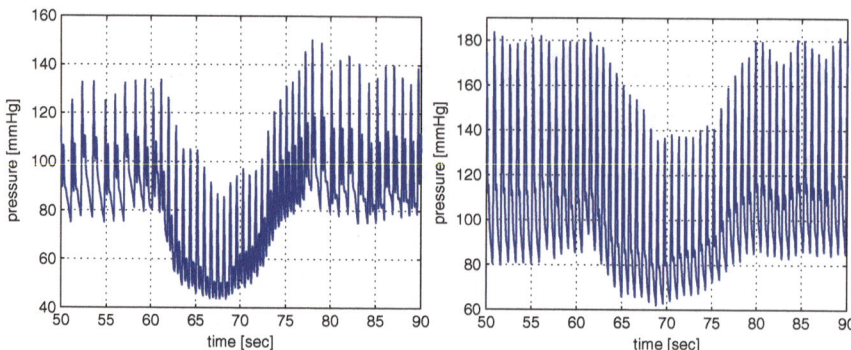

Fig. 10.1 Blood pressure [mmHg] versus time [s] during STS for a healthy young (*left*) and a hypertensive elderly (*right*) subject

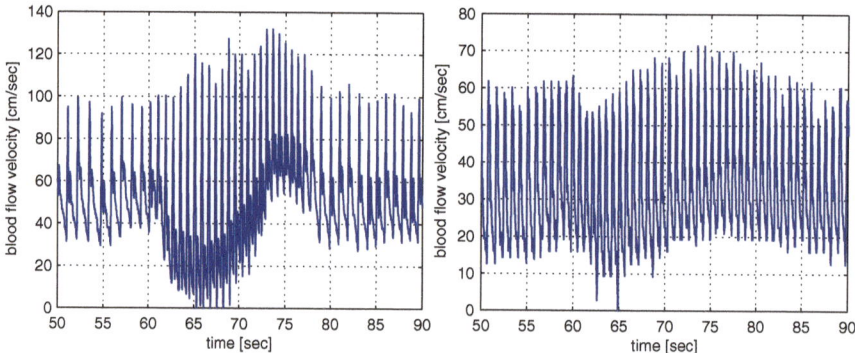

Fig. 10.2 Cerebral blood flow velocity [cm/s] versus time [s] during STS for a healthy young (*left*) and a hypertensive elderly (*right*) subject

in cardiac stroke volume, a decline in upper body arterial blood pressure (see Fig. 10.1), an increase in lower body arterial pressure, and an immediate decrease of blood flow to the brain (see Fig. 10.2). These changes are rapidly counteracted by short-term feedback control mechanisms working to reestablish blood flow and pressure. The main short-term feedback mechanisms involved are autonomic regulation (mainly baroreceptor feedback control) and cerebral autoregulation. The baroreceptor system controls blood pressure by regulating heart rate (see Fig. 10.3), vascular tone (resistance and compliance), and cardiac contractility in response to changes in blood pressure, while cerebral autoregulation controls cerebral blood flow by regulating cerebrovascular tone (resistance and compliance). These control mechanisms are complex and interact in ways that are not yet fully understood. Furthermore, it is believed that their functions are compromised by aging and disease. For example, mean blood flow velocity and pressure show similar dynamics for healthy young and hypertensive elderly subjects, while the pulse pressure (the width of pressure wave, the systolic minus the diastolic value) in response to

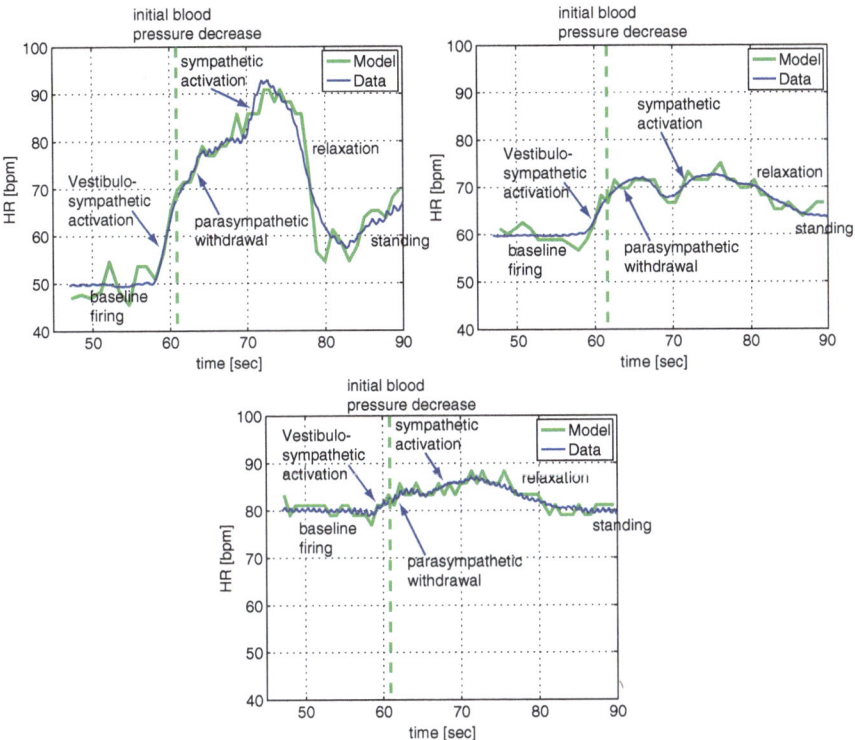

Fig. 10.3 Heart rate [bpm] versus time [s] for a healthy young (*top left*), a healthy elderly (*top right*) and a hypertensive elderly (*bottom*) subject (Adapted from [21])

standing up is increased for healthy young subjects but not for hypertensive elderly subjects (compare Figs. 10.1 and 10.2). It is likely that this widening of the pulse wave (or lack thereof) is a response to changes in the feedback control.

The baroreflex regulation acts through stimulation of baroreceptors located in the carotid and aortic walls. These receptors are stretch receptors, which sense changes in arterial blood pressure. A blood pressure drop leads to decreased firing of the afferent baroreceptor nerves, which gives rise to parasympathetic withdrawal and sympathetic activation. Parasympathetic withdrawal induces a fast increase in heart rate within 1–2 cardiac cycles, while sympathetic stimulation yields a delayed (within 6–8 cardiac cycles) increase in vascular tone (resistance and compliance), cardiac contractility, and a further increase in heart rate [8, 38]. At the same time cerebrovascular tone is modulated through metabolic vasoregulation (mediated by changes in CO_2) and myogenic autoregulation (regulation responding to local changes in blood pressure). It is not clear if autonomic control also regulates cerebrovascular tone. Some recent studies [46, 47] have indicated that cerebral vascular tone is modulated by stimulation of cholinergic nerves terminating in the cerebral vasculature.

In humans, autonomic and cerebral autoregulation is most often studied by analyzing dynamics during STS and HUT experiments [14, 37]. One difference between these two challenges is that HUT is often carried out using a slow-tilt lasting 5–10 s, while STS occurs rapidly over 1–5 s. As a result the regulatory response (in particular the parasympathetic portion of the response) to HUT is initiated before the subject is fully tilted thereby preventing excessive drop in blood pressure, while during STS blood is pooled to the legs fast leading to a much larger drop in pressure. This difference in the orthostatic motion impacts the loading of the baroreceptors. During HUT, gravitational forces act between the head and the torso as well as between the torso and the lower body, while STS mainly pools blood to the legs (in elderly a considerable amount of blood may be pooled in the legs during sitting). As a result HUT gives rise to an increased flow from the brain to the heart, which for a period of time may increase right atrial pressure stimulating pulmonary baroreceptors, which has potential to lower heart rate. During STS, there is no hydrostatic change between various regions in upper corpus, thus all pooling of blood will be in the legs. Therefore this motion does not give rise to the immediate decrease in heart rate upon standing.

Another difference is that HUT is almost entirely passive, while STS require muscle contraction in the legs activating muscle sympathetic nerves (MSNA), which may increase heart rate before the observed drop in blood pressure. Other explanation for the initial increase in heart rate include stimulation by the vestibular system, and/or by central command [11, 33–35, 45].

During HUT and STS experiments typical cardiovascular measurements include: arterial blood pressure, heart rate, and cerebral blood flow velocity measured using Transcranial Doppler ultrasound. These measurements are then used to assess short term autonomic (baroreflex) and cerebral autoregulation. Most common data analysis methods use some form of linear response models. For example, baroreflex sensitivity [9, 36] is often assessed using spectral transfer functions relating changes in systolic blood pressure to interbeat intervals. Problems with these methods are two-fold; first the methods are linear, second they are limited to analysis of the relationships between two signals. Another limitation is that these methods lack the ability to predict how changes in neural responses interact to maintain arterial blood pressure.

In this manuscript we discuss alternative approaches based on nonlinear dynamic mathematical models. Two models will be discussed in detail: an open loop model developed to predict STS and HUT regulation of heart rate and a closed loop model developed to predict blood flow and blood pressure baroreflex and cerebral autoregulation during STS. The heart rate model has been discussed in [17, 21, 22, 24, 25] and closed loop model has been discussed in [5, 6, 18–20, 30, 31]. Both models were developed to analyze patient specific data, i.e., our goal was to develop models that allow prediction of dynamic quantities including cerebral blood flow velocity, arterial blood pressure, and heart rate. To do so, we estimated a set of nominal parameters for each subject using allometric scaling laws and anthropometric data (height, weight, age gender). Using these nominal values patient specific values were obtained by predicting a subset of parameters minimizing (in a reliable way) the difference between computed and measured values of the observed quantities.

The subset of parameters identified using optimization have potential to function as biomarkers, which can reveal significant differences between healthy and diseased subjects (or even within a group of subjects). Not all optimized parameters will be different between groups, but those that do have potential to serve as biomarkers characterizing varying states of diseases. Finally, we show dynamics observed during longer time scales (>35 min) for a subject who experienced syncope. Note, at the onset of syncope blood pressure and heart rate decreases rapidly. We show (see Fig. 10.15) that our model is able predict a similar behavior for a suitable choice of parameters, i.e., when saturation in the vascular tone is reached.

10.2   Methods

10.2.1   Experimental Design

The studies reviewed here used STS and HUT data from a number of healthy, hypertensive, and elderly subjects. The healthy young and elderly subjects where not treated for any systemic disease and the hypertensive elderly subjects were diagnosed and treated for hypertension but had no history of more than one episode of syncope. Subjects with a history of diabetes, stroke and brain injury, renal liver and other systemic disorders were excluded. Instrumentation for all studies was done using similar protocols. Heart rate was measured using a three-lead electrocardiogram (ECG) (SpaceLab Medical Inc., Issaquah, WA). A transcranial Doppler system (MultiDop X4, DWL Neuroscan Inc. Sterling VA) was used to obtain continuous measurements of blood flow velocity in the middle cerebral artery. The data was acquired by insonating the artery through the temporal windows using a 2 MHz pulsed Doppler probe. The probe was positioned to record the maximal flow velocities and stabilized using a three-dimensional head frame positioning system. A photoplethysmographic device mounted on the middle finger of the non-dominant hand was used to obtain non-invasive beat-to-beat blood pressure (Finapres device, Ohmeda Monitoring Systems, Englewood, Colorado). To eliminate effects of gravity, the hand was held at the level of the right atrium and supported by a sling. All physiological signals were digitized at 500 Hz using Labview NINDAQ software (National Instruments, Austin, TX) and stored for offline analysis. Before data were analyzed they were down-sampled to 50 Hz.

- STS protocol: After instrumentation, subjects sat in a straight-backed chair with their legs elevated at $90°$ in front of them. After 5 min of stable recordings, the subjects were asked to stand up. Standing was defined as the moment both feet touched the floor, recorded by a force platform.
- HUT protocol: The subjects rested on their back in the supine position for 10 min. After resting in this position, the table was tilted to $70°$ for 10 min.
- Syncope protocol: Using the HUT protocol described above the subjects were tilted to $70°$ from supine to upright position and then kept upright until

presyncope (>35 min) at this time the subjects were returned to supine position, and the subjects regained consciousness immediately.

The STS and the open loop model data analyzed were collected from Lewis A. Lipsitz and Vera Novaks laboratories at Hebrew Senior Life and at the Beth Israel Deaconess Medical Center, Boston, MA. HUT data used for the syncope study was obtained by Jesper Mehlsen, Medical Director, Department of Physiology and Nuclear Medicine, Frederiksberg Hospital, University of Copenhagen, Denmark. All subjects provided informed consent approved by the Institutional Review Board at the participating institutions.

10.2.2 Modeling Strategy

The following protocol provides an outline of the methodology that has guided our work with the development of patient specific models:

- Data, knowledge and structure: Development of reliable patient specific physiological models require detailed knowledge of the underlying biological system combined with a clear definition of the outcomes that the model is supposed to predict or prescribe. Sufficient insight into both the physiology and available modeling techniques is crucial. In this spirit, structures in the data analyzed in this study were revealed using standard statistical tools combined with nonlinear optimization. Results of these analyses lead to knowledge of the system revealed through collaboration between mathematicians and physicians. Other methodologies that can be used for analyzing the structure of the system include filtering and generalized principal component analysis.
- Canonical models: The level of details incorporated in the models reflect the system analyzed. In order to identify and estimate patient specific parameters in an effective and reliable way the number of parameters has to be kept as low as possible so all unimportant elements are excluded (the principle of parsimony). The models will be based on first principles (e.g., conservation laws), and, whenever possible, the parameters shall have a physiological interpretation. Such models are denoted canonical models. The models discussed in this review were developed to study patient specific short-term regulation of heart rate, blood pressure, and cerebral blood flow velocity during STS and HUT. The models were formulated as dynamical systems. Steady state analysis of the model equations were considered. Afterward models were used for analysis of baseline dynamic behavior and later coupled with control models allowing prediction of dynamics during the orthostatic challenges (STS and HUT).
- Parameter identification and estimation: The main goal with the models presented here [2, 3, 5, 20–22, 27, 29–31] was to identify biomarkers (model parameters) that allow the models to display the dynamic behavior observed in the data. Estimation of these quantities requires solution of an inverse problem, where a set of parameters is estimated that minimize the error between measured

and computed quantities. For example, in the open loop heart rate model, a set of parameters minimizing the least squares error between computed and measured values of heart rate was identified, and in the closed loop model we minimized the least squares error between computed and measured values of blood flow velocity and arterial blood pressure. For a given model and a given set of data, it is likely that the model is insensitive to a subset of the model parameters, i.e., changes in that subset of parameters has a negligible impact on the solution. This set of parameters is called insensitive, and such parameters cannot be estimated via solution to the inverse problem. Second, among the subset of parameters that are sensitive it is likely that parameters are correlated. For example, two resistors in series could both be sensitive, but it is not possible to identify both resistors separately, knowing only the total potential drop and current, e.g., $R_T = R_1 + R_2 = 5$ have indefinitely many solutions for R_1 and R_2. The latter problem is addressed in some of our recent studies [5, 31] using a technique called subset selection. Once a set of sensitive uncorrelated parameters have been identified and estimated, statistical methods should be invoked to compare parameters within and between groups of individuals. In the models [17, 21, 31] we used analysis of variance (ANOVA) to show if the limited set of model parameters vary between healthy young and healthy elderly subjects. To understand patient specific behavior, model parameters and initial conditions (considered as a special kind of parameters) were estimated using literature data coupled with anthropometric (height, weight, age, gender) information, allometric scaling relations, as well as mean and steady state values extracted from the data.

- Optimization algorithms: To perform the required parameter estimations non-linear optimization algorithms was used. Two methodologies were used, gradient free methods: the Nelder–Mead method (a simplex method) [18, 20–22], genetic algorithms [7], and implicit filtering [7]; and gradient-based methods: the Newton method and the Levenberg–Marquardt method [5, 31]. The advantage of gradient-based methods is that sensitivities are computed as part of the optimization. A disadvantage of these methods is that the ODEs should be differentiated with respect to each of the parameters. Other methods that have received much attention recently include Kalman filtering [3], particle filters, functional differential analysis, and sequential Monte Carlo (SMC) methods.

- Validation of models: The process of solving the inverse problem estimating parameters that allow the models to predict data is one aspect of model validation, but this type of analysis should be combined with more general validation methods. For example, once a set of parameters have been estimated using one dataset, the models should be validated using other datasets not used for the parameter estimation. However, this type of validation often cannot be used for biological systems where inter and intra variations within and between groups of individuals are large. One way to validate such models is to use K-fold cross-validation [13] where a subset within one data-set is used for prediction of model parameters, while the remaining data are used for validation. Other validation methodologies include model reduction (discussed in [6]) and analysis of sub-mechanisms. Common for all of these validation methods is that if a model fails to be validated it should be adjusted. This process of iterative model development often generates important new insights into the underlying physiology [1, 12, 30, 40].

- Biomarkers: Patient specific model parameters estimated using reliable optimization techniques from well-validated models have potential to be used as biomarkers. This requires that confidence intervals for each identified parameters are small and that some certainty is achieved that the optimized parameters are not a result of optimization methods terminating at some local minimum. This type of methodology using models to estimate biomarkers can provide insight into quantities that cannot directly be measured experimentally. Examples include estimation of cerebrovascular resistance [17, 31], delay in baroreflex firing rate, baroreflex dampening, and baroreflex gain [21, 22]. To determine if biomarkers differ significantly between groups of subjects, and if biomarkers can be used to identify variant causes of a given illness statistical tests can be performed. For example, in a recent study [31] we showed that cerebrovascular resistance is increased with aging.

- Multiscale models: Frequently models contain several scales, one could have an overall system level model predicting overall pressure level coupled with a detailed 3D model describing wave propagation and blood flow dynamics in a given artery. Coupling of such models have received much attention recently, and it is important when models are coupled that the overall dynamics is preserved. For example if a given biomarker has been identified from a system level model, if modeled correctly, the detailed model should serve to refine prediction of the biomarker.

We emphasize that the variant steps discussed above are not all independent, and should merely be considered as components important to assess in development of subject specific models. Overall, the main components included are: choice of model, parameter identification, parameter estimation, model validation, and comparison of outcomes (biomarkers) among study populations.

10.2.3 Mathematical Models

Two models will be discussed: an open-loop model predicting baroreflex mediated changes in heart rate in response to changes in blood pressure (provided as an input to the model) and a closed-loop model predicting baroreflex and cerebral autoregulation of arterial (finger) blood pressure and cerebral blood flow velocity in response to changes in heart rate (provided as an input to the model). Both models have been used for "model based" data analysis using input quantities defined a priori to predict model parameters (biomarkers) that minimized the error between measured and computed values of output quantities. In addition to the biomarkers, this approach also allows prediction of internal quantities, which cannot be measured experimentally. We like to view our model based data analysis as an *in silico* investigation of the feedback mechanisms, allowing inaccessible parts to be made accessible (through the model predictions). For the open loop heart rate model, we used blood pressure data as an input to predict heart rate and the links made visible through the model include baroreflex firing rate, sympathetic and parasympathetic tone, and concentrations of acetylcholine and noradrenaline. For the closed loop model quantities predicted include cerebrovascular resistance, cerebral and systemic blood pressure. These model based quantities can be used to form hypotheses for how these "invisible" quantities may vary, and with more experiments, it may be possible to validate the various sub-models. Thus, with sufficient validation, we may view the model together with the outcomes as a method that can provide insight into an individuals control system like a fingerprint. Such biomarkers have potential to be relevant for treatment of several diseases such as hypertension, see [18, 20].

10.2.3.1   Open Loop Model

The overall function of the baroreceptor feedback mechanism is known. However, the underlying biochemical mechanistic processes are not fully understood and are difficult to investigate in-vivo. In the studies summarized here we used STS and HUT experiments to investigate the short-term baroreceptor feedback regulation of heart rate, detailed description of the model can be found in [21, 22, 24, 25, 30]. To investigate baroreflex regulation of heart rate we developed the model shown in Fig. 10.4. This model uses blood pressure as an input to predict heart rate using the following five steps:

- Mean blood pressure is used to predict afferent baroreflex firing rate.
- Sympathetic and parasympathetic outflows are computed from the baroreflex firing rate combined with stimulation by muscle sympathetic nerves, the vestibular system, and from central command.
- Concentrations of acetylcholine and noradrenaline are computed as functions of the sympathetic and parasympathetic outflow.
- Heart rate potential is computed from chemical concentrations.
- Heart rate is predicted from the heart rate potential.

The baroreflex model uses weighted mean blood pressure as an input to predict the afferent firing rate using a nonlinear differential equation of the form

$$\frac{dn_i}{dt} = k_i \frac{d\overline{p}}{dt} \frac{n(M-n)}{(M/2)^2} - \frac{n_i}{\tau_i}, \quad i = S, I, L, \tag{10.1}$$

$$n = n_S + n_I + n_L + N,$$

where n_i [1/s] denotes the firing-rate and k_i [1/mmHg] denotes the gain of the stimulus displayed by the neuron of type i (three types of neurons are included fast or short time-scale term S, intermediate I, and slow or long time-scale term L).

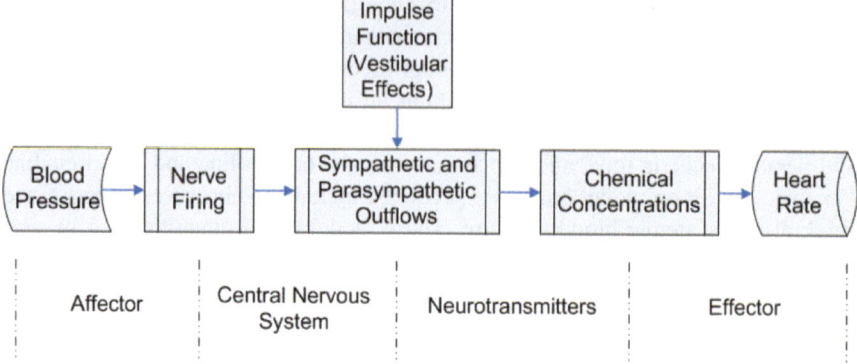

Fig. 10.4 Elements of the baroreceptor feedback chain controlling heart rate (Adapted from [21])

N [1/s] denotes the baseline firing rate, M [1/s] denotes the maximum firing rate, and n [1/s] denotes the integrated firing rate. This model accounts for hysteresis observed between increases and decreases in mean pressure [25]. If the change in pressure is negative (as opposed to positive), the two terms in (10.1) have the same sign leading to a larger net change in firing rate. The baseline firing-rate N cannot exceed the maximum firing-rate ($N < M$) and we assume that $N > M/2$. To enforce these bounds we have parameterized N using a sigmoidal function of the form

$$N = \frac{M}{2} + \frac{\eta^2}{1 + \eta^2} \left(M - \frac{N}{2} \right),$$

where η is the unknown parameter to be identified.

The baroreflex firing rate model depends on the weighted mean pressure [mmHg] computed as

$$\overline{p}(t) = \int_{-\infty}^{t} e^{-\alpha s} p(s) ds, \quad \text{or equivalently} \quad \frac{d\overline{p}}{dt} = \alpha(-\overline{p} + p),$$

where α is the weight. A large value of α [1/s] gives rise to a small weight of the past time (short memory), while a small value of α indicates larger weight (long memory). Thus, the weighted mean pressure $\overline{p}$ is a function of time, which oscillates with the same frequency as the instantaneous pressure p, but with smaller amplitude. For the studies analyzed here we let $\alpha = 1$. The afferent firing rate n is used for prediction of sympathetic T_{sym} and parasympathetic T_{par} outflows. Parasympathetic outflow is proportional to the firing rate, while the sympathetic outflow is inversely proportional to the firing rate and is dampened (with rate β) by the parasympathetic outflow. In addition, sympathetic outflow is modulated through activation via central command, the vestibular system, and via MSNA [11, 33–35, 45]. The latter is lumped into the contribution $u(t)$ defined using an impulse function. Thus,

$$T_{par} = \frac{n(t)}{M}, \quad T_{sym} = \frac{1 - n(t - \tau_d)/N + u(t)}{1 + \beta T_{par}}, \quad \text{where}$$

$$u(t) = -(b(t - t_m))^2 + u_0, \quad b = \sqrt{\frac{4u_0}{t_{per}^2}}, \quad \text{and} \quad t_m = t_{st} + \frac{t_{per}}{2},$$

where τ_d [s] denotes the delay of the sympathetic response, u_0 (dimensionless), t_{st} [s], and t_{per} [s] denote the magnitude and timing of the MSNA/central command/vestibular stimulation. Using the sympathetic and parasympathetic outflows, nondimensionalized concentrations of acetylcholine C_{ach} and noradrenaline C_{nor} were computed using the first order equation

$$\frac{dC_i}{dt} = \frac{-C_i + T_j}{\tau_i}, \quad (i, j) = \text{(nor, sym)} \quad \text{or} \quad (i, j) = \text{(ach,par)}. \tag{10.2}$$

Parameters in this equation include characteristic time scales for noradrenaline and acetylcholine, τ_{nor} and τ_{ach} [s]. In this equation we have lumped the long chain of biochemical reactions into a first order reaction equation and taken the accumulated release times τ_i to be equal to the average clearance and consumption time for the respective substances. The heart rate potential ϕ [beats] was computed using an integrate and fire model of the form

$$\frac{d\phi}{dt} = H_0 \left(1 + M_S C_{nor} - M_P C_{ach}\right),$$

where H_0 denotes intrinsic heart rate, which we predicted as a function of age ($H_0 = 118.1 - 0.57 \times$ age [10, 23]). The remaining parameters M_S and M_P represent the strength of the response to changes in the concentrations. To bound heart rate within physiological values, we constrained M_S and M_P in the interval [0,1]. This was done by introducing the parameters ζ_S and ζ_P that fulfill

$$M_S = \frac{\zeta_S^2}{1 + \zeta_S^2} \quad \text{and} \quad M_S = \frac{\zeta_P^2}{1 + \zeta_P^2}.$$

When ϕ reaches 1 it is reset to 0, and heart rate is computed as inverse of the interval $\phi = 0$ to $\phi = 1$, i.e.,

$$HR = 1/(t_{\phi=1} - t_{\phi=0}).$$

In summary, this model can be written in the form

$$\frac{d\overline{x}}{dt} = f(\overline{x}, \xi(t), \overline{p}(t); \theta), \tag{10.3}$$

$$\overline{x} = \{n_i, C_{ach}, C_{nor}, \phi\}, \quad i = S, I, L,$$

$$\theta = \{k_i, \tau_i, M, \eta, \tau_d, \beta, t_{st}, t_{per}, u_0, \tau_{ach}, \tau_{nor}, \zeta_S, \zeta_P\}, \quad i = S, I, L,$$

$$\xi = \{n, T_{sym}, T_{par}\},$$

where $\overline{x}(t)$ denotes the states, $\xi(t)$ the auxiliary equations, $\overline{p}(t)$ denotes the mean pressure (input from data), and θ denotes the model parameters. This model is validated against heart rate, i.e., with this model we seek to identify a set of parameters θ that minimize the least squares error

$$ J = r^{\mathsf{T}}r, \quad \text{where} \quad r = |y_c - y_d|. \tag{10.4} $$

The output y is the heart rate, i.e., $y_c = \mathrm{HR}_c = f(x, \theta)$ denotes computed values of heart rate and $y_d = \mathrm{HR}_d$ denote the heart rate data.

10.2.3.2 Closed Loop Model

The model discussed above, predicted regulation of heart rate as a function of blood pressure. However, data analyzed also include measurements of cerebral blood flow velocity. Therefore, in another series of studies [5, 18, 20, 22, 31] we developed a closed loop model predicting autonomic (baroreflex mediated) and cerebral autoregulation of arterial blood pressure and cerebral blood flow velocity using heart rate as an input. The most general form of the closed loop model is shown in Fig. 10.5. This model includes the systemic circulation, including the left heart (the atria and the ventricle) the aorta and vena cava, arteries and veins in the upper and lower torso as well as arteries and veins in the legs. The dotted lines on the figure indicate the diaphragm movement during respiration: the movement of the diaphragm change the trunk pressure and the transmural pressure of the vessels in the chest region. To model this we let

$$ p_{ext}(t) = \frac{A_u}{2}\left(\cos\left(\frac{2\pi t}{T_{insp}}\right) - 1\right) + B_u. \tag{10.5} $$

During inspiration ($t < T_{insp}$), the base level is given by B_u and the amplitude $A_u < 0$, for the remaining part of the cycle (5/3 times the inspiration period) the pressure is set at the base level. Similarly organs below the diaphragm, e.g., the liver, experiences changes in transmural pressure. However, the later blood pressure change is phase shifted by $180°$ compared to that in the chest region. This is imposed to the model by allowing the exterior pressure p_{ext} below diaphragm to vary as

$$ p_{ext}(t) = \frac{A_l}{2}\left(\cos\left(\frac{2\pi t}{T_{insp}}\right) - 1\right) + B_l, \tag{10.6} $$

where similar to the model discussed above A_l is the amplitude, B_l the base level value, and T_{insp} is the duration of inspiration. To shift the two signals by $180°$ we enforced that $A_l > 0$ (i.e., it has a sign opposite to A_u).

Finally, after the period of inspiration has ended we let $p_{ext}(t) = B_l$ during expiration. Typically, the length of the respiration cycle is $T_{resp} = 8/3 T_{insp}$.

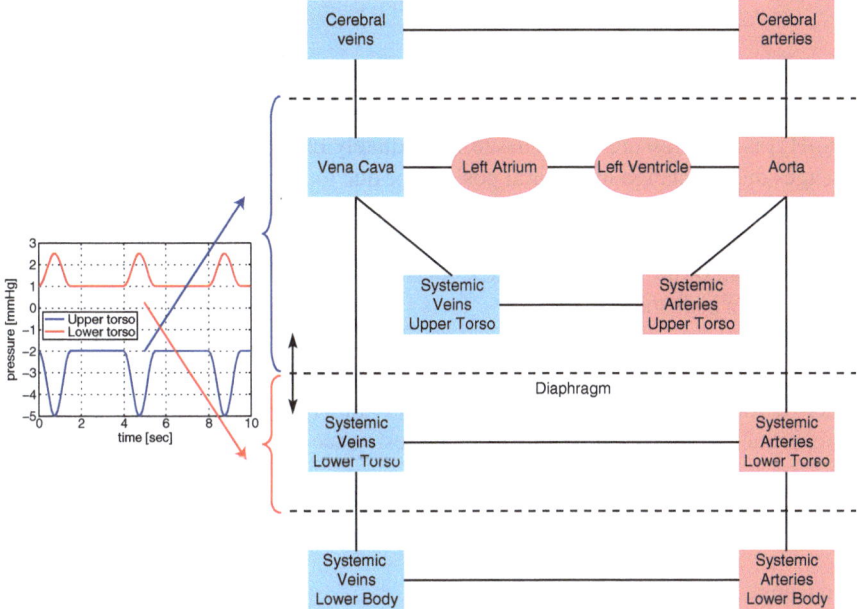

Fig. 10.5 Diagram of the closed loop model. Arteries and ventricle are *red* and veins are *blue*. The influences of the diaphragm movement on the exterior vessel pressure is shown to the *left*

In fact, this frequency and the depth of respiration are controlled. However, for the studies analyzed here, the subjects were asked to breathe to a metronome at a uniform depth, thus the model proposed above is adequate. If one has airflow data, it is possible to model exterior pressure directly as a function of airflow velocity, this approach was used in [4].

The closed loop model consists of three pars: A cardiovascular model predicting arterial blood flow and pressure in the various compartments; an autonomic regulation model predicting control of vascular tone (resistance and compliance), heart rate and cardiac contractility; and a cerebral autoregulation model predicting cerebrovascular resistance.

Cardiovascular Model. Each compartment in this model consists of a collection of arteries and veins of same caliber all with approximately the same pressure. Exceptions are the two compartments representing the left ventricle and atrium. Flow in this model is computed using an analogy to an electrical network with resistors and capacitors. Using this terminology, flow between compartments are analogous to current, pressure of each compartment is analogous to voltage, and compliance of each compartment is analogous to capacitance, while the resistance is the same in both formulations. Following this analogy the volume of each compartment is related to the pressure according to.

$$V - V_{unstr} = C(p - p_{ext}),   \tag{10.7}$$

where V [ml] is the total volume of the compartment, V_{unstr} [ml] is the unstressed volume, C [ml/mmHg] is the compliance, p [mmHg] is the blood pressure, and p_{ext} [mmHg] is the pressure of the tissue immediately outside the compartment which changes due to change in postural position or due to respiration. For compartments in the legs and brain we assumed that the external pressure p_{ext} is constant, while for the compartments in the upper and lower torso the external pressure is modulated by movement of the diaphragm. Flows between compartments are computed using Ohm's law which states that

$$q = \frac{p_{in} - p_{out}}{R},   \tag{10.8}$$

where q [ml/s] denotes the flow, p_{in} and p_{out} [mmHg] are the pressures of the two compartments, and R [ml/s mmHg] is the resistance to flow. Differentiating (10.7) gives

$$\frac{dV}{dt} = C\frac{d(p - p_{ext})}{dt} + (p - p_{ext})\frac{dC}{dt},$$

and using (10.8) we get

$$C\frac{dp}{dt} = C\frac{dp_{ext}}{dt} - (p - p_{ext})\frac{dC}{dt} + q_{in} - q_{out}.$$

A differential equation of this form can be derived for all arterial and venous compartments. Note, for steady state simulations we assumed that $p_{ext} = 0$ and C is constant. In general, p_{ext} should be modulated with respiration (e.g., as suggested in (10.5) and (10.6) and C should be controlled, or as discussed in several studies by Ursino et al. [42–44] it may be appropriate to model C using a nonlinear function of the stressed volume. For compartments representing the left heart (the left and right atrium) two different models have been analyzed. One model proposed by Ottesen [29] uses a generalized activation function to predict the pressure in the heart compartments; the other is a simple elastance model. The advantage of the elastance model is that it contains only four parameters, while the more accurate model by Ottesen contain 14 parameters. The Ottesen model predicts the left heart pressure as

$$p_{lh} = a\,(V(t) - b)^2 + (c(t)V(t) - d)\,f(\tilde{t})/f(t_p),   \tag{10.9}$$

$$f(\tilde{t}) = \begin{cases} P_p\dfrac{\tilde{t}(\beta - \tilde{t})^m}{n^n m^m(\beta/(m+n))^{m+n}} & \text{for } 0 \leq \tilde{t} \leq \beta, \\ 0 & \text{for } \beta \leq \tilde{t} \leq T_i, \end{cases}   \tag{10.10}$$

where a [mmHg/ml^2] is related to the elastance during relaxation, b [ml] represents the volume at zero diastolic pressure, $c(t)$ [mmHg/ml] represents contractility (note during steady state c is constant), and d [mmHg] is related to the volume-independent component of the developed pressure. In the activation function f, T_i [s] denotes the length of the i-th cardiac cycle, β [s] denotes the onset of relaxation, n and m characterize the contraction and relaxation phases and p_p [mmHg] is the peak value of the activation. The ability to vary heart rate is included in the pressure equation by scaling the time t_p [s] and peak values p_p of the activation function f, for details see [29]. An advantage of this model is that the ejection effect is easily incorporated, i.e., the fact that dynamical changes in ventricular volume due to ejection of blood affect the ventricular contractility. Thus the model is suitable when varying afterloads are considered. The corresponding elastance model is given by

$$p_{lh} = E(\tilde{t})(V(t) - V_d), \tag{10.11}$$

$$E(\tilde{t}) = \begin{cases} E_m + \dfrac{E_M - E_m}{2}\left(1 - \cos\left(\dfrac{\pi \tilde{t}}{T_M}\right)\right) & \text{for } 0 \leq t \leq T_M, \\[2ex] E_m + \dfrac{E_M - E_m}{2}\left(1 + \cos\left(\dfrac{\pi(\tilde{t} - T_M)}{T_R}\right)\right) & \text{for } T_M \leq \tilde{t} \leq T_M + T_R, \\[2ex] E_m & \text{for } T_M + T_R \leq \tilde{t} \leq T_i, \end{cases} \tag{10.12}$$

where V_d [ml] denotes the volume at zero end-systolic pressure. In the elastance function E, T_M and T_R [s] denote the time for maximum (systolic) elastance (T_M) and the remaining time to relaxation (T_R) and E_M and E_m [mmHg/ml] denotes the maximal (systolic, E_M) and minimal (E_m) diastolic elastance. This model accounts for varying heart rate by defining scaled parameters $T_{Mf} = T_M/T_i$ and $T_{Rf} = T_R/T_i$. As before T_i [s] denotes the length of the current cardiac cycle. For either of the two models, a differential equation for the heart compartments can be obtained from conservation of volume, i.e., we let

$$\frac{dV}{dt} = q_{in} - q_{out},$$

where as before the flows are computed using Ohm's law. It should be noted, that both the atrium and the ventricle are modeled using the same type of equations, but that parameters for the two heart chambers vary. Finally, the left ventricle cannot function without heart valves. In all studies summarized here, we used time varying resistances to represent the valves. These are defined such that a closed valve is represented by a high resistance $R_{valve,c}$ and an open valve is represented by a very low resistance $R_{valve,o}$. This can be done by defining valve resistances as

$$R_{valve} = \min\left(R_{valve,o} - e^{-k(p_{in} - p_{out})}, R_{valve,c}\right),$$

where k [1/mmHg] is a rate constant denoting the time it takes for the valve to close. Similar to the open loop model, the closed loop steady state model (i.e., no parameters are controlled C, R, c are constant parameters) shown in Fig. 10.5 can be represented by a system of differential equations of the form coupled with a set of auxiliary equations

$$\frac{d\overline{x}}{dt} = f(\overline{x}, \xi(t), T_i; \theta), \tag{10.13}$$

$$\overline{x} = \{p_i, p_{i,c}, p_{i,ut}, p_{i,lt}, p_{i,l}, V_{la}, V_{lv}\}, \quad i = a, v,$$

$$\theta = \{R, C, a, b, c, d, n, m, T_{mf}, T_{Mf}, E_M, E_m, T_i\},$$

$$\xi = \{R_{valve}, f, E, p_{lv}, p_{la}\},$$

where $\overline{x}(t)$ denotes the states, $\xi(t)$ denotes the auxiliary equations, T_i denote the length of each cardiac cycle (input from data), and θ denotes the model parameters.

This model is validated using measurements of heart rate (input), arterial blood pressure and blood flow velocity, i.e., with this model we seek to estimate a set of parameters θ that minimize the least squares error

$$J = r^T r, \quad \text{where} \quad r = |y_c - y_d|. \tag{10.14}$$

The output y is a vector concatenating blood pressure and blood flow velocity, i.e., $y_c = \{p_c(t_1), p_c(t_2), \ldots, p_c(t_N), v_c(t_1), v_c(t_2), \ldots, v_c(t_N)\} = f(x, \theta)$ denotes computed values and $y_d = \{p_c(t_1), p_c(t_2), \ldots, p_c(t_N), v_c(t_1), v_c(t_2), \ldots, v_c(t_N)\}$ denote the corresponding data.

Modeling Postural Changes. To allow the model to predict blood pressure and cerebral blood flow velocity dynamics during postural change from sitting to standing and head up tilt we incorporated changes in hydrostatic pressure to allow pooling of the blood in the legs. To do so we modified equations predicting flow to and from the upper to the lower body as

$$q = \frac{(p_{in} - \rho g h_{in}) - (p_{out} - \rho g h_{out})}{R},$$

$$h(t) = \frac{h_M}{1 + e^{-k(t - T_{up} - \delta)}},$$

where T_{up} [s] is the time at which the subject stands up, h_M [cm] is the maximum height needed for the mean arterial pressure to drop as indicated by the data, and δ [s] is the latency for the transition to upright position.

Modeling Autonomic Regulation. Only one of our previous studies [20] modeled autonomic regulation. In this study we assumed that cardiac contractility c [mmHg/ml] and systemic peripheral resistances R [ml/s mmHg] in the upper body and the legs were increased in response to the drop in arterial pressure, while compliance C [ml/mmHg] was decreased. Inspired by [27] we used a first order set-point equation to model this control,

$$\frac{dx}{dt} = \frac{-x + x_{ctr}(p_a)}{\tau},$$

$$x_{ctr}(p_a) = (x_M - x_m)\frac{\alpha^k}{p_a^k + \alpha^k} + x_m, \qquad x = R, c,$$

$$x_{ctr}(p_a) = (x_M - x_m)\frac{p_a^k}{p_a^k + \alpha^k} + x_m, \qquad x = C. \qquad (10.15)$$

In the above equation x_{ctr} is an increasing (for R, c)/decreasing (for C) sigmoidal function of arterial pressure. Using a sigmoidal function allows the system to display saturation beyond the limit of regulation. In this function x_M and x_m denote the upper and lower limit for the parameter controlled, the parameter α is set to ensure that $x(t)$ returns to the value of the controlled parameter found during steady state, and k denotes the steepness of the sigmoid. Finally, the parameter τ characterizes the time it takes for the control to reach its maximal effect. In addition to the active control we modeled resistances between arterial compartments using a sigmoidal equation similar to the one given in (10.15).

It should be noted, that this direct control as a function of pressure is significantly simpler than the more complex baroreflex model presented earlier, the disadvantage is that this is a purely empirical model not accounting for any of the physiological mechanisms, known to be involved in the baroreflex regulation. Another important point is that the true baroreflex model includes a delay, which is not accounted for in the simpler model discussed above. However, the two models could be combined to form a more complete model. For further discussion see Chap. 11.

Modeling Cerebral Regulation. Most models in the literature tempting to predict cerebral regulation are derived from models proposed by Ursino et al. [42, 43]. However, it is not clear if this model includes all possible mechanisms believed to be involved in cerebral regulation. Many factors have been proposed to play a role, including responses to changes in CO_2 (often denoted as cerebral vasoregulation), responses to myogenic regulation (this is sometimes what is understood by the term cerebral autoregulation). Some recent studies have also indicated that a portion of cerebral regulation stems from neurogenic regulation. Initially, we attempted to model cerebral regulation using model similar to the set-point function proposed in (10.15) including only the "myogenic" aspects of the regulation. However, using this type of equation did not enable prediction of observed variation in cerebral blood flow velocity. Instead we used an open-loop control model formulated using a piecewise linear function with unknown coefficients to obtain a representative function that describes the time-varying response of the cerebrovascular resistance. To obtain such a function, we parameterized the cerebrovascular resistance using the piecewise linear function of the form

$$R(t) = \sum_{i=1}^{N} \gamma_i H_i(t), \qquad (10.16)$$

$$H_i(t) = \begin{cases} \dfrac{t - t_{i-1}}{t_i - t_{i-1}} & \text{for } t_{i-1} \leq t \leq t_i, \\ \dfrac{t_{i+1} - t}{t_{i+1} - t_i} & \text{for } t_i \leq t \leq t_{i+1}, \\ 0 & \text{otherwise,} \end{cases}$$

where γ_i [ml/s mmHg] are the unknown coefficients, which should be estimated together with the other parameters. Following the parameter estimation we used the "predicted time-varying response" to propose a cerebral regulation model. The most promising model analyzed had the form

$$R = R_{met} + R_{myo} + R_{neu}, \qquad (10.17)$$

$$\frac{dR_{met}}{dt} = \frac{-R_{met} + \tilde{R}_{met}(q_c)}{\tau_{met}},$$

$$\frac{dR_{myo}}{dt} = \frac{-R_{myo} + \tilde{R}_{myo}(p_{ac})}{\tau_{myo}},$$

$$\frac{dR_{neu}}{dt} = \frac{-R_{neu} + k_{neu} C_{ach}(p_a)}{\tau_{neu}},$$

where R_{met} [ml/s mmHg] is the contribution from the metabolic regulation, R_{myo} [ml/s mmHg] is the contribution from myogenic regulation, and R_{neu} [ml/s mmHg] is the contribution from neurally mediated control. Note, R_{met} is modeled as a function of cerebral flow q_c [ml/s], while myogenic contribution is modeled as a function of cerebral arterial pressure p_{ca} [mmHg]. Finally, the neurogenic contribution is modeled as a function of arterial pressure p_a [mmHg]. It is believed that a potential neurogenic contribution to cerebral vasoregulation is both cholinergic and adrenergic in nature. Therefore we let the set-point equation use concentration of acetylcholine C_{ach} (dimensionless). The concentration of acetylcholine was computed similar to the open loop heart rate model, see Eq. (10.2). For the metabolic and myogenic contributions the control functions $\tilde{R}$ were sigmoidal functions similar to the one given in (10.15). Results of the two control models (10.17) and (10.16) are shown in Fig. 10.6. It should be noted that if we omitted any part of the proposed control function we were not able to reproduce the dynamic found from the piecewise linear function. One thing should be kept in mind is that both the spline model (10.16) and the differential equations model (10.17) has 26 parameters. Ideally, it would be better if the cerebral regulation model contained fewer parameters.

Fig. 10.6 Cerebrovascular resistance predicted using a piecewise linear function (10.16) (*blue dashed line*) versus the control function (*red*) proposed in (10.17)

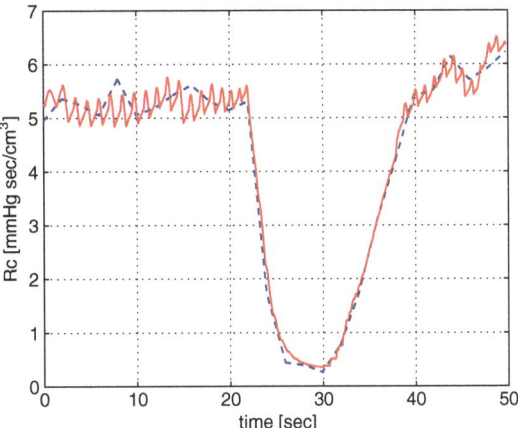

10.2.4 Parameter Estimation and Model Validation

Many real life processes and systems can be modeled using non-linear ODE's or PDE's. A frequent difficulty in biomedical applications is that the model equations often have a large number of unknown parameters. For some systems it is possible to determine model parameters directly from the experimentally setup. However, in many cases it is difficult or even impossible to measure biomarkers, while other quantities can be measured. For example, during the STS and HUT experiments, typical measurements include arterial pressure and cerebral flow velocity.

To solve the inverse problem, one can invoke optimization techniques to estimate a set of model parameters that minimize the least squares error between computed and measured quantities. Whether the parameters for the mathematical model can be estimated assuming sufficient and error-free data is subject to an *a priori* identifiability analysis. Two aspects typically have to be investigated: First using sensitivity analysis we are able to split model parameters in two sets include: "sensitive" and "insensitive" parameters. Sensitive parameters are characterized as parameters where variation in the parameter values invoke a significant change in the model output (see Fig. 10.7), while a change of insensitive parameters has a negligible impact on the model output. Second, among sensitive parameters, correlations can be present. This type of model dependencies can be predicted accurately for linear models, but for nonlinear models it is more difficult to analyze the system. Once a set of identifiable parameters have been identified, these parameters can be estimated using nonlinear optimization techniques. Following the approach put forward in [5,6,31] we describe each of the three components in detail.

Sensitivity Analysis. Sensitivities are computed with respect to output vector y (heart rate for the open loop model summarized in (10.3) and blood pressure and blood flow velocity for the closed loop model summarized in (10.13)). For both of

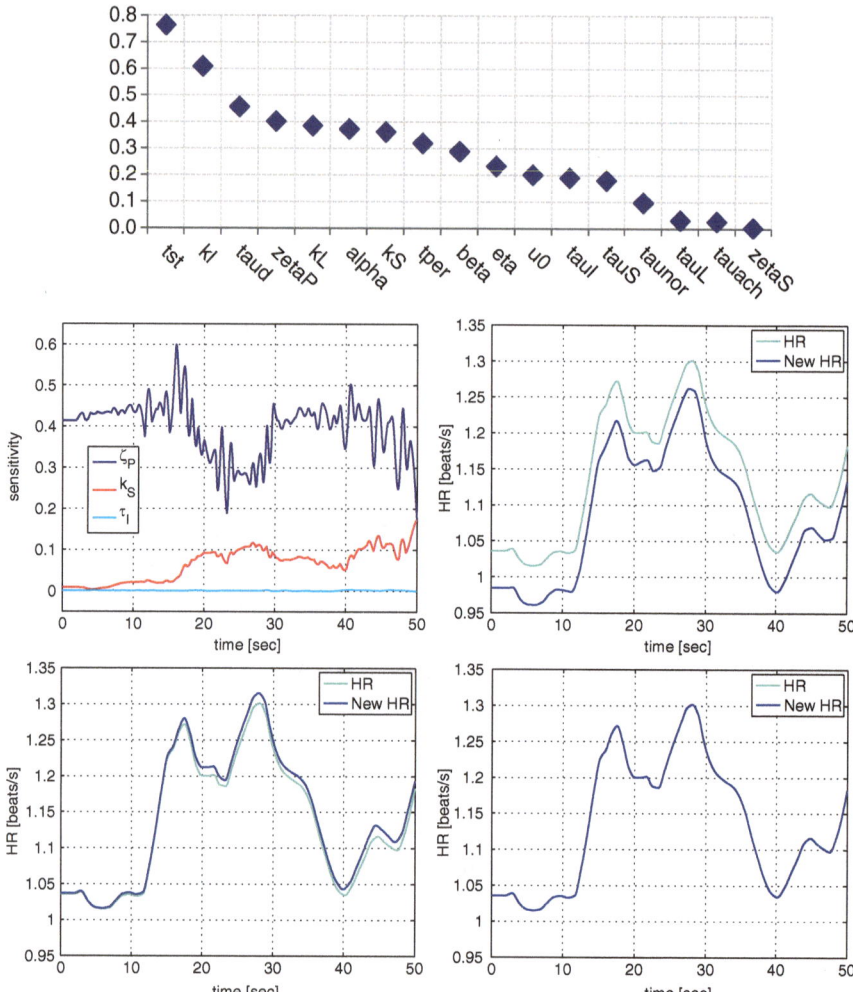

Fig. 10.7 Sensitivity ranking for the open loop heart rate model. Sensitivities are computed with respect to heart rate for a healthy young subject. A weighted 2-norm was used to obtain the ranked sensitivities. *Bottom panels* show time varying sensitivities for a sensitive, an intermediate, and an insensitive parameter as well as the corresponding heart rates computed while varying the parameter in question (Modified from [22])

these models, the nominal parameter values range several orders of magnitude, e.g., for the small closed loop model analyzed in [31] the parameter $E_m \approx 0.05$, while $C_{vs} \approx 36$. To compute sensitivities more accurately, we scaled the parameters by the natural logarithm, i.e., the model input to the optimizer is given by $\tilde{\theta} = \ln(\theta)$.

Using the scaled parameters sensitivities can be computed as the change in the output variables with respect to the parameters, the absolute sensitivity is defined by

$$S_{i,k}(t,\tilde{\theta}) = \left. \frac{\partial y_k(t,\theta)}{\partial \tilde{\theta}_i} \right|_{\theta_0}, \tag{10.18}$$

where θ_0 denotes the nominal values for the parameters. Even though parameters are scaled, quantities compared may still have different units (e.g., pressure and velocity in the closed loop model), thus it is appropriate to analyze relative sensitivities defined as

$$S_{ik}(t,\tilde{\theta}) = \left. \frac{\partial y_k(t,\theta)}{\partial \tilde{\theta}_i} \frac{\tilde{\theta}_i}{y_k(t,\theta)} \right|_{\theta_0}, \quad \tilde{\theta}_i, y_k \neq 0. \tag{10.19}$$

Note, for the open loop model $S_{i,k}$ is computed with respect to one output, heart rate, while for the closed loop model both pressure and velocity are concatenated together to provide the output vector of length $2N$. As discussed in [4,6] the sensitivities can either be found using automatic differentiation, by setting up a set of analytical equations, or using finite differences. The finite difference approximation of the sensitivities is less accurate, but for most practical applications including those studied here, finite difference approximations provides sufficient accuracy. This should be compared to the likelihood of introducing errors during analytical sensitivity calculations, in particular for models that have many parameters, e.g., if a model has two outputs and 21 parameters the system $2 \times 21 = 42$ sensitivity equations should be derived. An alternative approach is to use automatic differentiation to derive sensitivity equations, but these methods are computationally ineffective in particular if the differential equations are solved using Matlab. More efficient packages exist in C and Fortran, but these have not been analyzed for this study.

Using finite differences, the derivatives in the sensitivity equations can be computed using the forward difference approximation

$$\frac{\partial y_k}{\partial \tilde{\theta}_i} \approx \frac{y_k(t,\tilde{\theta}+h e_i) - y_k(t,\tilde{\theta})}{h},$$

where

$$e_i = \left[0\ldots 0 \overset{i}{\hat{1}} 0 \ldots 0 \right]^{\mathrm{T}}$$

is the unit vector in the i-th component direction. To rank the parameters from the most to the least sensitive, we used a scaled 2-norm to get the total sensitivity, S_i, to the i-th parameter

$$S_i = \left(\frac{1}{2N} \sum_{j=1}^{2N} S_{i,k}^2 \right)^{1/2}.$$

Figures 10.7 and 10.8 show examples of the sensitivity analysis. Figure 10.7 shows ranking and time varying sensitivities for the open loop model using data from a healthy young subject. Note the change in solution arising from varying the

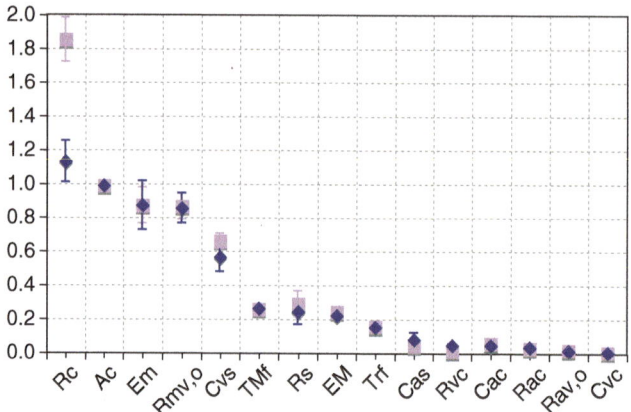

Fig. 10.8 Sensitivity ranking for a five compartment (systemic arteries and veins, cerebral arteries and veins, and the left ventricle) closed loop model. Sensitivities are computed with respect to arterial pressure and cerebral blood flow velocity. A weighted 2-norm was used to obtain an average sensitivity over the entire time series (Adapted from [32])

insensitive parameter τ_I barely changes the model solution, while a significant change in heart rate is observed when the sensitive parameter ζ_P is varied. Figure 10.8 shows the average ranking using data from 16 healthy young and elderly subjects computed from the closed loop model. The specific closed loop model used for these calculations included four vascular and one ventricular compartments, the model had 21 parameters.

It should be noted that the classical sensitivity analysis described above is a local analysis, and thus sensitivities depend on the values of the parameters. In this study the goal is to use sensitivity analysis to rank parameters in order of sensitivity and use this ranking in conjunction with results from subset selection to identify a set of parameters that can be estimated for all subjects. This is done prior to actual parameter estimations, thus the sensitivity ranking was computed using nominal parameter values.

In summary, results from the sensitivity analysis showed that both models include both sensitive and insensitive parameters. Sensitive parameters can often be estimated using optimization techniques, while insensitive parameters are difficult to estimate since a small change in the parameter value gives rise to a small change in the solution. This can become problematic when models are used to extract biological information from estimated parameter values as done in the studies summarized here. However, it should be noted that identifiability is a mathematical notion. For biological implications the precise values of parameters are not always important as long as they have certain characteristics, e.g., like being positive. An alternative method is to use generalized sensitivity analysis [41], which provide more insight into the dynamics of the model.

Subset Selection. Subset selection can be approached using a number of methods as described in [32]. Below we outline the method used in [31]. In this study subset selection analyzes the Jacobian matrix ($r' = dr/d\theta$) computed from the residual vector r (see Eqs. (10.3) and (10.13)). The entry at row i and column j of the Jacobian is $\partial r_i/\partial \theta_j$. The Jacobian, singular value decomposition $r' = U\Sigma V^\mathsf{T}$ is used to obtain a numerical rank for r'. This numerical rank is then used to determine ρ parameters that can be identified given the model output y defined in (10.3) and (10.13). QR decomposition is used to determine the ρ identifiable parameters to which our system is sensitive as *a group*. This differs from sensitivity analysis, which finds parameters to which our system is *individually* sensitive. To estimate the number of uncorrelated parameters we used an error estimate in our computation of the Jacobian as a lower bound on acceptable singular values. For example, in the studies analyzed here we used Matlab's differential equations solver ODE15S with an absolute error tolerance of 10^{-6}, i.e., the error of the numerical model solution is of order 10^{-6} and the error in the Jacobian matrix is approximately $\sqrt{10^{-6}} = 10^{-3}$. Consequently, singular values should not be smaller than 10^{-3}. Since the error of the Jacobian is an approximation, the smallest singular value that we accept is 10^{-2}. Once the number of identifiable parameters has been determined, we find the most dominant parameters by performing a QR decomposition with column pivoting on the most dominant right singular vectors. The process begins by choosing the most sensitive parameter in a way similar but not identical to the sensitivity analysis of the previous section, the column with largest 2-norm is chosen. The algorithm chooses additional parameters in a way that keeps the condition number of the chosen columns small. Below we summarize subset selection method as an algorithm.

Subset selection algorithm:

1. Given an initial parameter estimate, θ_0, compute the Jacobian, $r'(\theta_0)$ and the singular value decomposition $r' = U\Sigma V^\mathsf{T}$, where Σ is a diagonal matrix containing the singular values of r' in decreasing order, and V is an orthogonal matrix of right singular vectors.
2. Determine ρ, the numerical rank of r'. This can be done by determining a smallest allowable singular value.
3. Partition the matrix of eigenvectors in the form $V = [V_\rho \, V_{n-\rho}]$.
4. Determine a permutation matrix P by constructing a QR decomposition with column pivoting, for V_ρ^T. That is, determine P such that

$$V_\rho^\mathsf{T} P = QR,$$

where Q is an orthogonal matrix and the first ρ columns of R form an upper triangular matrix with diagonal elements in decreasing order.[1]

---

[1]We note, that depending on the QR factorization algorithm used a different permutation vector may be obtained. The importance of this step is to obtain a parameter set that gives rise to a well-conditioned Jacobian, that can be inverted allowing unique estimation of the selected

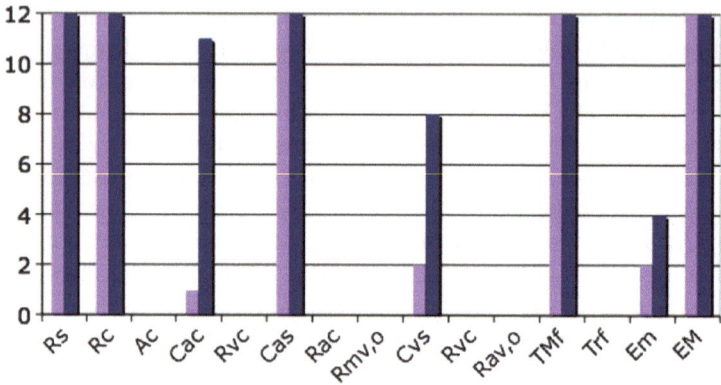

Fig. 10.9 Subsets computed for 16 healthy young (*dark blue*) and 16 healthy elderly (*light purple*) subjects using a five compartment closed loop model combined with arterial blood pressure and cerebral blood flow velocity data. (Adopted from [32])

5. Use P to reorder the parameter vector θ_0 according to $\hat{\theta}_0 = P^{\mathsf{T}}\theta_0$.
6. Make the partition $\hat{\theta}_0 = [\hat{\theta}_{0,\rho}\ \hat{\theta}_{0,n-\rho}]$ where $\hat{\theta}_{0,\rho}$ contains the first ρ elements of $\hat{\theta}_0$. Fix $\hat{\theta}_{n-\rho}$ at the a priori estimate $\hat{\theta}_{0,n-\rho}$.
7. Compute the new estimate of the parameter vector $\hat{\theta}$ by solving the reduced-order minimization problem

$$\hat{\theta} = \arg\min_{\theta} J(\theta), \quad \text{with } \hat{\theta}_{n-\rho} \text{ fixed at nominal values } \hat{\theta}_{0,n-\rho}.$$

Figure 10.9 shows possible subsets of identifiable parameters from the same model that we used to show results of the sensitivity analysis. The model used for this study included four vascular and one ventricular compartments, the model had 21 parameters, and out of these only four could be estimated reliably given arterial blood pressure and cerebral blood flow velocity data measured during sitting. It should be noted that this model did not include any control mechanisms.

Optimization Techniques. Sensitivity analysis and subset selection are examples of methods that can be used to address the question of parameter identifiability, other methods include Kalman filtering and principal component analysis. Common for these methods is that they identify a set of parameters and then using nonlinear optimization techniques it is possible to estimate the parameters in question. To do so we formulate a least squares problem (LQP) and seek to estimate a set of parameters that minimize the least squares error between measured and

---

model parameters. The exact parameter set obtained should include parameters that make sense to estimate in terms of the physiological system studied. In other words, the sub-set of parameters may not be unique, thus care must be taken in analyzing the exact parameter values chosen. Subset selection is further discussed in Chaps. 2, 3 and 11.

computed quantities (as discussed in Eqs. (10.4) and (10.14). In general, least squares problems (LQP) have several minima, because the problem has more than one solution or the error-function can have several stationary points that do not correspond to the lowest value of the fitness landscape. Local search methods, like the Levenberg–Marquardt method, easily get trapped in one of the local minima rather than finding the global minimum. To explore the whole search space one needs global search methods and if possible a physiological range for realistic values. Unfortunately, these methods converge very slowly once near a minimum. In contrast, gradient-based methods are efficient optimizers for nonlinear LQP's once a sufficiently good initial guess for the parameter values is available. Thus a recommendable strategy is to use the solutions from the global search as initial guesses for local optimization. In this way, one reduces the chance of missing the global minimum and the determination of all the minima is precise and fast. In the problems analyzed and discussed here we used the Nelder–Mead method (a gradient free global optimization method) to estimate parameters for the open loop model and a gradient based method (a Levenberg–Marquart method with Trust Regions) for the closed loop model. Both methods worked well, but the Nelder–Mead method was significantly slower, while the Levenberg–Marquart method required that we estimated initial parameter estimates carefully.

In addition to estimation of model parameters, another important question is model validation. The studies analyzed here do not address this question. Methods typically used include K-fold cross validation [13], which use a part of the time series data for parameter estimation and another part for model validation.

10.3   Discussion

Below we summarize and discuss result reported in [5, 6, 18–22, 24, 25, 30, 31]. We divide the presentation into three subsections; open loop model, closed loop model, and syncope.

10.3.1   Open Loop Model

Results with the open loop model have been reported in [7, 21, 22, 24, 25] results were obtained with the STS and HUT protocol. In [21] we analyzed STS data from three groups of subjects including healthy young, healthy elderly, and hypertensive elderly healthy young subjects, and in [22] we analyzed both STS and HUT data from five young subjects. For both studies all model parameters were predicted using the Nelder–Mead optimization method. Results showed that standard deviations were typically high, however, for both studies we were able to detect interesting differences between the groups of subjects. In [21] (results shown in Figs. 10.3 and 10.10) we observed that the hysteresis loop changed significantly

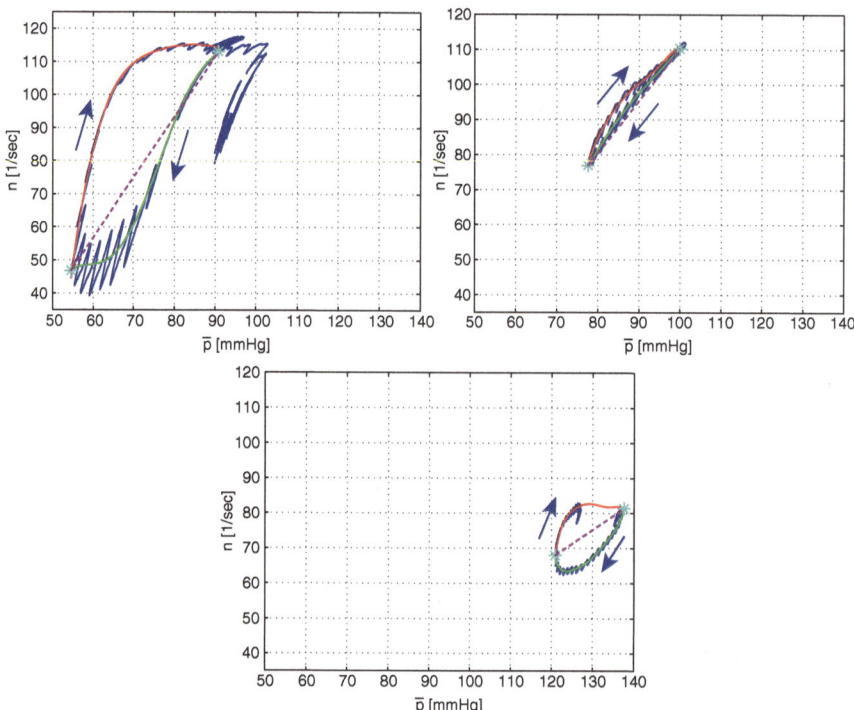

Fig. 10.10 Example hysteresis depicting changes in baroreflex firing rate (n) as a function of mean blood pressure ($\overline{p}$). Results are shown for a healthy young subject (*left*), a healthy elderly subject (*center*), and a hypertensive elderly subject (*right*). *Dashed lines* through the *curves* were used to determine the overall slope, and *arrows* indicate responses to change in pressure. Baroreceptor hysteresis is an important feature that demonstrates how baroreflex controls adapt in response to differences in vascular compliance that occur with aging (Adapted from [21])

with age: for health young subjects the hysteresis loop was significantly larger and the slope higher than for any of the other two groups, see Fig. 10.10. The healthy elderly subjects had a very narrow loop and mean pressures were higher than for the young subjects. Finally for the hypertensive subjects we noticed that for most subjects the hysteresis loop was not closed, indicating that within the time frame included in the experiments the pressure does not return to the value obtained during sitting. This may indicate that part of the regulation is not working as expected. In addition, comparison of parameters between groups revealed that parameters k_I, k_L, β, τ_d, τ_{ach}, and M_P changed significantly between groups. These results were obtained using ANOVA analysis using 20 data sets for each group. Data from ten subjects (two experiments per subject) were analyzed. It should be noted that these results were obtained without any applying any model and parameter reduction techniques. Seen from a physiological point of view observing differences in the given parameters are reasonable, reduction in k_I and k_L indicate that with age and hypertension, the firing rate sensitivity to changes in pressure is reduced, increase of β and τ_d, indicate that with age and hypertension, the delay in onset of sympathetic response is increased and that parasympathetic dampening of the sympathetic response is attenuated, finally τ_{ach} increase with age, but decrease with hypertension. An age related increase in τ_{ach} suggests that with age it takes longer for the parasympathetic response to reach its maximum effect, while the decrease with hypertension, may be compensating for the fact that the vessels are significantly stiffer. Finally M_P is increased with both age and hypertension indicating that parasympathetic regulation plays a more important role than subsequent sympathetic stimulation of heart rate. Results from [22] (see Fig. 10.11) showed that there were significant differences between results from the STS and HUT procedures. Comparing the two tests showed a much larger increase in heart rate during HUT than during STS and a more significant drop in blood pressure during STS than during HUT, leading to more pronounced changes in firing rate and sympathetic/parasympathetic tone. Another noticeable difference is the change in the area of the hysteresis loop: the loop is significantly wider during STS than during HUT. Finally, we noticed that during HUT heart rate decrease during the initial preparation to tilt (before any blood pressure drop was observed), while during STS heart rate increased before the subject changed posture (the latter observation was also found in our first study [21]). This initial drop in heart rate is associated with a slight increase in blood pressure (compare panels A and E). This may be due to a short increase in venous return due to hydrostatic pressure difference imposed between the heart and the head. On the other hand the increase in heart rate immediately before standing may, as explained earlier, be due to vestibular and muscle sympathetic activation [11, 33–35, 45].

In addition to analysis of the results, in [21] we used sensitivity analysis (see Figs. 10.7 and 10.8) to rank model parameters from the most to the least sensitive. Results of this analysis showed that some parameters are very insensitive including τ_I, k_I, ζ_S, and τ_{ach}. Results varying a sensitive parameter, an intermediate parameter and an insensitive parameter are showed together with the ranking. One aspect not done for this study is to analyze if any of the sensitive parameters are correlated, more analysis is needed to investigate potential correlations. Such correlations are likely to exist, some initial attempts to study those have been done by Fowler [7] comparing results obtained with Nelder–Mead with both implicit filtering and using a genetic algorithm to optimize model parameters.

10.3.2 Closed Loop Model

Closed loop cardiovascular models were investigated in [4, 6, 19, 20, 31]. All of these manuscripts discussed a model similar to the one shown in Fig. 10.5. In [19] and [20] we analyzed a model including all but two compartments (this model did not separate arteries and veins in the upper and lower torso). Instead the model included a compartment representing arteries in the hand. This model used heart rate data as an input to predict blood flow velocity and cerebral blood pressure

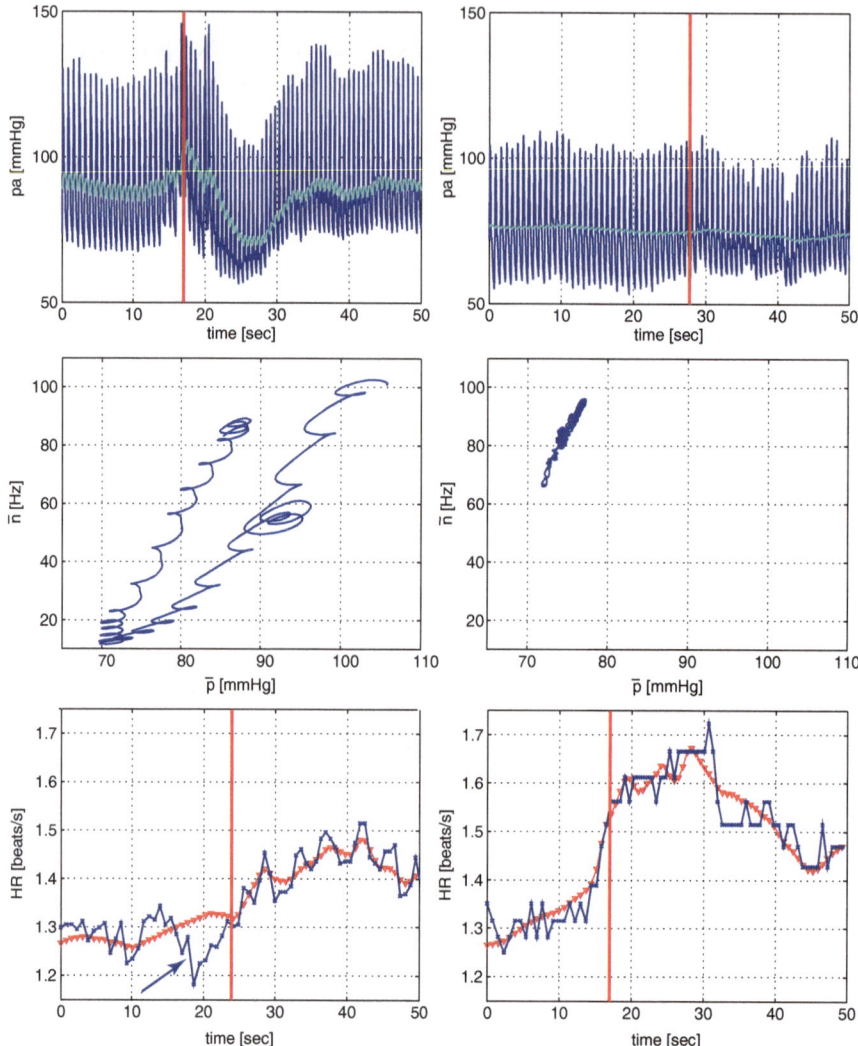

Fig. 10.11 Blood pressure [mmHg], firing rate [Hz], and heart rate [bps] versus time [s] for STS (*left panel*) and for HUT experiment (*right panel*). *Red curves* are model results and *blue curves* are data (Adapted from [22])

during STS. In this study all model parameters were optimized and the model was validated against data from one healthy young subject. In [6] we used sensitivity analysis to show that cerebral blood flow velocity and arterial blood pressure data could be predicted using a simpler model not including the left atrium and the finger compartment, and when analyzing data during steady state (i.e., only modeling dynamics during sitting), we showed that arteries and veins in the upper and lower body could be lumped together. Again, this model only used data from one healthy

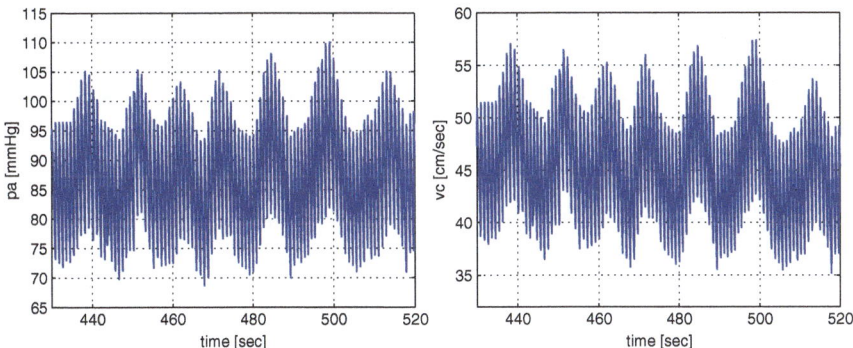

Fig. 10.12 Model generated arterial pressure (*left panel*) and cerebral blood velocity (*right panel*) when respiratory effects are included. The resulting *curves* are realistic and in accordance with measurements

young subject. Furthermore, it should be noted that none of these model included a description of the diaphragm movement. Diaphragm movement was included in [4] (see Fig. 10.12), where we used airflow data to predict the movement. This study was carried out using a cardiovascular–respiratory model, that included both the systemic and pulmonary circulations. Note, this model was not designed to predict changes during STS, thus again fewer compartments were included separating the systemic circulation between upper and lower body. The latter model was developed to study dynamics during a CO_2 challenge for a subject with congestive heart failure. Finally in [31] we developed a methodology for identifying a small subset of parameters that can be estimated given data during sitting. To do so, we used an even simpler model including only five compartments: the left ventricle, systemic arteries and veins, and cerebral arteries and veins. Again this model used heart rate as an input to predict dynamics of cerebral blood flow velocity and arterial blood pressure. Since this model was significantly simpler, and because we used subset selection, we were able to predict dynamics for two groups of subjects including 16 healthy young subjects and 16 healthy elderly subjects. Results of this comparison showed that both cerebral resistance and compliance were modulated by aging. Results also showed that the total resistance was increased and that T_M, f time for systolic pressure was increased. All closed loop models were able to predict both blood pressure and cerebral blood flow velocity as shown in Fig. 10.13, however, only models including subset selection and sensitivity analysis could be analyzed against several data sets. The more advanced models (including more compartments) simply had to many parameters, thus making it very tedious to achieve good predictions of the data. In future work, we plan to apply subset selection to the more complex models, and with this method we anticipate that even the more complex models can be validated against multiple datasets, allowing identification and prediction of biomarkers.

The results shown in Fig. 10.13 were obtained with the complex control model developed in [19, 20]. In this figure we show how the closed loop compartmental

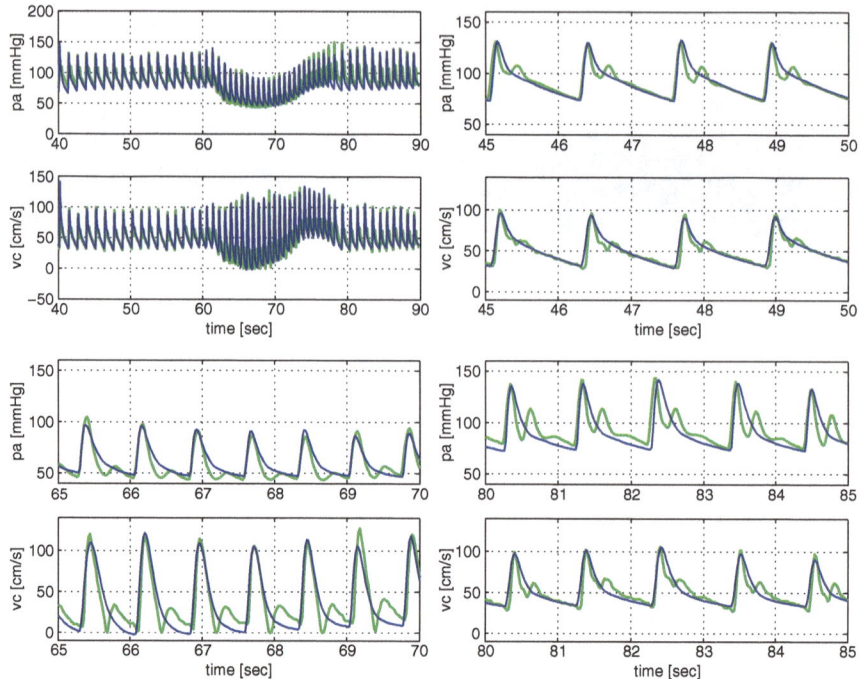

Fig. 10.13 Pressure [mmHg] (first and third row) and flow velocity [cm/s] (second and fourth row) during STS including respiratory mechanical effects (*data green, model blue*). (Modified from [20])

model allows prediction of arterial blood pressure (p_a) and cerebral blood flow velocity (v_c) during postural change from sitting to standing. The figures shows the overall estimation of these quantities and three "zooms" showing the prediction during sitting (for $t \in [45 : 50]$ s), during the transition (for $t \in [65 : 70]$ s), and during standing (for $t \in [80 : 85]$ s). Note, while the model was able to predict the amplitude and approximate shape of the waveform, this type of model cannot predict the wave reflections observed in the data. To predict these, it is necessary to use a fluid dynamics model, that more realistically allows modeling of the wave propagation [16, 39].

Another important aim with these models was to analyze internal dynamics of the states not measured. Results of this analysis is shown in Fig. 10.14. The first figure shows prediction of arterial blood pressure and cerebral blood flow velocity modeling the movement of the diaphragm as a function of measured airflow velocity. This result was obtained using a more elaborate closed loop compartmental model accounting for both vascular movement of blood and respiratory dynamics of CO_2 and O_2, this model was described in [4, 5]. The second figure (Fig. 10.14) shows not only effect on arterial pressure but on the internal states as well. It should be noted that respiration also affects the overall flow distribution and favors some branches at the expense of others and moreover Starlings law of the heart follows as

Fig. 10.14 *Upper panel* shows atria (*blue*), ventricular (*red*), aortic (*black*), and respiratory (*green*) pressure [mmHg] versus time [s] when respiration is included. *Lower panel* shows average flow through the heart (cardiac output) [ml] as a function of mean pressure, results show impact of changing respiration amplitude (*blue*) and depth (*red*, mean), respectively

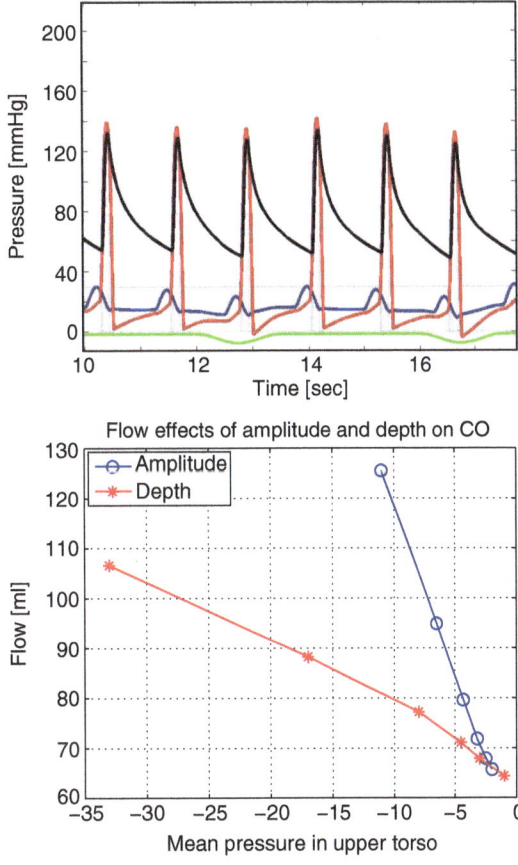

a result. Furthermore, including mechanical coupling with respiration also affect model parameter values. Especially, parameters related to the ventricles and the control mechanisms are sensitive to effect of respiration. Surprisingly the impact of mechanical movement giving rise to respiration is strong. This is illustrated in Fig. 10.14, where the movement of the diaphragm during respiration is seen to be significant.

10.3.3   Syncope

Control mechanisms of the cardiovascular system play an important role in adaptation to postural changes during everyday activities. Syncope (meaning a pause in music) is the medical term for temporary loss of consciousness or fainting. Syncope is a common and significant medical problem that accounts for 3–6 % of emergency rooms visits and hospital admissions every year and in some cases may hallmark a significant underlying morbidity. Syncope is multifactorial and can be triggered from many different inputs into the central and peripheral autonomic systems. It often manifest as an abrupt loss of muscle tone with/without falling accompanied by a decline in blood pressure, blood flow to the brain, and heart rate, the latter may lead to temporary cardiac arrest. This presentation may have clinical prodromes and changes in autonomic nervous system firing for minutes before the actual loss of consciousness occurs. Syncope is considered to be a reflex mechanism, protecting the vital organs (mainly the brain) from lack of perfusion.

The term syncope is related to a broad range of problems, and mechanisms related to triggering of syncope are multifactorial and different mechanisms can present in the same patients on various occasion not well understood. The most common type of syncope is neurally mediated syncope, which is characterized by peripheral vasodilation and a decrease in blood pressure (hypotension) along with a slowing of heart rate (bradycardia) or increasing heart rate (tachycardia). The result is temporary insufficient blood flow to the brain. The event is usually initiated by a withdrawal of peripheral sympathetic tone in upright posture, release of vaso-constrictors or active vasodilatation, accompanied by a decline in blood pressure. Cardio-acceleration and central vasodilatation are compensatory mechanisms that may fail if blood pressure declines further. Bradycardia or slow heart rate typically occurs later when blood pressure falls below certain threshold. In the model of HUT syncope appear approximately after 30 min as a sudden incident as a result of a crash in the control system where the effect of the controls saturates.

In the example discussed here the reduction in flow is initiated by the sudden drop in blood pressure observed during HUT, followed by compromised autonomic and cerebral autoregulation. In contrast to the heart rate regulation a unifying description of pressure changes shows that multiple simultaneous control mechanisms may be important in order to understand the experiments.

There are many reasons to use the model to analyze questions related to syncope, most importantly, it is not well known what mechanisms triggers syncope. A number of theories have been put forward to explain the phenomena [15]. How each of these theories impact pressure, flow, and HR dynamics could be studied using the proposed models.

To model dynamics during syncope we used the model illustrated in Fig. 10.5 modified to include intermediate control mechanisms essential for describing dynamics of HUT experiments over longer time scales (>35 min). To do so it was important to account for dynamics of the venous pump, functioning to prevent pooling of flow in the extremities due to effects of gravity. This type of dynamics was not included in the original closed loop model, which was developed to study dynamics over a short time scale (<2 min). To model this we included a fluid shift compartment at the level of the legs. Through this compartment fluid was continuously transferred from the cardiovascular system to the extravascular environment. Consequently a combination of control mechanisms including arterial resistance and venous compliance regulation, fluid shift and dead volumes were included in this study. In addition, we incorporated a slight tension of the diaphragm

Fig. 10.15 *Upper panel*
shows arterial pulsatile
pressure [mmHg] versus time
[s], *center panel* shows HR
[bpm], and *bottom panel*
shows results from a
simulation. Results are shown
during a HUT experiment.
Note that the control
mechanisms reach their
saturation causing the
syncope

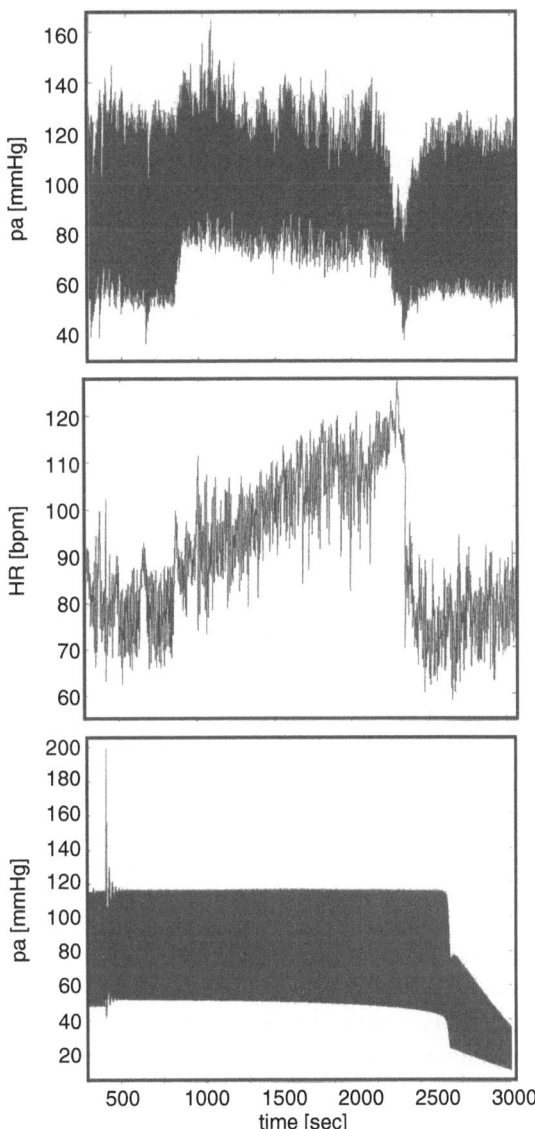

The resulting model nicely describes the HUT experiments as shown in Fig. 10.15.
In particular notice, how well the model reproduce data. The crash happens in the
model when the arterial compliance regulation reaches its saturation. Thus the heart
rate cannot manage to regulate the system alone, venous return declines and the
system breaks down resulting in syncope.

10.4   Conclusion

Above we have discussed a number of methodologies that can be used to develop patient specific models that allow prediction of arterial blood pressure, cerebral blood flow velocity, and heart rate during STS and HUT. Two model approaches were discussed open-loop models that use blood pressure as an input to predict heart rate, and a closed loop model that allow prediction of both blood pressure and blood flow velocity, using heart rate as an input. We showed how these models could be used to predict patient specific data by solving the inverse problem estimating a set of model parameters that allow prediction of patient specific dynamics. Two analysis techniques were investigated, and prove to be essential to obtain good results: sensitivity analysis, which allow us to rank model parameters from the most to the least sensitive, and subset selection, which allow us to identify correlation among model parameters. The sensitivity ranking can be used to identify insensitive parameters that cannot be estimated given data, and allowed us to identify essential components of the model. Subset selection allowed us to predict a subset of uncorrelated parameters among the sensitive parameters, which can be reliably be estimated using the available data. Finally, we show how these models can be used to model clinical events (syncope), and using this example, we have shown that this type of modeling approach has potential to be used for better understanding of underlying mechanisms triggering syncope. Another important outcome is that the type of models developed here allow prediction of internal states, which cannot be measured experimentally.

However, the models themselves have a number of limitations. First, the sparsity of data used for prediction of dynamics in the closed loop models discussed above, is probably the main limitation in obtaining good patient specific models. As discussed earlier, only 4–6 parameters can be estimated reliably given these data. Another limitation, not of the models, but of the patient specific analysis is that without measurements of cardiac output it is difficult to get flow distributions correct. Many different estimates of cardiac output can be distributed to obtain the correct arterial pressure and cerebral blood flow velocity, and as discussed in [31], the models analyzed here typically underestimates cardiac output. Consequently, one important suggestion we can put forward for researchers interested in using the proposed modeling tools discussed here is to incorporate some measurements of cardiac output.

Another important lesson we learned (discussed in [6]), is that for any given application, it is essential to include as few compartments as necessary to model the proposed dynamics, and then be conservative in reducing the number of parameters estimated to predict patient specific dynamics.

Finally, if one is interested in studying dynamics of the wave form, the ODE compartmental models should be coupled with a fluid dynamics model (e.g., as done in [16, 39]) allowing accurate prediction of the reflected waves.

Acknowledgements The authors wish to thank graduate students involved with this project including Laura Ellwein, Department of Biomedical Engineering, Marquette University, Scott Pope, SAS Corp, Raleigh, NC, April Alston, Department of Mathematics, NCSU, and numerous undergraduate students participating in research and REU program at NCSU. Furthermore, authors would like to thank Hien Tran and Tim Kelley, Department of Mathematics, NCSU. The work by Mette Olufsen was supported in part by NSF-DMS grant # 0616597 and NSF-OISE grant # 0437037. V. Novak, director of the SAFE Laboratory at BIMDC was supported by NIH-NIA Harvard Older American Independence Center 2P60 AG08812-11A1, Core B.

References

1. Allen, L.J.S.: An Introduction to Mathematical Biology. Pearson Education, Up Sadddle River, NJ (2003)
2. Danielsen, M.: Modeling of feedback mechanisms which control the heart function in a view to an implementation in cardiovascular models. PhD Thesis, Roskilde University, Denmark. Text No 358 (1998)
3. DeVault, K., Gremaud, P., Zhao, P., Vernieres, G., Novak, V., Olufsen, M.S.: Blood flow in the circle of Willis: Modeling and calibration. SIAM Int. J. Multiscale Model. Simul. **7** 888–909 (2008)
4. Ellwein, L.M.: Cardiovascular and Respiratory Regulation, Modeling and Parameter Estimation. PhD Thesis, Applied Mathematics, NC State University, Raleigh, NC (2009)
5. Ellwein, L.M., Pope, S.R., Xie, A., Batzel, J.J., Kelley, C.T., Olufsen, M.S.: Patient-specific modeling of cardiovascular and respiratory dynamics during hypercapnia. Math Biosci, in press (2012). doi: 10.1016/j.mbs.2012.09.003
6. Ellwein, L.M., Tran, H.T., Zapata, C., Novak, V., Olufsen, M.S.: Sensitivity analysis and model assessment: Mathematical models for arterial blood flow and blood pressure. J. Cardiovasc. Eng. **8**, 94–108 (2008)
7. Fowler, K.R., Gray, G.A., Olufsen, M.S.: Modeling heart rate regulation – Part II: Parameter identification. J. Cardiovasc. Eng. **8**, 109–119 (2008)
8. Guyton, A.C., Hall, J.E.: Textbook of Medical Physiology, 9th edn. WB Saunders, Philadelphia, PE (1996)
9. Johnson, P., Shore, A., Potter, J.F., Panerai, R.B., James, M.: Baroreflex sensitivity measured by spectral and sequence analysis in cerebrovascular disease – methodological considerations Clin. Auton. Res. **16**, 270–275 (2006)
10. Jose, A., Collison, D.: The normal range and determinants of the intrinsic heart rate in man. Cardiovasc. Res. **4**, 160–167 (1970)
11. Kaufmann, H., Biaggioni, I., Voustianiouk, A., Diedrich, A., Costa, F., Clarke, R., Gizzi, M., Raphan, T., Cohen, B.: Vestibular control of sympathetic activity: An otolith-sympathetic relfex in humans. Exp. Brain Res. **143**, 463–469 (2002)
12. Kyrylov, V., Severyanova, L.A., Vieira, A.: Modeling robust oscillatory behaviour of the hypothalamic-pituitary-adrenal axis. IEEE Trans. Biomed. Eng. **52**, 1977–1983 (2005)
13. Liang, K.H., Krus, D.J., Webb, J.M.: K-fold crossvalidation in canonical analysis. Multivariate Behav. Res. **30**, 539–546 (1995)
14. Low, P.A., Bannister, R.G.: Multiple System Atrophy and Pure Autonomic Failure. In: Low, P.A. (eds.) Clinical Autonomic Disorders, pp. 555–575. Lippincott-Raven, Philadelphia, PE (1997)
15. Mosqueda-Garcia, R., Furlan, R., Tank, J., Fernandez-Violante, R.: The elusive pathophysiology of neurally mediated syncope. Circulation **102**, 2898–2906 (2000)
16. Olufsen, M.S.: Structured tree outflow condition for blood flow in larger systemic arteries. Am. J. Physiol. **276**, H257–H268 (1999)

17. Olufsen, M.S., Nadim, A., Lipsitz, L.A.: Dynamics of cerebral blood flow regulation explained using a lumped parameter model. Am. J. Physiol. **282**, R611–R622 (2002)
18. Olufsen, M.S., Nadim, A.: On deriving lumped models for blood flow and pressure in the systemic arteries. Math. Biosci. Eng. **1**, 61–80 (2004)
19. Olufsen, M.S., Tran, H.T., Ottesen, J.T.: Modeling cerebral blood flow control during posture change from sitting to standing. Cardiovasc. Eng. **4**(1), 47–58 (2004)
20. Olufsen, M.S., Ottesen, J.T., Tran, H.T., Ellwein, L.M., Lipsitz, L.A., Novak, V.: Blood pressure and blood flow variation during postural change from sitting to standing: Model development and validation. J. Appl. Physiol. **99**, 1523–1537 (2005)
21. Olufsen, M.S., Tran, H.T., Ottesen, J.T., Lipsitz, L.A., Novak, V.: Modeling baroreflex regulation of heart rate during orthostatic stress. Am. J. Physiol. **291**, R1355–R1368 (2006)
22. Olufsen, M.S., Alston, A.V., Tran, H.T., Ottesen, J.T., Novak, V.: Modeling heart rate regulation, Part I: Sit-to-stand versus head-up tilt. J. Cardiovasc. Eng. **8**, 73–87 (2008)
23. Opthof, T.: The normal range and determinants of the intrinsic heart rate in man. Cardiovasc. Res. **45**, 173–176 (2000)
24. Ottesen, J.T.: Modeling of the baroreflex-feedback mechanism with time-delay. J. Math. Biol. **36**, 41–63 (1997)
25. Ottesen, J.T.: Nonlinearity of baroreceptor nerves. Surv. Math. Ind. **7**, 187–201, (1997)
26. Ottesen, J.T.: General Compartmental Models of the Cardiovascular System in Mathematical Modelling in Medicine, pp. 121–138. IOS, Amsterdam (2000)
27. Ottesen, J.T.: Modeling the dynamical baroreflex-feedback control. Math. Comp. Mod. **31**, 167–173 (2000)
28. Ottesen, J.T.: Valveless pumping in a fluid-filled closed elastic tube-system: one-dimensional theory with experimental validation. J. Math. Biol. **46**, 309–332 (2003)
29. Ottesen, J.T., Danielsen, M.: Modeling ventricular contraction with heart rate changes. J. Theor. Biol. **222**, 337–346 (2003)
30. Ottesen, J.T., Olufsen, M.S., Larsen, J.: Applied Mathematical Models in Human Physiology. IAM, Philadelphia, PA (2004)
31. Pope, S., Ellwein, L.M., Zapata, C.L., Novak, V., Kelley, C.T., Olufsen, M.S.: Estimation and identification of parameters in a lumped cerebrovascular model. Math. Biosci. Engl. **6**, 93–115 (2009)
32. Pope, S.R.: Parameter identification in lumped compartment Cardiorespiratory models. PhD Thesis, Applied Mathematics, NC State University, Raleigh, NC (2009)
33. Ray, C.: Interaction of the vestibular system and baroreflexes on sympathetic nerve activity in humans. Am. J. Physiol. **279**, H2399–H2404 (2000)
34. Ray, C., Monahan, K.: The vestibulosympathetic reflex in humans: Neural interactions between cardiovascular reflexes. Clin. Exp. Pharmacol. Physiol. **29**, 98–102 (2002)
35. Ray, C., Carter, J.: Vestibular activation of sympathetic nerve activity. Acta Physiol. Scand. **177**, 313–319 (2003)
36. Robbe, H.W.J., Mulder, L.J.M., Ruddel, H., Langewitz, W.A., Weldman, J.B.P., Mulder, G.: Assessment of baroreceptor reflex sensitivity by means of spectral analysis. Hypertension **10**, 538–543 (1987)
37. Robertson, D., Low, P.A., Polinsky, R.J.: Primer on the Autonomic Nervous System, 2nd edn. Academic, Boston, MA (2005)
38. Smith, J.J., Kampine, J.T.: Circulatory Physiology, the Essentials, 3rd edn. Williams and Wilkins, Baltimore, MD (1990)
39. Steele, B., Olufsen, M.S., Taylor, C.: Fractal network model for simulating abdominal and lower extremity blood flow during rest and exercise conditions. Comp. Meth. Biomech. Biomed. Eng. **10**, 39–51 (2007)
40. Strogatz, S.H.: Nonlinear Dynamics and Chaos. Perseus Books Publishing, LLC (1994)
41. Thomaseth, K., Cobelli, C.: Generalized sensitivity functions in physiological system identification. Ann. Biomed. Eng. **27**, 607–616 (1999)
42. Ursino, M., Lodi, C.A.: A simple mathematical model of the interaction between intracranial pressure and cerebral hemodynamics. J. Appl. Physiol. **82**, 1256–1269 (1997)

43. Ursino, M., Magosso, E.: Short-term autonomic control of cardiovascular function: a mini-review with the help of mathematical models. J. Integr. Neurosci. **2**, 219–247 (2003)
44. Ursino, M., Magosso, E.: Short-term autonomic control of the cardio-respiratory system: A summary with the help of a comprehensive mathematical model. IEEE Eng. Med. Biol. Soc. **1**, 354–358 (2006)
45. Wilson, T.D., Cotter, L.A., Draper, J.A., Misra, S.P., Rice, C.D., Cass, S.P., Yates, B.J.: Vestibular inputs elicit patterned changes in limb blood flow in conscious cats. J. Physiol. **575**, 671–684 (2006)
46. Zhang, R., Zuckerman, J.H., Iwasaki, K., Wilson, T.E., Crandall, C.G., Levine, B.D.: Autonomic neural control of dynamic cerebral autoregulation in humans. Circulation **106**, 1814–1820 (2002)
47. Zhang, R., Levine, B.D.: Autonomic ganglionic blockade does not prevent reduction in cerebral blood flow velocity during orthostasis in humans. Stroke **38**, 1238–1244 (2007)

Chapter 11
Parameter Estimation of a Model for Baroreflex Control of Unstressed Volume

Karl Thomaseth, Jerry J. Batzel, Mostafa Bachar, and Raffaello Furlan

Abstract The baroreflex involves a number of control pathways. In this chapter we consider in greater detail the role of the control of unstressed volume mobilization. We also consider an alternative approach for choosing parameters most likely to be estimable and we apply this method to a model incorporating the control of unstressed volume and compare to data.

11.1 Introduction

The baroreflex represents the primary cardiovascular system (CVS) short-term global control response mechanism. The baroreflex acts to stabilize blood pressure during stresses that alter this pressure. The baroreflex control response includes varying heart rate H and heart muscle contractility S, systemic resistance R_s, and vascular unstressed volume V_u (and perhaps vascular compliance c). Increasing

K. Thomaseth (✉)
Institute of Biomedical Engineering, National Research Council (ISIB-CNR),
Corso Stati Uniti 4, 35127 Padova, Italy
e-mail: karl.thomaseth@isib.cnr.it

J.J. Batzel
Institute for Mathematics and Scientific Computing, University of Graz and Institute of Physiology, Medical University of Graz, A 8010 Graz, Austria
e-mail: jerry.batzel@uni-graz.at

M. Bachar
Department of Mathematics, College of Sciences, King Saud University, Riyadh, Saudi Arabia
e-mail: mbachar@ksu.edu.sa

R. Furlan
Internal Medicine IV, Humanitas Clinical and Research Center, University of Milan,
Rozzano (Milan), Italy
e-mail: raffaello.furlan@unimi.it

J.J. Batzel et al. (eds.), *Mathematical Modeling and Validation in Physiology*,
Lecture Notes in Mathematics 2064, DOI 10.1007/978-3-642-32882-4_11,
© Springer-Verlag Berlin Heidelberg 2013

H, S, and R_s will act to raise pressure as will a reduction in unstressed volume which increases effective blood volume as outlined below.

The baroreflex control of vascular resistance involves contraction or dilation of small arterioles. Increasing vascular contraction in the arterioles will increase systemic resistance which will act to support blood pressure. This contraction can be supplemented or overridden by local mechanisms that adjust local blood flow to respond to local metabolic activity.

The baroreflex also can vary venous vascular volume in a way that affects so-called unstressed volume. Unstressed blood volume V_u is the blood volume that fills a vascular element before causing distension of the vascular walls (filling volume). Any pressure inducing additional volume will stretch the vascular walls to accommodate the additional volume. This additional volume generated by stretching the vascular walls is termed stressed volume and the pressure generating the distension is termed dynamic pressure (this pressure is the pressure involved in determining blood flow). When the baroreflex reduces unstressed volume reservoirs, this implies that more blood is added to the dynamic circulation helping to support blood pressures.

As mentioned above, in response to blood pressure change, the baroreflex (in conjunction with local self-regulatory mechanisms) varies the levels of V_u (and venous vascular compliance), H, S, and R_s, allowing for a complex blending of control responses to a variety of CVS stresses. Given the complexity of interactions via the various baroreflex control pathways, modeling can, together with specialized data, provide important insight into this key cardiovascular control mechanism. The material presented in Chap. 10 examines a number of issues related to cardiovascular control during orthostatic stress. This chapter focuses in particular on the role of the control of V_u.

11.2 Stressed and Unstressed Vascular Volume

Unstressed volume V_u represents reservoirs of blood which can be accessed (mobilized) by control mechanisms to support blood pressure when blood volume is lost or otherwise removed from dynamic circulation. Approximately 25–30% of total blood volume is V_u mobilizable by baroreflex sympathetic nerve activation [11,16]. Mobilization of V_u helps to maintain mean arterial pressure (MAP), despite the central hypovolemia (low dynamic blood volume) induced by head-up-tilt (HUT) or lower body negative pressure (LBNP), both of which induce CVS stresses similar to orthostatic stress (stress due to blood pooling in lower extremities during upright posture). Further discussion on orthostatic stress is given in Sect. 10.1. In addition, the control of V_u can be an important control component when the system is responding to blood loss such as occurs during hemorrhage [6].

Figure 11.1 indicates the relation between compliance, pressure, stressed and unstressed volume. As volume is introduced above the filling volume of a vascular element (this filling volume as mentioned above is V_u), pressure induces a stretching

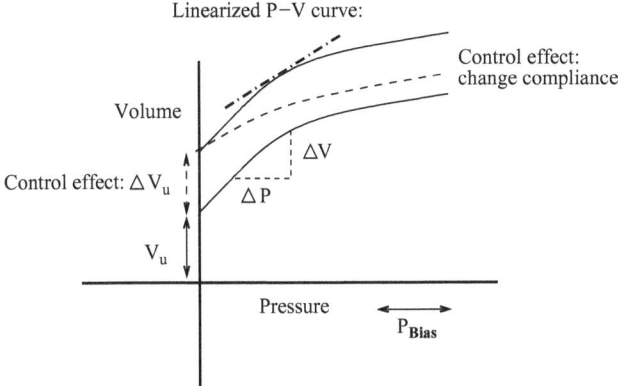

Fig. 11.1 The pressure–volume relationship of a generic blood compartment illustrates the concepts of stressed and unstressed volume and indicates the effects of variations in unstressed volume ΔV_u and bias pressure P_{bias} on capacitance and compliance. The compliance is the local slope $c = \Delta V / \Delta P$ at a particular pressure–volume combination. Control variations in V_u will shift the volume-pressure curve upwards or downwards while a control change in compliance will steepen or flatten the curve. One can see that a linearized approximation defines compliance around a given operating point (P,V). A change in P_{bias} implies a left-right shift in the pressure–volume curve. For example, an external negative pressure which stretches the vascular walls will shift the curve leftward and imply a higher unstressed volume from which the pressure–volume curve begins but with a difference level of compliance due to the stretch

to accommodate an additional (stressed) volume in the vascular element. The volume added due to stretch depends on the elastic characteristics of the walls responding to the pressure. The term compliance is the derivative of the functional relationship of the (stressed) volume to the pressure inducing this volume. The compliance actually varies depending on the level of stretch in the vascular walls (as elastic properties change with stretch) [18]. However, over narrow pressure ranges, compliance is typically assumed constant or piecewise constant. The sum of stressed and unstressed volume divided by the distending pressure will be termed the capacitance of the vascular element (a number of definitions of capacitance exist).

11.3   Model Structure

The above described physiological complexities imply that mathematical modeling is necessary to study quantitatively the interaction of the various factors and mechanisms involved in short-term CVS control. In particular, our purpose in this chapter is to consider the parameter estimation process using an example of patient-specific data. We describe a different subset selection approach for determining parameters to estimate. This method can be compared to the approaches discussed in Chaps. 2, 3, and 10.

Given the focus on the parameter estimation aspect, we present here only an overview of the model structure that we employ to study the baroreflex control during an HUT or LBNP test. This model includes features of unstressed volume control. Variations of the model applied in this paper have been used to model orthostatic stress (LBNP or HUT) and blood loss due to hemorrhage [5, 6, 12]. Details of the full model can be found in Appendix A.1 of [1] available at www. uni-graz.at/imawww/reports/index.html .

The model includes ten compartments representing various body tissues compartments as well as 11 additional state variables associated with control mechanisms, and plasma-interstitial fluid exchange. For the purpose of parameter identification, the mathematical model equations and the corresponding parameter-sensitivity equations were generated symbolically using a specialized software tool [19]. The blood compartments were expressed directly in terms of compartmental blood volumes rather than blood pressures as done in [1]. This equivalent representation simplifies the associated mass balance relations, which become plain expressions of flows between compartments that are independent on time derivatives of compliances, unstressed volumes or bias pressures. The instantaneous blood pressure–volume relation for each compartment is expressed, as shown in Fig. 11.1, by a piece-wise linear function. The effect of control on unstressed volume, at a given fixed compartmental volume, affects compartmental pressures like a wedge moving vertically. Variations of unstressed volume can therefore, in principle, produce instantaneous pressure variations without violating continuity conditions imposed upon total volume.

11.3.1 Mass Balance Equations

The generic form for mass balance relations depends upon the interplay of several model variables such as P which represents the pressure, c the compliance, V the volume of a compartment, F the flow between vascular compartments, R the resistance to flow between compartments, and other model variables and parameters. The concept of model parameter adopted in the following is rather flexible. A parameter may refer in a first instance to adjustable coefficients that remain constant during a simulation run, in contrast to fixed constants such as $\pi = 3.14\ldots$, and become in a second instance time-varying functions that may be either user-defined, e.g., model inputs, or be functions of other variables in the system.

Compartments and modeled control relations are depicted in Fig. 11.2. Subscripts reference the compartments in this block diagram in a straight forward way using the symbols "as", "per", "up", "ren", "spl", "leg", "avc", "vc", "ap", and "vp". For example, "ren" refers to the renal compartment and "avc" refers to the abdominal vena cava (see Table 11.2). For each compartment, variations in compliance, local resistance, and V_u can be induced using various formats of baroreflex control mechanisms (or local mechanisms) as described below.

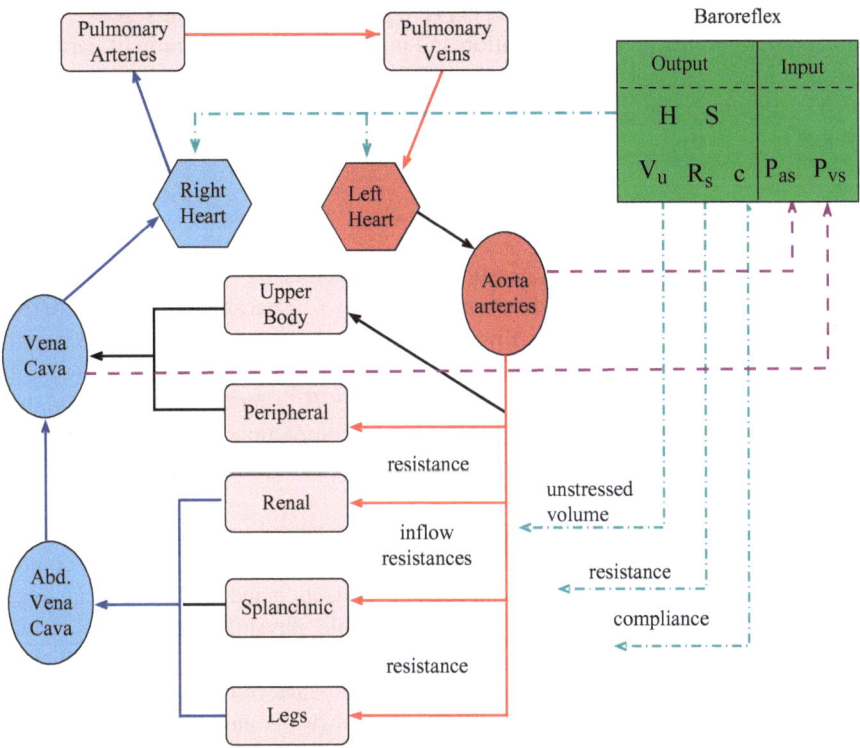

Fig. 11.2 Model block diagram of the circulatory system, representing, except hexagonal elements, blood compartments described as state variables, flows between compartments (*continuous lines*), afferent sensory signals (*dashed*) and efferent (*dash-dotted*) baroreflex control signals. The baroreflex control, based of afferent signals related to systemic arterial P_{as} and venous P_{vs} pressures, determines heart rate H, myocardial contractility S, unstressed volume V_u, systemic resistance R_s, and compliance c

The standard form of the volume dynamics of compartment "comp" has the mass balance relation given by

$$\frac{dV_{comp}}{dt} = F_{in,comp} - F_{out,comp} , \qquad (11.1)$$

where $F_{in,comp}$ represents the natural circulating blood flow into the compartment as well as additions to the compartment via external routes, e.g., blood infusion, and $F_{out,comp}$ the natural circulating flow out of the compartment including blood loss from the compartment via vascular flows, e.g., hemorrhage could also be included as a term. By viewing the overall CVS, including cumulative blood volume loss/gain, as a closed system the total volume of blood in the system becomes a constant. This would allow, in principle, to consider the dynamics of a reduced (by one) number of compartments, and to calculate the volume of one compartment as the difference between total blood in the system and sum of the remaining compartmental volumes.

While this strategy has been adopted in [1], it has been dropped in this study for sake of simplicity in the model formulation and in the symbolic derivation of parameter sensitivities.

The instantaneous pressure of a generic vascular compartment "comp" in the model is given by

$$P_{\text{comp}} = \frac{1}{c} \lfloor V_{\text{comp}} - V_{\text{u,comp}} \rfloor + P_{\text{bias,comp}}, \tag{11.2}$$

where c is compliance, $V_{\text{u,comp}}$ is unstressed volume, $\lfloor V_{\text{comp}} - V_{\text{u,comp}} \rfloor$ is non-negatively constrained stressed volume, and $P_{\text{bias,comp}}$ is any additional external (orthostatic, positive or negative) pressure to compartment "comp". More specifically, $P_{\text{bias,comp}}$ reflects transmural pressure viewed here, in contrast with [1], as outside minus inside pressure so that a positive P_{bias} term represents a higher outside pressure that will eventually decrease the compartmental volume. Similarly, a negative term represents a lower outside pressure that causes vascular volume to increase.

It is important to stress that P_{bias} is the main external input available for non-invasive experimentation aimed to infer upon the function of the baroreflex control system in humans. This is possible through the collection and model-based analysis of variations in heart rate and other measurable physiological cardiovascular variables following arterial and venous pressure changes, which can be elicited through external lower body positive or negative pressure, or following gravitational stress due to orthostasis stretching the lower limb walls and generating in effect an additional unstressed volume contribution to total volume. An open problem remains the attribution of the correct extent of bias pressure elicited at various blood compartments during different perturbation experiments.

The generic expression of blood flow entering (most) compartments due to differences with arterial pressure is given, according to Ohm's law, by

$$F_{\text{in,comp}} = \frac{P_{\text{as}} - P_{\text{comp}}}{R_{\text{in,comp}}}, \tag{11.3}$$

where $R_{\text{in,comp}}$ is arterial vascular resistance of compartment "comp". Similarly blood flow leaving (most) compartments towards a generic venous pool "v" are given by

$$F_{\text{out,comp}} = \left\lfloor \frac{P_{\text{comp}} - P_{\text{v}}}{R_{\text{out,comp}}} \right\rfloor, \tag{11.4}$$

where $R_{\text{out,comp}}$ is venous vascular resistance of compartment "comp", and $\lfloor \cdot \rfloor$ represents non-negative constrained blood flow in presence of venous vascular valves.

Mean artero-venous pressure differences are sustained by left and right heart cardiac outputs separately, which are of course identical on average. Left and right heart cardiac outputs are modeled as the product of heart rate H and the respective stroke volumes which depend upon the respective ventricular contractilities and the corresponding pre- and after-loads [1]. It must be underlined that the CVS model does not describe pulsatile blood flow but only variability of cardiovascular parameters averaged over single heart beats.

11.3.2 Control Equations

Control response depends on sensory input to the baroreflex reflecting systemic arterial pressure P_{as} and systemic venous pressure represented in the model by P_{vc} (vena cava pressure) as depicted in Fig. 11.2. Assumptions on the distribution of control effects to various compartments can be found in [1]. We will apply the same control presented in the Chap. 10. The generic form of a baroreflex feedback control loop is implemented by

$$\frac{dx}{dt} = \frac{-x(t) + x_{ctr}(\bar{P})}{\tau}, \qquad x_{ctr} = (x_{max} - x_{min})\frac{\alpha_{xP}^{\beta}}{\alpha_{xP}^{\beta} + \bar{P}^{\beta}} + x_{min}, \quad (11.5)$$

where $x(t)$ is the control, $\bar{P}$ is a current pressure, τ is a time constant that characterizes the time it takes (delay effect) for the control variable to obtain its full effect. The expression x_{ctr} is a set-point function. It reflects the observed baroreflex characteristic of a decreasing or increasing sigmoidal relation of control variable (decreasing for heart rate and resistance, increasing for V_u) in response to level of blood pressure. Here x_{min} and x_{max} are the minimum and maximum values for the controlled parameter x, respectively. The quantity α_{xP} is the resting nominal pressure referencing a midpoint in the control level. Also, β helps to determine the steepness of the sigmoid and hence is connected to the characteristic gain. Further details on the development of this control can also be found in [1, 14, 15]. The final system steady state need not be exactly α_{xp}. The above equation is decreasing in $\bar{P}$ and hence can be employed for heart rate and resistance control. Reversing the maximum and minimum value positions generates an increasing function which is appropriate for unstressed volume control. Note also that the choice of α_{xp} adjusts the relative position of the control value between the maximum and minimum values in steady state. We assume a central position for heart rate and resistance while assuming unstressed volume is near the maximum values (which implies that the control responds primarily during volume reductions). Other formulations of control such as given in [2, 22] can easily be incorporated as well.

11.3.3 Control Responses: Unstressed Volume and Systemic Resistance

As mentioned above, complete details of the full model can be found in [1]. We summarize here the implementation of the unstressed volume and resistance controls:

- Each vascular compartment includes a degree of unstressed volume.
- Baroreflex changes in systemic resistance R_s are distributed among relevant compartments inflow resistances to the compartments depicted in Fig. 11.2. Note that the change in resistance (ΔR_s) is a variable representing sympathetic drive to vary R_s by some amount. In principle ΔR_s could grow very high but changes locally will be constrained by autoregulation through parameters that restrict the increase to local resistance that would block a minimum blood flow.
- A similar partition is implemented for ΔV_u. Unstressed volume is distributed among several compartments but changes are assumed to be implemented only in certain compartments namely in the renal and splanchnic compartments.
- H only enters one equation at one place so no division is necessary.

11.4   Data

The data used in this paper were collected from HUT tests [9]. One data set is applied to parameter estimation. Additional representative research and typical experimental design for such tests can be found in [8, 9, 13]. Figures 11.3 and 11.4 illustrate the characteristics of data that was collected. Measurements were taken for systolic and diastolic blood pressure, from which mean pressure is calculated. Heart rate was calculated from observed RR intervals. In addition central venous pressure (CVP) was measured invasively and muscle sympathetic nerve activity was measured to provide assessment of sympathetic response to orthostatic stress. Respiratory movement was measured to allow for more accurate assessment of heart rate and blood pressure variability and assess respiratory activity modulating sympathetic neural traffic. Several points should be made:

- Raw data: Arterial pressure and RR intervals were collected essentially continuously. The data was collected using the Finapres system which monitors RR intervals between heart beats and which employs a finger cuff (calibrated by the typical arm cuff) to monitor blood pressure. Central venous pressure was measured invasively with sensor transducers placed in venous return pathways to the right heart (median or basilic vein). Other hemodynamic quantities could also be monitored such as stroke volume and systemic resistance but these variables are estimated using internal modeling strategies by the Finapres. These values are most useful for following dynamic changes and were not used as part of the estimation process. The data, as can be seen from Fig. 11.3 includes noise and artifacts.
- Processed data: This data was derived by removing artifacts and calculating a moving average of measured values to smooth the data as depicted in Fig. 11.4.

As a result of artifacts, data were used beginning at 900 s near the start of the HUT. The data was followed for about 15 min as discussed below. A number of other non-invasive but tricky measurements are possible, including Doppler measurement of blood flow velocity to estimate cardiac output and NIRS to monitor regional blood flows. These measurements could enhance the estimation process.

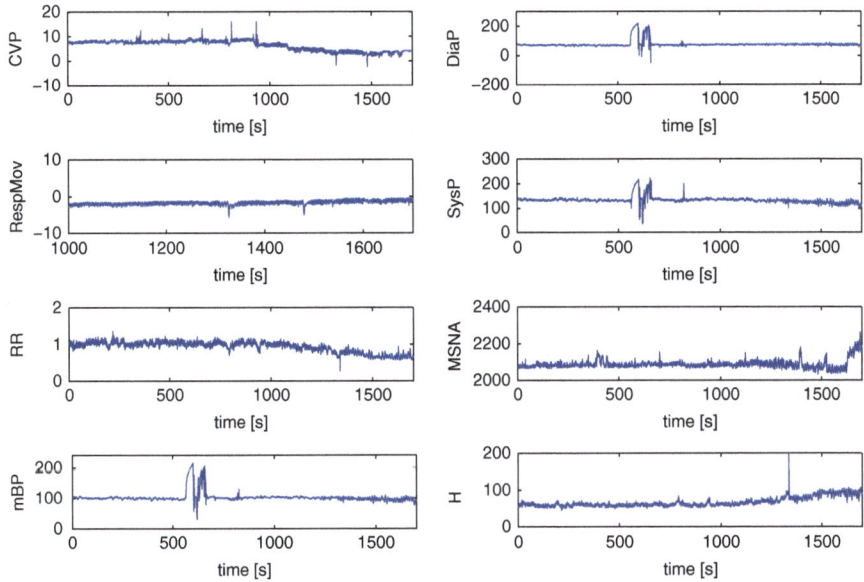

Fig. 11.3 Raw data measured for HUT: heart rate H, central venous pressure CVP, mean arterial pressure mBP, systolic pressure sysP, diastolic pressure diaP, respiratory movement RespMov, muscle sympathetic nerve activity MSNA

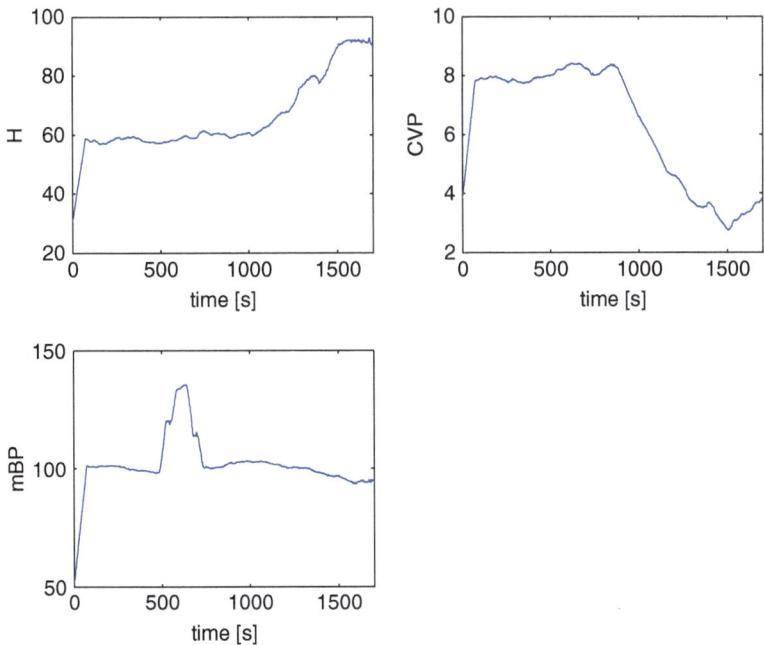

Fig. 11.4 Processed data of measured heart rate H, central venous pressure CVP, and mean arterial pressure mBP

11.5   Model Identification

The CVS model is described by a system of 21 nonlinear differential equations
that define the dynamics of compartmental blood volumes and of auxiliary state
variables. The number of (potentially) adjustable parameters is 114, and it is evident
that not all parameters are identifiable from the adopted input–output experiment. In
particular, the model outputs considered for parameter estimation are heart rate H,
systolic pressure P_{as} and central venous pressure (vena cava) P_{vc}. The measured
outputs coincide therefore with three state variables of the system, which is however
irrelevant for identification purposes.

The perturbation experiment consisted of a HUT test with stepwise increments of
the inclination angle of a tilt table, starting from the horizontal resting condition, and
with the patient in supine position. To approximate the pressure bias P_{bias} provoked
during the HUT perturbation test, the model input to the CVS was expressed,
in a first attempt, as a staircase, piece-wise constant function with increments of
10 [mmHg] every 3 min beginning at 15 min. The input bias pressure was assumed
to cause an equal decrease of P_{bias} in the leg compartment, and a partial (30 %)
decrease in the splanchnic and abdominal vena cava regions. The pressure bias
chosen to correspond to the degree of HUT (from 0 ° up to about 65 °) was based on
typical conversion correspondences between LBNP pressure and HUT degree found
in the literature (e.g., [10]).

The above staircase input was used during an early stage of model identification,
but did not provide satisfactory results because model outputs exhibited, unlike the
data, rapid transients in coincidence with the changes in bias pressure during HUT.
The second, and more successful, attempt for describing the model input consisted
in a continuously varying bias pressure with constant slope of 10/3 mmHg per min.

With either model input representation the single input multiple output model
resulted clearly unidentifiable according to the criteria described below, and a
model order reduction by subspace selection for parameter identification appeared
necessary which was implemented as follows.

11.6   Sensitivity Identifiability: A Subset Selection Approach

Parameter identification plays a central role in physiological systems modeling
for validating modeling hypotheses against experimental data and, in general, for
solving the inverse problem in practical applications. Either global or local identifiability is a mandatory requirement for estimating with some degree of confidence
model parameters from input–output experiments. The most restrictive requirement
is global *a priori* identifiability, which is a structural property of a model that is in
general ascertainable only for particular classes of models of reduced complexity.
On the contrary, local *a posteriori* identifiability can be thought of as the least
restrictive requirement for estimated parameters to optimize locally, yet uniquely
the cost function associated with the adopted fitting criterion, e.g., weighted nonlinear least squares or maximum likelihood. For continuously differentiable cost

functions the local optimum is characterized by vanishing gradients with respect to parameters calculated at the optimal solution, and the optimizing parameter vector is uniquely defined, according to the inverse function theorem, if the Jacobian matrix is non-singular. Since the Jacobian matrix depends generally upon the sensitivities with respect to parameters of the measured model outputs taken at discrete sampling times, the requirements for local identifiability can be expressed in terms of the properties of the parameter-sensitivity matrix of measured outputs.

Various different strategies exist to overcome lack of local identifiability which include modifications of the cost function, such as in Bayesian inference by including prior information on parameters; reduction of the dimension of the vector of estimated parameters down to an identifiable subset of parameters; or through linear transformations of the parameter space and subsequent selection of a reduced rank subspace with a smaller number of actually estimated parameters. In this paper, we apply this latter approach of reduced rank subspace selection for parameter identification based on singular value decomposition.

11.6.1 Parameter Identification Framework

For the purpose of model parameter estimation using non-linear weighted least squares (NLWLS) we consider a generic model described by a system of non-linear ordinary differential equations

$$\dot{\mathbf{x}}(t) = f(\mathbf{x}(t), \mathbf{p}, \mathbf{u}(t), t), \qquad (11.6)$$

where $\mathbf{x}(t) \in \mathbb{R}^{n_x}$ is the state trajectory with initial condition $\mathbf{x}(0) = \mathbf{x}_0(\mathbf{p}), \mathbf{p} \in \mathbb{R}^{n_p}$ is the parameter vector and $\mathbf{u}(t) \in \mathbb{R}^{n_u}$ is the input vector. The measurable output vector is, generally, given by a system of non-linear functions

$$\mathbf{y}(t, \mathbf{p}) = \mathbf{g}(\mathbf{x}(t), \mathbf{p}, t), \qquad (11.7)$$

where $\mathbf{y}(t, \mathbf{p}) \in \mathbb{R}^{n_y}$ is expressed explicitly as a function of parameter vector $\mathbf{p}$, because $\mathbf{x}(t)$ is itself a function of $\mathbf{p}$ according to Eq. (11.6). The dependence of $\mathbf{y}(t, \mathbf{p})$ upon a known input $\mathbf{u}(t)$ is tacit. In the present study $n_x = 21, n_p = 114, n_u = 1, n_y = 3$, and $\mathbf{g}$ is linear.

Parameter identification is based on noisy measurements, taken over a finite horizon at discrete time points $\{t_j, j = 1, \ldots, N\}$, and given by

$$\mathbf{z}(t_j) = \mathbf{y}(t_j, \mathbf{p}^*) + \mathbf{e}(t_j), \qquad (11.8)$$

where $\mathbf{e}(t_j)$ is assumed, for simplicity, zero-mean uncorrelated white noise with known diagonal covariance matrix, and $\mathbf{p}^*$ represents the true parameter vector that generated the particular set of observed data. With the given hypotheses about measurement noise, Eq. (11.8) can be expressed in terms of the scalar components

$$\mathbf{z}_i(t_j) = \mathbf{y}_i(t_j, \mathbf{p}^*) + \mathbf{e}_i(t_j); \quad i = 1, \ldots, n_y. \qquad (11.9)$$

Is is worth stressing that $\{\mathbf{z}_i(t_j), i = 1, \ldots, n_y; j = 1, \ldots, N\}$ represent experimental data, while $\mathbf{y}_i(t_j, \mathbf{p})$ represent the i-th simulated model output at time t_j calculated for a particular value of parameter vector $\mathbf{p}$. Moreover, by recognizing that $\mathbf{p}^*$ will remain largely unidentifiable the role of $\mathbf{p}^*$ in Eqs. (11.8) or (11.9) is considered of minor importance. A more practical approach is to assign initial values, $\mathbf{p}_0$, on the basis of prior knowledge and hypotheses about the CVS, and to improve the quality of model predictions by fitting the model outputs to available data through adjustments of a reduced subset of parameter vector $\mathbf{p}$. Any prior information available on parameters, such as positivity constraints or bounds, can be included into the model equations. In this study we constrained parameters to be positive by means of the log-transformation, which consists of replacing a generic positively constrained parameter, $p > 0$, with $e^{\ln p}$, where the unbounded $\ln p$ replaces p in the list of parameters. This transformation has several advantages, including increased robustness of numerical simulation, and implicit parameter scaling in the calculation of sensitivities.

Irrespective of non-linear parameter transformations, the cost function used for NLWLS is the weighted sum of squares given by

$$WSS(\mathbf{p}) = \frac{1}{2} \sum_{i=1}^{n_y} \sum_{j=1}^{N} w_{ij} \left(\mathbf{z}_i(t_j) - \mathbf{y}_i(t_j, \mathbf{p}) \right)^2, \tag{11.10}$$

where the weights w_{ij} are usually taken as the reciprocal of measurement noise variance of output $\mathbf{y}_i$ sampled at time t_j, but can be also used to exclude some dubious data point by letting $w_{ij} = 0$ or to fit primarily one particular model output to the related data by increasing the corresponding weights. Equation (11.10) can be expressed more concisely as

$$WSS(\mathbf{p}) = \frac{1}{2}(\mathbf{Z} - \mathbf{Y}(\mathbf{p}))^{\mathsf{T}} \mathbf{W}(\mathbf{Z} - \mathbf{Y}(\mathbf{p})), \tag{11.11}$$

where $\mathbf{Z}$ and $\mathbf{Y}(\mathbf{p})$ represent the vectors of sequential measurements and model outputs, respectively, e.g., $\mathbf{Z} = [\mathbf{z}_1(t_1), \ldots, \mathbf{z}_1(t_N), \mathbf{z}_2(t_1), \ldots, \mathbf{z}_2(t_N), \ldots, \mathbf{z}_{n_y}(t_1), \ldots, \mathbf{z}_{n_y}(t_N)]^{\mathsf{T}}$, and $\mathbf{W}$ is the diagonal weighing matrix.

Given the above notation, well known properties and results are derived in the following. The NLWLS problem yields the parameter estimates defined as

$$\hat{\mathbf{p}} = \arg \min_{\mathbf{p}} WSS(\mathbf{p}). \tag{11.12}$$

The optimal solution is characterized by the optimality condition

$$\nabla_{\mathbf{p}} WSS(\mathbf{p})\big|_{\mathbf{p}=\hat{\mathbf{p}}} = \mathbf{S}(\hat{\mathbf{p}})^{\mathsf{T}} \mathbf{W}(\mathbf{Z} - \mathbf{Y}(\hat{\mathbf{p}})) = \mathbf{0}, \tag{11.13}$$

where $\mathbf{S}(\mathbf{p})$ is the sensitivity matrix of the model outputs $\mathbf{Y}(\mathbf{p})$ (see below). Moreover, the local behavior of the cost function (11.13) around the optimum

$\hat{\mathbf{p}}$ is characterized by its Hessian matrix which must be positive definite in order to uniquely characterize the local optimal solution $\hat{\mathbf{p}}$. This is equivalent to the concept of local identifiability of $\hat{\mathbf{p}}$. With some abuse of notation the Hessian matrix of the *WSS* cost function becomes

$$\nabla_{\mathbf{p}^2}^2 WSS(\mathbf{p})\Big|_{\mathbf{p}=\hat{\mathbf{p}}} = \mathbf{S}(\hat{\mathbf{p}})^\mathsf{T}\mathbf{W}\,\mathbf{S}(\hat{\mathbf{p}}) - \nabla_{\mathbf{p}^2}^2\mathbf{Y}(\hat{\mathbf{p}})\mathbf{W}(\mathbf{Z} - \mathbf{Y}(\hat{\mathbf{p}})) \simeq \overline{\mathbf{S}}(\hat{\mathbf{p}})^\mathsf{T}\overline{\mathbf{S}}(\hat{\mathbf{p}}) \geq 0,$$

(11.14)

where $\overline{\mathbf{S}} = \mathbf{W}^{\frac{1}{2}}\,\mathbf{S}$ is the weighted sensitivity matrix. The right hand side approximation is justified if either the weighted Hessians of the model outputs at various times ($\nabla_{\mathbf{p}^2}^2\mathbf{Y}(\hat{\mathbf{p}})\mathbf{W}^{\frac{1}{2}}$) are small, i.e., quasi-linear behavior with small curvature, or if the weighted estimation residuals $\mathbf{W}^{\frac{1}{2}}(\mathbf{Z} - \mathbf{Y}(\hat{\mathbf{p}}))$ are small, or both. Even if we assume a priori that one of these simplifying assumptions is valid, the Hessian $\nabla_{\mathbf{p}^2}^2 WSS(\hat{\mathbf{p}})$ is only guaranteed to be positive semidefinite. Only if the weighted sensitivity matrix $\overline{\mathbf{S}}(\hat{\mathbf{p}})$ has full rank the Hessian becomes positive definite. This observation is equivalent to the fact that model parameters are locally identifiable only if the sensitivity matrix of the measured outputs has full rank.

11.6.1.1 Calculation of Model Sensitivities

Given the model differential equations (11.6), the matrix $\nabla_{\mathbf{p}}\mathbf{x}(t)$ defines the sensitivity of the state trajectory with respect to parameter variations, or equivalently $d\mathbf{x}(t) \simeq \nabla_{\mathbf{p}}\mathbf{x}(t) \cdot d\mathbf{p}$. The i-th column of $\nabla_{\mathbf{p}}\mathbf{x}(t)$, which will be indicated as $\mathbf{x}_{\mathbf{p}_i}(t)$, represents the sensitivity at time t of the state vector with respect to the i-th component of parameter vector $\mathbf{p}$. This sensitivity vector is the solution of the following dynamic equations

$$\dot{\mathbf{x}}_{\mathbf{p}_i}(t) = \nabla_x f(\mathbf{x}(t), \mathbf{p}, \mathbf{u}(t), t) \cdot \mathbf{x}_{\mathbf{p}_i}(t) + \nabla_{\mathbf{p}_i}\mathbf{f}(\mathbf{x}(t), \mathbf{p}, \mathbf{u}(t), t) \qquad (11.15)$$

with initial conditions $\mathbf{x}_{\mathbf{p}_i}(0) = \partial\mathbf{x}(0)/\partial\mathbf{p}_i$. The matrix $\nabla_x\mathbf{f}(\mathbf{x}(t), \mathbf{p}, \mathbf{u}(t), t)$ represents the Jacobian of the dynamic system equations with respect to the state, which needs to be determined only once for all parameters.

Similarly, with reference to the output equations (11.7) we define the output sensitivity matrix $\nabla_{\mathbf{p}}\mathbf{y}(t)$, such that $d\mathbf{y}(t) = \nabla_{\mathbf{p}}\mathbf{y}(t) \cdot d\mathbf{p}$, and whose i-th column $\mathbf{y}_{\mathbf{p}_i}(t)$, represents the sensitivity of the output trajectory with respect to the i-th element of $\mathbf{p}$. It is defined as

$$\mathbf{y}_{\mathbf{p}_i}(t) = \nabla_x\mathbf{g}(\mathbf{x}(t), \mathbf{p}, t) \cdot \mathbf{x}_{\mathbf{p}_i}(t) + \nabla_{\mathbf{p}_i}\mathbf{g}(\mathbf{x}(t), \mathbf{p}, t). \qquad (11.16)$$

The implementation of the above approach is thus based on analytic derivation of model equations rather than on numerical differentiation of output trajectories using parameter perturbations and finite differences. This is a so-called algorithmic differentiation method in which sensitivities are computed from symbolic derivatives of the same computer code used for calculating model outputs. The derivatives of model outputs with respect to parameters are therefore "correct" even if the sensitivities are small in the order of roundoff errors, and are robust with respect to changes in numerical integration step size, which can cause large errors with numerical differentiation.

The implementation of the above equations (11.15) and (11.16) requires symbolic differentiation of the model's differential and output equations with respect to state variables and parameters, and needs the generation of computer code for the numerical solution of the extended system of model equations. This task can be fully automated using computer algebra software or using ad hoc symbolic differentiation as implemented in [19]. In the present study the total number of differential equation used to simulate the system dynamics (11.6) and the sensitivity differential equations (11.15) was $n_x \cdot (n_p + 1) = 2{,}415$, and the number of system outputs and their sensitivities was $n_y \cdot (n_p + 1) = 345$. The numerical simulation using a variable step 4/5-th order Runge–Kutta–Fehlberg method was surprisingly time efficient, most likely thanks to the common subexpression elimination capabilities of the optimizing compiler used (GNU Fortran (GCC) 4.2.3). Simulations, graphics and optimization algorithms were carried out within the statistical software package R (http://www.R-project.org/).

11.6.2 Reduced Rank Subspace Selection Using Singular Value Decomposition

The widely used singular value decomposition (SVD) approach for reduced rank subset selection is presented within the context of iterative, restricted step, Gauss–Newton method used to minimize the weighted sum of squares function (11.11). In particular, given at the k-th iteration the parameter vector $\mathbf{p}_k$, the Gauss–Newton iteration moves into the opposite direction to the gradient of cost function (11.11), taking into account the local curvature of the cost function using the approximation of the Hessian introduced in (11.14). In particular, the direction in which to move the parameter vector is calculated by solving the normal equations

$$\left[\overline{\mathbf{S}}(\mathbf{p}_k)^\mathsf{T} \overline{\mathbf{S}}(\mathbf{p}_k) \right] d\mathbf{p}_k = \overline{\mathbf{S}}(\mathbf{p}_k)^\mathsf{T} \overline{\mathbf{E}}(\mathbf{p}_k), \qquad (11.17)$$

where $\overline{\mathbf{E}}(\mathbf{p}_k) = \mathbf{W}^{\frac{1}{2}}(\mathbf{Z} - \mathbf{Y}(\mathbf{p}_k))$ are current weighted residuals. The same weighing matrix $\mathbf{W}^{\frac{1}{2}}$ is thus used to normalize the rows of the sensitivity matrix $\mathbf{S}(\mathbf{p}_k)$ as well as the current prediction errors.

The actual restricted step taken in direction $d\mathbf{p}_k$ determines the new vector of parameters

$$\mathbf{p}_{k+1} = \mathbf{p}_k + \alpha_k \, d\mathbf{p}_k, \qquad (11.18)$$

where $0 < \alpha_k \leq 1$ is chosen such that $WSS(\mathbf{p}_{k+1}) < WSS(\mathbf{p}_k)$. This latter inequality can be satisfied for some $\alpha_k > 0$, if Eq. (11.17) has a unique solution, that is if $\overline{\mathbf{S}}(\mathbf{p}_k)$ has full rank. This could be obtained through left multiplication of the weighted residuals by the pseudoinverse $\overline{\mathbf{S}}(\mathbf{p}_k)^+ = \left[\overline{\mathbf{S}}(\mathbf{p}_k)^\mathsf{T}\overline{\mathbf{S}}(\mathbf{p}_k)\right]^{-1}\overline{\mathbf{S}}(\mathbf{p}_k)^\mathsf{T}$. However, taken for granted that the model parameter vector $\mathbf{p}_k$ is locally unidentifiable, the sensitivity matrix $\overline{\mathbf{S}}(\mathbf{p}_k)$ is rank deficient, and (11.17) has not a unique solution.

Singular value decomposition (SVD) is a dependable approach to determine the pseudoinverse of a matrix and is based on the following factorization

$$\overline{\mathbf{S}}(\mathbf{p}_k) = \mathbf{U}_k \Sigma_k \mathbf{V}_k^T, \tag{11.19}$$

where $\mathbf{U}_k \in \mathbb{R}^{n_y \cdot N \times n_y \cdot N}$ and $\mathbf{V}_k \in \mathbb{R}^{n_p \times n_p}$ are the orthonormal eigenvector matrices of $\overline{\mathbf{S}}(\mathbf{p}_k)\overline{\mathbf{S}}(\mathbf{p}_k)^T$, and $\overline{\mathbf{S}}(\mathbf{p}_k)^\mathsf{T}\overline{\mathbf{S}}(\mathbf{p}_k)$, respectively, and $\Sigma_k \in \mathbb{R}^{n_y \cdot N \times n_p}$ is diagonal (referring to the top $n_p \times n_p$ submatrix) with sorted *singular values* $\sigma_1 \geq \sigma_2 \geq \ldots \geq \sigma_{n_p} \geq 0$, which are also the square roots of the eigenvalues of the positive-semidefinite matrix $\overline{\mathbf{S}}(\mathbf{p}_k)^\mathsf{T}\overline{\mathbf{S}}(\mathbf{p}_k)$.

By hypothesis, $\overline{\mathbf{S}}(\mathbf{p}_k)$ is rank deficient and the effective rank $r < n_p$ is characterized, in theory, by $\sigma_{r+1} = 0$ and, in practice, by $\sigma_{r+1}/\sigma_1 \approx 0$. This justifies the approximation of $\overline{\mathbf{S}}(\mathbf{p}_k)$ by

$$\overline{\mathbf{S}}(\mathbf{p}_k) \approx \mathbf{U}_k \tilde{\Sigma}_k \mathbf{V}_k^T \tag{11.20}$$

where $\tilde{\Sigma}_k$ has only the first r positive singular values and others are zero. Because of roundoff errors the rank r is rarely defined exactly through $\sigma_{r+1} = 0$, and in practice it is linked to the largest singular value such that $\sigma_{r+1} < \delta \sigma_1 \leq \sigma_r$, for a chosen $\delta > 0$, usually as a function of machine precision. For the aims of this study we are interested in a solution of the normal equations (11.17) using the approximation (11.20), with a numerical rank r_k calculated for a particular threshold δ_k. This yields the (approximate) pseudoinverse of $\overline{\mathbf{S}}(\mathbf{p}_k)$ given by

$$\overline{\mathbf{S}}(\mathbf{p}_k)^+ \approx \mathbf{V}_k \tilde{\Sigma}_k^+ \mathbf{U}_k^T, \tag{11.21}$$

where $\tilde{\Sigma}_k^+$ has the reciprocals of the first r_k diagonal elements of $[\tilde{\Sigma}_k]^\mathsf{T}$ and zero otherwise. The (approximate) solution of (11.17) is then given by

$$d\mathbf{p}_k = \overline{\mathbf{S}}(\mathbf{p}_k)^+ \overline{\mathbf{E}}(\mathbf{p}_k) = \mathbf{V}_k \tilde{\Sigma}_k^+ \mathbf{U}_k^T \overline{\mathbf{E}}(\mathbf{p}_k), \tag{11.22}$$

which represents a practically feasible approach for computing the search direction in the Gauss–Newton algorithm. Equation (11.22) bears the interpretation that the direction of parameter variations $d\mathbf{p}_k$ is a linear combination of the first r columns of $\mathbf{V}_k$ (right singular eigenvectors of $\overline{\mathbf{S}}(\mathbf{p}_k)$), with coefficients proportional to the reciprocal of the corresponding singular values multiplied by the projection of the weighted prediction errors $\overline{\mathbf{E}}(\mathbf{p}_k)$ onto the first r columns of $\mathbf{U}_k$ (left singular eigenvectors of $\overline{\mathbf{S}}(\mathbf{p}_k)$). In formula

$$d\mathbf{p}_k = \sum_{i=1}^{r_k} \left(\frac{1}{\sigma_i}\mathbf{U}_{ki}^T \overline{\mathbf{E}}(\mathbf{p}_k)\right)\mathbf{V}_{ki}. \tag{11.23}$$

11.6.3 Effective Dimensions of Estimated Parameter Vectors

The selection of a reduced number of orthonormal right singular eigenvectors of $\bar{\mathbf{S}}(\mathbf{p}_k)$ for representing search directions in the original parameter space has an intuitive interpretation in terms of restrictions of step size along certain directions in the parameter space. This can be seen by writing Eq. (11.23) as

$$d\mathbf{p}_k = \sum_{i=1}^{r_k} \theta_i \mathbf{V}_{ki} = \mathbf{V}_k \begin{bmatrix} \theta_{r_k} \\ 0 \end{bmatrix}. \qquad (11.24)$$

with θ_{r_k} representing the first r_k components of the transformed parameter vector $\theta = \mathbf{V}_k^\mathsf{T} d\mathbf{p}$, where $d\mathbf{p}$ represents the search direction towards the "true" parameter vector to be approximated by $d\mathbf{p}_k$. By considering a generic j-th scalar component of $d\mathbf{p}$, its value is thus "spread out" onto the values of θ with coefficients equal to the j-th column of $\mathbf{V}_k^\mathsf{T}$, i.e., the j-th row of $\mathbf{V}_k$. The j-th component of the approximating vector $d\mathbf{p}_k$ is then the sum of the first r_k components of θ multiplied by the j-th row of $\mathbf{V}_k$. The sum of the first r_k squared row elements of $\mathbf{V}_k$ represent therefore the fraction of the "true" parameter variations that are accounted for by $d\mathbf{p}_k$ in Eq. (11.24). These are actual fractions $\in [0, 1]$ because $\mathbf{V}_k$ is orthonormal.

The choice of an effective rank r_k limits therefore the search dimension in the transformed parameter space, i.e., of θ_{r_k}, as well as the step size of the individual search directions in the original parameter space. Unlike other parameter selection approaches, e.g., based on QR factorization with pivoting, there is thus no clear-cut interpretation for the dimension of the effectively estimated parameter vector, because there is no strict limitation upon the number of (original) model parameters that may vary during the estimation process. Such limitations are rather imposed implicitly by fractional scaling factors associated with each estimated parameter. These factors can be used, especially after sorting in decreasing order, to assess the relative importance of the various parameters in the model identification process. Parameters having scaling factors close to unity may be interpreted as fully identifiable from the experimental data, while others with small factors as essentially fixed. To assess the most sensitive parameters one may restrict the attention to parameters having scaling factors above a certain level, e.g., 5 %. In contrast, insensitive model parameters that are characterized by relatively small norms of the corresponding columns in the sensitivity matrix, are mapped to right singular eigenvectors of $\bar{\mathbf{S}}(\mathbf{p}_k)$ with small singular values and are therefore expected to be invariant also for the original parametrization.

A drawback of the above interpretation framework is that the SVD of the sensitivity matrix $\bar{\mathbf{S}}(\mathbf{p}_k)$ changes generally at each iteration and the effective rank determined with a given threshold level δ_k may vary as well. The above analysis may therefore yield different interpretations if carried out at different points of the estimation procedure, e.g., with initial parameter values versus final solution.

A further source of uncertainty in the evaluation of the relevance of individual model parameters in the model fitting procedure is related to parameter scaling, which has a direct effect on the magnitude of model output sensitivities with respect to parameters. If applicable, the systematic use of log-transformation of parameters reduces the influence of parameter scaling providing implicitly model output sensitivities with respect to fractional changes of parameters.

11.6.3.1 Implementation and Practical Issues

The dimension of the reduced rank subspace of parameter variations has been defined as a function of the threshold δ_k, i.e., $r_k = r_k(\delta_k)$, rather than as an arbitrarily chosen fixed number. This provides at least theoretically for $\delta \to 0$ a consistent estimate of the "true" dimension, r_{max}, of the identifiable subspace. In practice r_{max} and the associated minimum threshold δ_{min} can be derived from the log-plot of singular values that typically exhibit an abrupt decline in correspondence to r_{max}. For the purpose of parameter identification such a numerical rank can however exceed to a large extent the number of parameter components that can be effectively identified from the data. In fact, given a subspace of parameter variations defined by the first r_k eigenvectors of $\mathbf{V}_k$, a reduction in δ_k affecting r_k will add new orthogonal components to the search direction (see Eq. (11.23)). This may be beneficial for improving the solution of the NLWLS problem as long as the new added components do not interfere too much with, or even overwhelm, the previous components.

To clarify let us assume that $\mathbf{p}_k$ satisfies the optimality conditions (11.13) such that

$$\overline{\mathbf{S}}(\mathbf{p}_k)^{\mathsf{T}}\overline{\mathbf{E}}(\mathbf{p}_k) = \mathbf{V}_k \tilde{\Sigma}_k^{\mathsf{T}} \mathbf{U}_k^{\mathsf{T}} \overline{\mathbf{E}}(\mathbf{p}_k) = \mathbf{0}, \tag{11.25}$$

which means that the weighted residuals are orthogonal to the first r_k columns of $\mathbf{U}_k$. New components added by reducing δ_k maintain the optimality of $\mathbf{p}_k$ and a new minimum is searched in new orthogonal directions. On the contrary, if $\mathbf{p}_k$ is far away from the optimum $\hat{\mathbf{p}}$, the projection of $\overline{\mathbf{E}}(\mathbf{p}_k)$ onto $\mathbf{U}_k$ may yield large coefficients in (11.23) for all components of $\mathbf{V}_k$. In such a case the norm of $d\mathbf{p}_k$ increases inversely proportionally to the singular values σ_i and adding too many components may cause convergence problems due to non-linearities of the optimization problem.

An iterative procedure that was found effective for finding optimal solutions of the NLWLS problem was to fix a decreasing sequence for δ_K, where K is an outer iteration counter, e.g., $\delta_K \in \{10^{-2}, 10^{-3}, 10^{-4}, \ldots\}$, and iterating, with inner iteration counter k, the restricted step Gauss–Newton algorithm described above with $\delta_k = \delta_K$ until convergence to $\mathbf{p}_K$. This latter was used as initial parameter value for the restart of the algorithm with updated K. To avoid over-parameterization and over-fitting of the experimental data, the outer iteration was stopped manually after subjective evaluation of the goodness of fit and improvements of the cost function achieved between two outer iterations.

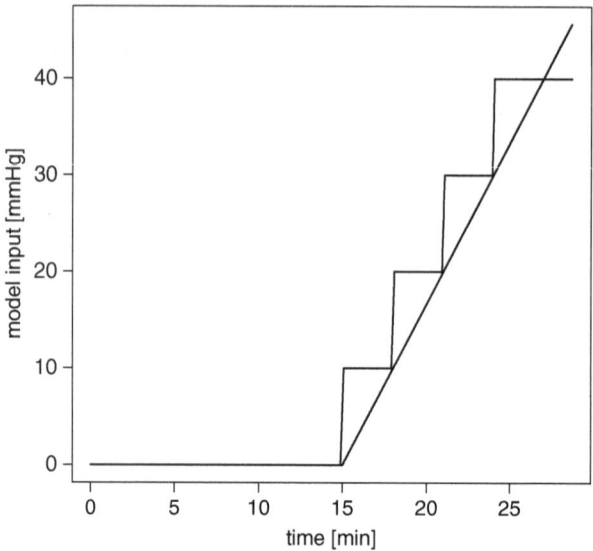

Fig. 11.5 Stepwise and linearly increasing model inputs used to simulate the HUT and separately used for parameter identification

11.7 Results

The parameter identification procedure was initially applied to the CVS model with a stepwise increasing function, according to the experimental procedure. The model showed however some difficulties in describing the smooth decline in central venous pressure observed experimentally, which did not appear to be affected by the step changes in tilt table angle. For this reason the parameter identification procedure was repeated also with a linear increase of HUT perturbation, as depicted in Fig. 11.5.

Experimental data and final model predictions obtained with the smallest tolerance level $\delta_K = 10^{-4}$ and with the stepwise and linearly increasing (ramp) test inputs are shown in Figs. 11.6 and 11.7, respectively. The final model fits obtained with the two input representations were nearly equivalent as regards the predictions of heart rate and mean blood pressure. In contrast, the prediction of central venous pressure was less accurate with the step input, probably due to inadequate modeling assumptions about the distributed effect on the CVS of the perturbation input.

With both input representations the model failed to predict the raise of blood pressure at the beginning of the observation interval, which occurred even before the beginning of the test starting at 15 min. These kinds of random short-term blood pressure, as well as heart rate, fluctuations are normal in healthy subjects and are associated with spontaneous and evoked sympathetic and parasympathetic autonomic activity. An advanced model description should probably include stochastic terms in the dynamic model equations.

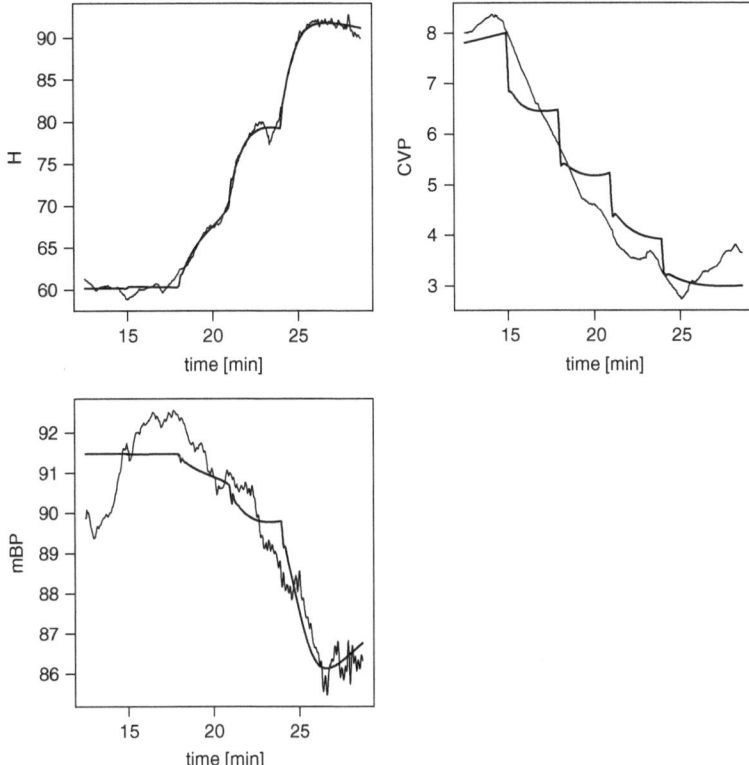

Fig. 11.6 Experimental data (*thin lines*) and final best fit model predictions (*thick lines*) obtained with the stepwise increasing HUT test input

For each model input representation, the final model fit was obtained by applying the iterative parameter estimation procedure as described previously with three levels of δ_K. The model fit improved, as expected, with decreasing δ_K and increasing effective rank (Table 11.1), but did not substantially improve with further reduction of δ_K. Table 11.1 shows that the ramp input representation definitely outperformed the step input representation only with the smallest δ_K considered, achieving a smaller WSS with a smaller effective rank.

A detailed picture of the relationship between effective rank and different levels of δ is given in Fig. 11.8 for the final model with parameter estimates obtained with the ramp input representation with $\delta_K = 10^{-4}$. It can be observed that the final effective rank $r_K = 13$ used for parameter estimation is only a small portion of the "numerical" rank of the sensitivity matrix, which lies between 57 and 74 (Fig. 11.8). Tables 11.2–11.4 provide units for states and certain parameters.

As regards the effective dimensions of the estimated parameter vectors, Tables 11.5 and 11.6 report the fractions of estimated parameter variability for the stepwise and continuously changing HUT test input, respectively. The scaling factors are sorted in decreasing order and only values above 5 % are reported.

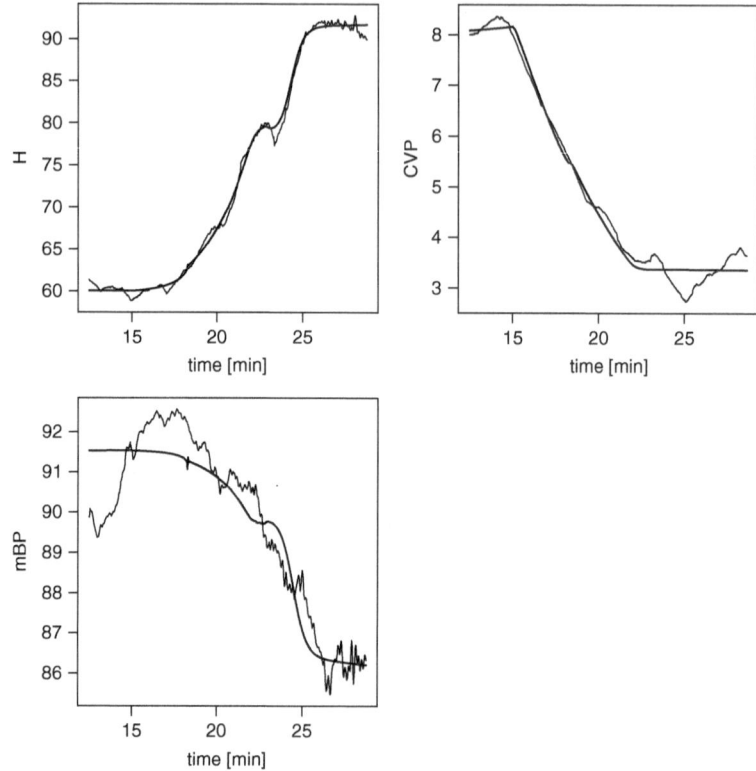

Fig. 11.7 Experimental data (*thin lines*) and final best fit model predictions (*thick lines*) obtained with the linearly increasing HUT test input

Table 11.1 Final weighted sum of squared residuals and effective rank at different tolerance levels with step and ramp input

	Step input		Ramp input	
δ	WSS	Rank	WSS	Rank
10^{-2}	5813.8	6	3425.5	7
10^{-3}	1490.7	11	1667.9	11
10^{-4}	659.8	14	646.3	13

Despite some differences found between the two model input representations, some parameters, mainly related to the control of heart rate and vascular resistances, are ranked highest in both tables. On the contrary, physiologically relevant parameters, such as those related to the control of unstressed volume, appear to play a significant role in parameter estimation only for one or the other input representation.

Tables 11.7–11.11 provide the parameter estimates (Initial value *Init*) and the estimated values for the step and continuous rise in HUT. Parameter symbols in Tables 11.5–11.11 are generated by the special software tool described in [19] which constructs code from model equations. This code format is easily translated

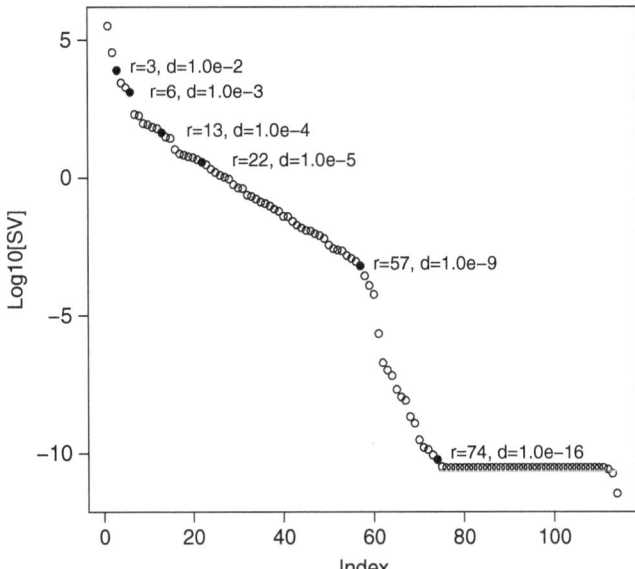

Fig. 11.8 Singular values and effective rank evaluated with different hypothetical tolerances δ using final results obtained with the linearly increasing HUT test input with $\delta = 10^{-4}$, at which a rank $= 13$ was determined. Between rank $= 57$ ($\delta = 10^{-9}$) and rank $= 74$ ($\delta = 10^{-16}$) the singular values drop rapidly to virtually zero values

to the symbols provided in [1], Fig. 1, and the generic equations provided here. Underscores denote subscripts. For the control parameters unstressed volume and resistance, the middle symbols c,v, and k, refer to the nominal value (starting value before perturbation), the maximal sustainable value, and the proportion of overall change, respectively. The latter two refer to the fact that the model builds in a constraint for the proportion of total unstressed volume and a constraint on minimal compartment blood flow (i.e., a constraint on inflow resistance).

11.8 Sensitivity Identifiability: A General Strategy

The above presented method for selecting parameters to be estimated represents one approach to refine the parameter estimation process. Other approaches are discussed in Chaps. 1, 2 and 10. The coordination of model structure with data availability represents a key step in overall model development and validation. The goal is to match model and data in such a way as to improve the robustness and accuracy of the parameter estimation process. Chapter 1 discusses in detail the overall issue of model validation, and how analysis of model identifiability with respect to available data fits into this process.

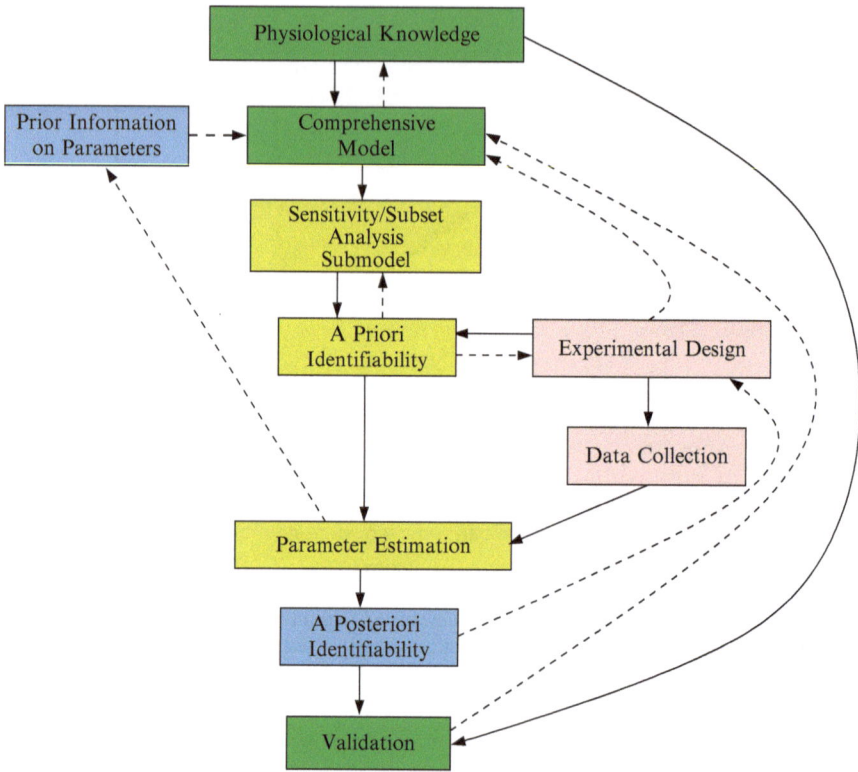

Fig. 11.9 Flow chart for the parameter estimation process

A conceptual iterative scheme employing sensitivity analysis and subset selection is depicted in Fig. 11.9. This figure illustrates that model design and validation involves a process of refinement in which information on available data guides reasonable model reduction and how analysis of model structure can guide experimental design.

The major steps in this iterative process include the following components:

- Once a model has been constructed which incorporates an appropriate degree of physiological detail for the task of the model, classical sensitivity analysis (analysis of how a given model output changes in response to small change in a given parameter) and subset selection can be applied to analyzed the identification problem. This analysis is referred to as sensitivity identifiability analysis as described in [4, 17] but broadened here to include subset selection and generalized sensitivity analysis. This analysis can provide guidance on reasonable model reduction leading to combinations of parameters to estimate that can be identified given the available model output (we will refer to this as a priori identifiability [3]). *A posteriori* identifiability refers to an assessment of the reliability of the estimates given the quality of data.

- However, subset selection and classical sensitivity analysis can be used to not only detect parameters to estimate but also can be employed to assess the value of adding new (and perhaps expensive or invasive measurements (see, e.g., [7]). Conversely, the application of generalized sensitivity analysis (as described in Chap. 1) can provide some guidance on the design of the experiment and how to carry out data collection to improve the parameter estimation process.
- The iterative application of these tools (and decisions based on the information provided) is indicated by the dashed lines showing how one step or aspect of the process can influence others. For example the double arrows between a priori identifiability and experimental design indicates how information on either aspect can shape the other. Generalized sensitivity (Chap. 1) comes into play here. Notice also that experimental design can lead to new information that changes model design. In addition, the parameter estimation process itself can be repeated leading to improved initial guesses for the parameters.
- The final stage of model validation can be carried out by tuning the model parameters to subsets of given data and testing if the model with these parameters can adequately predict observed behavior when perturbations, conditions, or some of the parameters are varied to represent a new situation. For examples and further discussion see [20, 21].

The following observations are made in regards to the method described here in relation to the above model validation protocol:

- The presented method has proved to be a robust approach for (partial) parameter estimation which was a necessary prerequisite for increasing our confidence in the model's capabilities and weaknesses (validation). The results of this study suggest that there is likely a misspecification of the effect of external bias pressure on the various compartments (a temporary workaround has been the use of a linearly increasing HUT test input, which markedly improved the prediction of central venous pressure) and that random variability of cardiovascular parameters may contribute to large modeling errors, especially during resting conditions.
- The newly proposed index for quantifying the fraction of estimated parameter variability within the reduced rank subset selection method provides a means for assessing which parameters are estimable with a particular experiment design and which are not, giving the basis for modifying the experiment design, especially for improving the estimation of poorly estimated parameters. Such kind of evaluation can be based on virtual experiments carried out through simulation studies. In this regard, the availability of a robust parameter estimation approach allows the refinement of prior information on parameter values, improving the quality of simulations.

Appendix

Table 11.2 The state variables and other variables related to the vascular submodel

Variable	Meaning	Unit
P_{as}	Pressure in the arterial systemic compartment	mmHg
P_{vc}	Pressure in the venous systemic (or vena cava) compartment	mmHg
P_{ap}	Pressure in the arterial pulmonary compartment	mmHg
P_{vp}	Pressure in the venous pulmonary compartment	mmHg
P_{up}	Pressure in the upper body tissue compartment	mmHg
P_{ren}	Pressure in the renal compartment	mmHg
P_{spl}	Pressure in the splanchnic body tissue compartment	mmHg
P_{leg}	Pressure in the lower body tissue compartment	mmHg
P_{avc}	Pressure in the abdominal vena cava	mmHg
P_{per}	Pressure in the peripheral (skin) compartment	mmHg
S_{ℓ}	Contractility of the left ventricle	mmHg
S_{r}	Contractility of the right ventricle	mmHg
R_{s}	Peripheral resistance in the systemic circuit	mmHg min/l
H	Heart rate	min^{-1}
V_{u}	Unstressed blood volume	l
$V_{str,\ell}$	Stroke volume of the left ventricle	l
$V_{str,r}$	Stroke volume of the right ventricle	l
$V_{diast,\ell}$	End-diastolic volume of the left ventricle	l
$V_{diast,r}$	End-diastolic volume of the right ventricle	l
$V_{syst,\ell}$	End-systolic volume of the left ventricle	l
$V_{syst,r}$	End-systolic volume of the right ventricle	l
Q_{ℓ}	Cardiac output of the left ventricle	l/min
Q_{r}	Cardiac output of the right ventricle	l/min

Table 11.3 The state variables and other variables of the fluid exchange submodel. Full details of all fluid sub model parameters and values can be found in [1]

Variable	Meaning	Unit
V_{blood}	Total volume of blood	liters
V_{RBC}	Volume of red blood cells	liters
V_I	Interstitial fluid volume	liters
V_P	Plasma volume	liters
dV_{PI}	Interstitial to plasma flow rate	liters/min
dV_{Inj}	Transfusion flow rate	liters/min
dV_{Hem}	Hemorrhage flow rate	liters/min
Q_{Lymph}	Lymph fluid flow	liters/min
Hct	Hematocrit	%
π	Colloidal pressure	mmHg
p	Fluid pressure	mmHg
m_P	Protein mass	liters

Table 11.4 Compliance and example resistance parameters of the basic model. Each compartment has a compliance, an unstressed volume, and inflow and outflow resistances and maximal values for resistances reflecting a required baseline blood flow at all times. Full details of all model parameter and values can be found in [1]

Parameter	Meaning	Units
c_as	Compliance of the arterial systemic compartment	l/mmHg
c_vc	Compliance of the vena cava compartment	l/mmHg
c_avc	Compliance of the abdominal vena cava compartment	l/mmHg
c_ap	Compliance of the arterial pulmonary compartment	l/mmHg
c_vp	Compliance of the venous pulmonary compartment	l/mmHg
c_leg	Compliance of the lower body tissue compartment	l/mmHg
c_spl	Compliance of the splanchnic compartment	l/mmHg
c_per	Compliance of the peripheral compartment	l/mmHg
c_ren	Compliance of the renal compartment	l/mmHg
c_myup	Compliance of the upper body tissue compartment	l/mmHg
c_subl	Compliance of the relaxed left ventricle	l/mmHg
c_subr	Compliance of the relaxed right ventricle	l/mmHg
R_p	Resistance in the peripheral region of the pulmonary circuit	mmHg min/l
R_ell	Inflow resistance of the left ventricle	mmHg min/l
R_r	Inflow resistance of the right ventricle	mmHg min/l
kappa_comp_i	Compartment i proportion of total unstressed volume or total change in resistance	Scaling factor
C_comp_i	Compartment i maximal proportional increase in normal compartment resistance	Scaling factor

Table 11.5 Fraction of estimated parameter variability with stepwise increasing HUT input. The symbols follow the structure of the block diagram and the parameters in the control functions. The middle symbol values z, c, and k refer to the control equations and stand for nominal value (starting value before perturbation), the maximal sustainable value, and proportion of overall change, respectively. The latter two refer to the fact that the model builds in a constraint for the proportion of total unstressed volume and a constraint on minimal compartment blood flow (i.e., a constraint on inflow resistance). The symbols for heart contractility (depending on h) submodel (l and r left and right respectively) and fluid (f) submodel parameter values can be referenced to [1]. The symbol ic refers to initial condition.

Parameter	Weight
ctrl_h_max	1
ctrl_h_alpha	1
ctrl_r_alpha	0.9996
ctrl_h_beta	0.9907
hkappa	0.9547
ic_p_spl	0.9422
r_c_up	0.8627
c_z_spl	0.8123

(continued)

Table 11.5 (continued)

Parameter	Weight
r_z_spl_out	0.6688
r_z_up_in	0.5488
halpha_l	0.4283
r_c_spl	0.3202
r_z_spl_in	0.3202
halpha_r	0.2841
hc_l	0.2742
ctrl_r_beta	0.2677
ic_m_p	0.2636
hbeta_l	0.2612
r_k_up	0.2505
hbeta_r	0.2437
hc_r	0.2407
c_z_vp	0.2365
hr_r	0.2329
hgamma_l	0.1833
ctrl_r_delta_max	0.1585
r_z_leg_in	0.1223
r_z_ren_in	0.1083
r_c_ren	0.104
ctrl_vu_min	0.09396
c_z_leg	0.07332
f_k_l	0.07317
ic_v_i	0.07242
r_z_per_in	0.06991
hgamma_r	0.06444
ctrl_r_tau	0.05647
f_f	0.05124

Table 11.6 Fraction of estimated parameter variability with continuously increasing HUT input. The symbols follow the structure of the block diagram and the parameters in the control functions. The middle symbol values z, c, and k refer to the control equations and stand for nominal value (starting value before perturbation), the maximal sustainable value, and proportion of overall change, respectively. The latter two refer to the fact that the model builds in a constraint for the proportion of total unstressed volume and a constraint on minimal compartment blood flow (i.e., a constraint on inflow resistance). The symbols for heart contractility (depending on h) submodel (l and r left and right respectively) and fluid (f) submodel parameter values can be referenced to [1]

Parameter	Weight
ctrl_h_alpha	0.9999
ctrl_h_max	0.9984
ctrl_r_alpha	0.9963

(continued)

Table 11.6 (continued)

Parameter	Weight
ctrl_vu_alpha	0.9771
hkappa	0.9172
r_c_up	0.7892
ctrl_vu_max	0.7804
r_c_leg	0.5564
r_c_per	0.5294
hr_r	0.4888
hbeta_l	0.4778
halpha_l	0.4608
r_z_leg_in	0.3949
hbeta_r	0.3882
halpha_r	0.3527
hc_r	0.3311
ctrl_h_beta	0.2961
r_z_up_in	0.2804
ctrl_r_beta	0.2195
f_f	0.1814
r_k_leg	0.1546
vu_z_spl	0.131
r_z_leg_out	0.1197
r_z_per_in	0.1128
c_z_leg	0.1033
r_z_p	0.1026
ctrl_r_delta_max	0.09437
ic_p_leg	0.08659
r_z_spl_in	0.07741
f_k_f	0.07493
r_k_per	0.05907
r_c_ren	0.05879
hr_l	0.05238
r_k_spl	0.05185

Table 11.7 Initial guesses of control (ctr) and compliance (c) parameter values and estimates obtained with different HUT inputs. In this
table r represents resistance, h heart rate, and vu unstressed volume.
The symbol z represents the nominal value which for compliances is the
same as any bound on the compliance since compliance is not changed
by the control. The other symbols are as they appear in the control
equations in the text. Init represents the starting guess

Parameter	Init	Step input	Ramp input
c_z_ap	0.0043	0.006725	0.004027
c_z_as	0.002	0.002543	0.002232
c_z_avc	0.025	0.01201	0.01693
c_z_leg	0.019	0.007762	0.03768
c_z_per	0.008	0.008515	0.007012

(continued)

Table 11.7 (continued)

Parameter	Init	Step input	Ramp input
c_z_ren	0.015	0.01302	0.01025
c_z_spl	0.055	0.5403	0.01535
c_z_up	0.008	0.01053	0.005669
c_z_vc	0.017	0.01903	0.01158
c_z_vp	0.0084	0.02433	0.009189
ctrl_h_alpha	100	91.97	91.83
ctrl_h_beta	3	65.24	65.24
ctrl_h_max	105	92.87	92.86
ctrl_h_min	50	13.88	19.35
ctrl_h_tau	1	0.1007	0.1966
ctrl_r_alpha	90	91.83	92.16
ctrl_r_beta	7	72.88	23.52
ctrl_r_delta_max	600	1803	1326
ctrl_r_delta_min	800	2386	1620
ctrl_r_tau	2	1.338	0.1395
ctrl_vu_alpha	5	10.83	3.28
ctrl_vu_beta	7	9.847	54.98
ctrl_vu_max	3.4	2.421	5.987
ctrl_vu_min	2.53	2.712	0.9727
ctrl_vu_tau	30	32.69	38.56

Table 11.8 Initial guesses (Init) of heart contractility (depending on h) submodel (l and r left and right respectively) and fluid (f) submodel parameter values and estimates obtained with different HUT inputs. Init represents the starting guess. See [1] for details on the fluid model

Parameter	Init	Step input	Ramp input
f_a	0.05	0.06068	0.04939
f_b_p	17.5	14.86	17.77
f_c_p	6.5	6.065	6.538
f_d	0.8	0.5782	0.7983
f_e_p	1.2	1.266	1.194
f_f	1.5	0.7147	1.405
f_g	0.02	0.02049	0.02003
f_h	0.63	0.6435	0.6318
f_j	0.02	0.02478	0.01975
f_k_1	0.21	0.3217	0.2068
f_k_2	0.0016	0.001864	0.001618
f_k_3	9e-06	9.349e-06	9.121e-06
f_k_f	0.0057	0.004665	0.006019
f_m_i	210	184.7	211.7
f_p_off	13.2	7.991	13.84
halpha_l	89.47	280.5	72.67
halpha_r	28.46	9.742	15.26
hbeta_l	68.71	41.58	82.5

(continued)

Table 11.8 (continued)

Parameter	Init	Step input	Ramp input
hbeta_r	1.666	3.535	2.969
hc_l	0.01289	0.007541	0.02959
hc_r	0.06077	0.1221	0.08517
hgamma_l	37.33	21.91	38.34
hgamma_r	11.88	19.08	12.2
hkappa	0.5	0.6178	0.4215
hr_l	11.35	10.55	24.42
hr_r	4.158	4.147	12.29

Table 11.9 Modeling initial conditions as parameter values to be estimated and estimates obtained with different HUT inputs. Init represents the starting guess

Parameter	Init	Step input	Ramp input
ic_delta_r_s	1e-05	1e-05	1e-05
ic_h	60	59.12	34.99
ic_m_p	210	556.5	218.4
ic_p_ap	12.89	13.34	13.25
ic_p_as	87.6	97.4	102.3
ic_p_avc	9.111	9.955	9.696
ic_p_leg	15.15	16.1	24.24
ic_p_per	12.89	13.72	13.45
ic_p_ren	14.46	16.89	15.32
ic_p_spl	13.55	20.02	13.62
ic_p_up	12.89	13.74	13.37
ic_p_vc	8.485	9.343	8.813
ic_p_vp	11.43	12.24	12.05
ic_s_l	53.76	53.85	54.2
ic_s_r	4.099	4.077	4.159
ic_sigma_l	1e-05	1e-05	1e-05
ic_sigma_r	1e-05	1e-05	1e-05
ic_v_i	16.9	29.56	16.38
ic_v_rbc	2.2	3.146	2.461

Table 11.10 Initial guess of resistance parameter values and estimates obtained with different HUT inputs. The middle symbols z, c, and k refer to the control equations and stand for nominal value, maximal sustainable value, and proportion of overall change, respectively, which represent the starting value before control and the resistance bounds to allow for a minimal blood flow (See [1]). Init represents the starting guess

Parameter	Init	Step input	Ramp input
r_c_leg	1.7	2.595	2.402
r_c_per	1.5	1.828	1.606
r_c_ren	1.5	1.322	1.567

(continued)

Table 11.10 (continued)

Parameter	Init	Step input	Ramp input
r_c_spl	1.8	0.8942	4.522
r_c_up	0.9	0.4144	0.8642
r_k_leg	0.2	0.1322	0.2179
r_k_per	0.1	0.07967	0.07747
r_k_ren	0.4	1.212	0.2828
r_k_spl	0.4	1.505	1.991
r_k_up	0.1	0.1493	0.1507
r_z_avc	0.167	0.1701	0.1394
r_z_leg_in	60	206	45.96
r_z_leg_out	5	7.46	7.581
r_z_p	1.965	2.394	1.096
r_z_per_in	65	181.7	36.55
r_z_per_out	3.833	3.951	4.002
r_z_ren_in	68.33	27.43	142.8
r_z_ren_out	5	3.605	4.992
r_z_spl_in	50	38.46	78.44
r_z_spl_out	3	2.421	1.413
r_z_up_in	65	109	19.59
r_z_up_out	3.833	4.609	2.929

Table 11.11 Initial guess of unstressed volume parameter values and estimates obtained with different HUT inputs. The middle symbols z and k refer to the control equations and stand for nominal value and proportion of overall change, respectively, which represent the starting value before control and a cap on the proportion that can be changed (See [1]). Init represents the starting guess

Parameter	Init	Step input	Ramp input
vu_k_ap	1e-05	1e-05	1e-05
vu_k_as	1e-05	1e-05	1e-05
vu_k_avc	1e-05	1e-05	1e-05
vu_k_leg	1e-05	1e-05	1e-05
vu_k_per	1e-05	1e-05	1e-05
vu_k_ren	0.103	0.1044	0.1145
vu_k_spl	0.897	0.9604	6.196
vu_k_up	1e-05	1e-05	1e-05
vu_k_vc	1e-05	1e-05	1e-05
vu_k_vp	1e-05	1e-05	1e-05
vu_z_ap	0.09	0.08926	0.09375
vu_z_as	0.715	0.6705	1.045
vu_z_avc	0.25	0.2443	0.2812
vu_z_leg	0.35	0.339	0.4143
vu_z_per	0.05	0.04977	0.05114
vu_z_ren	0.15	0.1479	0.1607
vu_z_spl	1.3	1.16	3.028

(continued)

Table 11.11 (continued)

Parameter	Init	Step input	Ramp input
vu_z_up	0.65	0.613	0.9123
vu_z_vc	0.085	0.08434	0.08834
vu_z_vp	0.49	0.4687	0.6259

Acknowledgements This research was partially funded by FWF (Austria) under project P18778-N13.

References

1. Batzel, J.J., Fürtinger, S., Bachar, M., Fink, M., Kappel, F.: Sensitivity identifiability of a baroreflex control system model. Tech. Rep. IMA03-09, Institute for Mathematics and Scientific Computing, University of Graz (2009 (journal submission))
2. Cavalcanti, S., Cavani, S., Ciandrini, A., Avanzolini, G.: Mathematical modeling of arterial pressure response to hemodialysis-induced hypovolemia. Comput. Biol. Med. **36**, 128–144 (2006)
3. Cobelli, C., Carson, E.R., Finkelstein, L., Leaning, M.S.: Validation of simple and complex models in physiology and medicine. Am. J. Physiol. Regul. Integr. Comp. Physiol. **246**(2) R259–R266 (1984)
4. Cobelli, C., DiStefano 3rd, J.J.: Parameter and structural identifiability concepts and ambiguities: A critical review and analysis. Am. J. Physiol. **239**(1), R7–R24 (1980)
5. Fink, M., Batzel, J.J., Kappel, F.: An optimal control approach to modeling the cardiovascular-respiratory system: An application to orthostatic stress. Cardiovasc. Eng. **4**(1), 27–38 (2004)
6. Fink, M., Batzel, J.J., Kappel, F.: Modeling the human cardiovascular-respiratory control response to blood volume loss due to hemorrhage. In: Commault, C., Marchand, N. (eds.) Positive Systems: Lecture Notes in Control and Information Sciences, vol. 341, pp. 145–152. Springer, Berlin Heidelberg (2006)
7. Fink, M., Batzel, J.J., Tran, H.: A respiratory system model: parameter estimation and sensitivity analysis. Cardiovasc. Eng. **8**(2), 120–134 (2008)
8. Furlan, R., Porta, A., Costa, F., Tank, J., Baker, L., Schiavi, R., Robertson, D., Malliani, A., Mosqueda-Garcia, R.: Oscillatory patterns in sympathetic neural discharge and cardiovascular variables during orthostatic stimulus. Circulation **101**(8), 886–892 (2000)
9. Furlan, R., Jacob, G., Palazzolo, L., Rimoldi, A., Diedrich, A., Harris, P.A., Porta, A., Malliani, A., Mosqueda-Garcia, R., Robertson, D.: Sequential modulation of cardiac autonomic control induced by cardiopulmonary and arterial baroreflex mechanisms. Circulation **104**(24), 2932–2937 (2001)
10. Goswami, N., Loeppky, J.A., Hinghofer-Szalkay, H.: LBNP: past protocols and technical considerations for experimental design. Aviat. Space Environ. Med. **79**(5), 459–471 (2008)
11. Janssens, U., Graf, J.: Volume status and central venous pressure. Anaesthesist **58**(5), 513–519 (2009)
12. Kappel, F., Fink, M., Batzel, J.J.: Aspects of control of the cardiovascular-respiratory system during orthostatic stress induced by lower body negative pressure. Math. Biosci. **206**(2), 273–308 (2007)
13. Mosqueda-Garcia, R., Furlan, R., Fernandez-Violante, R., Desai, T., Snell, M., Jarai, Z., Ananthram, V., Robertson, R.M., Robertson, D.: Sympathetic and baroreceptor reflex function in neurally mediated syncope evoked by tilt. J. Clin. Invest. **99**(11), 2736–2744 (1997)

14. Olufsen, M.S., Ottesen, J.T., Tran, H.T.: Modeling cerebral blood flow control during posture change from sitting to standing. Cardiovasc. Eng. **4**(1), 47–58 (2004)
15. Olufsen, M.S., Ottesen, J.T., Tran, H.T., Ellwein, L.M., Lipsitz, L.A., Novak, V.: Blood pressure and blood flow variation during postural change from sitting to standing: Model development and validation. J. Appl. Physiol. **99**(4), 1523–1537 (2005)
16. Pang, C.C.: Autonomic control of the venous system in health and disease: effects of drugs. Pharmacol. Ther. **90**(2-3), 179–230 (2001)
17. Reid, J.G.: Structural identifiability in linear time-invariant systems. IEEE Trans. Automat. Contr. **22**, 242–246 (1977)
18. Risk, M.R., Lirofonis, V., Armentano, R.L., Freeman, R.: A biphasic model of limb venous compliance: a comparison with linear and exponential models. J. Appl. Physiol. **95**, 1207–1215 (2003)
19. Thomaseth, K.: Multidisciplinary modelling of biomedical systems. Comput. Meth. Programs Biomed. **71**(3), 189–201 (2003)
20. Thomaseth, K., Cobelli, C.: Generalized sensitivity functions in physiological system identification. Ann. Biomed. Eng. **27**(5), 607–616 (1999)
21. Thomaseth, K., Cobelli, C.: Analysis of information content of pharmakokinetic data using generalized sensitivity functions. In: Proceedings of the 22nd Annual EMBS International Conference of the IEEE, vol. 1, pp. 435–437 (2000)
22. Ursino, M., Antonucci, M., Belardinelli, E.: Role of active changes in venous capacity by the carotid baroreflex: analysis with a mathematical model. Am. J. Physiol. **267**, H2531–H2546 (1994)

Glossary

Autonomic regulation Regulation by the autonomic nervous system. This system regulates a number of cardiovascular and other quantities. Cardiovascular quantities regulated include heart rate, cardiac contractility, resistance, compliance, and unstressed volume. Cardiovascular autonomic regulation is mediated by the baroreflex system, via the sympathetic and parasympathetic systems.

Baroreflex regulation Regulation mediated via stimulation of the baroreceptors which are stretch sensors located in the blood vessels of most mammals. In humans high-pressure receptors can be found in the aortic arch and the carotid sinuses, and low-pressure receptors can be found in large systemic veins, in pulmonary arteries, and in the walls of the right atrium. In response to stretch due to pressure change, baroreceptors induce changes in sympathetic and parasympathetic outflow, which in turn give rise to changes in heart rate, cardiac contractility, vascular tone (primarily through high-pressure receptors) and unstressed volume and blood volume (primarily through low-pressure) receptors.

Blood gases Gases transported in the blood; typically used to refer to the blood gases oxygen and carbon dioxide involved in respiration.

Cerebral autoregulation Autoregulation occurring in the brain. The brain autoregulates to maintain blood flow despite changes in blood pressure. It is believed that autoregulation is primarily mediated via myogenic and metabolic regulation.

Cheyne-Stokes breathing A respiratory abnormality characterized by alternate periods of hyperventilation and apnea.

Closed loop model A model where the outputs of the model are incorporated in a feedback loop to influence system response or function (compare open loop).

Euvolemia Normal amount of intravascular volume for a given body size.

Exercise hyperpnea Increase in respiratory ventilation in proportion to metabolic demand without essential deviation of blood CO_2 tension during muscular exercise.

J.J. Batzel et al. (eds.), *Mathematical Modeling and Validation in Physiology*,
Lecture Notes in Mathematics 2064, DOI 10.1007/978-3-642-32882-4,
© Springer-Verlag Berlin Heidelberg 2013

It is in contrast to chemoreflex in which ventilation increases in proportion to change in blood CO_2 tension.

Head-up tilt A laboratory test to examine the cardiovascular response to orthostasis. The test passively tilts a subject from supine to approximately $70°$ angle.

Homeostasis The property of a system to regulate its internal environment in order to maintain a stable and constant condition.

Hypovolemia Reduced amount of intravascular volume for a given body size.

Ill-conditioned estimation Parameter estimation problem in which the numerical values of the parameter estimates depend strongly on small changes in input data.

Inverse problem A general framework for using observed measurements to infer information about the system under investigation.

Kalman filter A set of mathematical equations for estimating the state of a process, taking explicitly into account measurement errors and taking measurement data into account incrementally (on-line estimation).

Metabolic regulation Regulation of local blood flow mediated via enzymes regulating metabolic pathways in response to changes in the cell's environment, level of metabolism, or signals from other cells. In the cerebral vasculature, for example, an increase in tissue metabolism in an activated nerve area will induce an increase in local blood flow, providing an increase in oxygen and nutrients to the tissues. A subsequent reduction in activity results in blood vessels constriction with local blood flow returning to normal.

Model validation The process of testing the correctness of a model. This involves many steps including testing for reasonable model behavior in a variety of conditions with all state variables producing consistent and physiologically reasonable outputs for reasonable parameter values. In essence, one searches for the conditions under which the model behaves appropriately. This domain of validity will depend on (simplifying) model assumptions which in turn depend on the goal for which the model is developed. Often, a narrower meaning is given to validation which involves using portions of available data for parameter estimation and then using the estimated parameters with the remaining data to verify that the model can predict additional observed responses. One method used for model validation includes K-fold cross-validation.

Multiscale modeling Model design containing multiple scales. These can be multiple scales in time or in space. Examples include models coupling 3D spatial and 1D spatial components, models with both higher dimensions and zero-D models, or models including several scales in time, e.g., a model having some components that resolve dynamics over seconds and other components that resolve over hours.

Myogenic regulation The myogenic response (also called the Bayliss effect) regulates contraction of smooth muscles in response to stretch. In the cerebral vasculature, this effect is mainly seen in the arterioles, where an increase in blood pressure causes distension of the arterioles, in response the smooth muscles are constricted. This effect is mediated via calcium signaling and acts to smooth changes induced by perturbations in blood pressure.

Open loop model A model formulation in which the outputs of the model are not used to influence system response. Only arbitrary independently chosen model inputs can be applied to the model to influence the system response.

Orthostatic intolerance Inability to maintain an upright posture without developing cardiorespiratory signs of impending loss of consciousness (such as dizziness) and potentially leading to actual syncope.

Parameter estimation The process of obtaining values for a set of parameters given a model and available data. Typically, the process is carried out using optimization techniques that usually minimize the least squares cost (the sum of squared errors between the observed and computed quantities). Several problems can arise for parameters that are insensitive (i.e., a change in a given parameter does not affect the observed output). In addition parameters may be functionally dependent which can make their unique estimation impossible.

Parameter identifiability This term indicates that the parameters of a model can be identified from the knowledge of model behavior for certain scenarios assuming perfect data reflecting such scenarios.

Parasympathetic regulation A part of the autonomic nervous system. It is activated via the baroreceptors and function via regulation of acetylcholine. In the cardiovascular system parasympathetic regulation mainly impacts heart rate. Parasympathetic regulation of heart rate is mediated via the Vagus nerve. Parasympathetic regulation is almost immediate and the effect on heart rate can be observed within one cardiac cycle from the sensed change in blood pressure.

Periodic breathing Waxing and waning of breathing in a visually cyclic (but not truly periodic) manner.

Physiological respiratory dead space Total wasted ventilation in non-perfused, under-perfused, or poorly-mixed alveolar units.

Respiration The term respiration is used in several ways. It can refer at the cellular level to the process of generating energy including aerobic respiration (using oxygen and generating carbon dioxide) and anaerobic (not using oxygen). It can also refer to the total process of transporting and exchanging oxygen and carbon dioxide between the environment and the cell in the process of generating energy. Hence, terms exist for internal respiration (exchange of blood gases between the cell-extracellular fluid boundary) and external respiration (exchange at the alveolar-capillary boundary) to distinguish individual steps in the process. This term is often used interchangeably with ventilation which is really only one aspect of respiration (see ventilation). Cellular respiration restricted to the aerobic process represents intercellular exchange of oxygen and carbon dioxide (producing as well energy and water).

Sensitivity analysis A study of how the variation in the parameters of a mathematical model will impact a particular measured state variable.

Shunt (right-to-left) Portion of blood flow bypassing the lung unit, representing a perfused but non-ventilated gas exchange unit. The shunt brings venous blood with high CO_2 tension to the arterial blood, hence elevating arterial CO_2 tension.

Sit-to-stand A physiological stress test (similar to head-up tilt) often used in the clinic to test the cardiovascular control system. During sit-to-stand test clinicians often measure dynamics (beat-to-beat) values for blood pressure, heart rate, blood flow, and quantities related to respiration including expiratory concentration of O_2 and CO_2, as well as airflow.

Sleep apnea Pathological state characterized by cessation of breathing which occurs repetitively during sleep.

Subset selection Determination of a subset of model parameters that can be estimated reliably from experimental data.

Sympathetic regulation A part of the autonomic nervous system. It is activated via the baroreceptors and function via regulation of noradrenaline (or norephinephrine). In the cardiovascular system, sympathetic regulation impact heart rate, cardiac contractility, and vascular tone. Regulation of these quantities is mediated by the ganglionic nervous system, and it typically takes 5–10 s from the sensed change in blood pressure to the actual control of the effector organs.

Transcranial Doppler A test that measures the velocity of blood flow through the brain's blood vessels. Typically, flow is measured in the middle cerebral arteries, though the test also allows for measurement in anterior and posterior cerebral arteries.

Unscented transformation A practical method to estimate the statistics of a random variable undergoing a nonlinear transformation.

Vascular tone This refers to the degree of constriction experienced by a blood vessel relative to its maximally dilated state. All arteries and under basal conditions exhibit some degree of smooth muscle contraction that determines the diameter, and hence tone, of the vessel.

Ventilation The movement of air into and out of the lungs (the process of breathing). Minute ventilation is a measure in liters per minute. Tidal volume is a single inhalation or exhalation measured in liters per breath. Compare to the term Respiration.

Index

J.J. Batzel et al. (eds.), *Mathematical Modeling and Validation in Physiology*,
Lecture Notes in Mathematics 2064, DOI 10.1007/978-3-642-32882-4,
© Springer-Verlag Berlin Heidelberg 2013

LECTURE NOTES IN MATHEMATICS  Springer

Edited by J.-M. Morel, B. Teissier; P.K. Maini

Editorial Policy (for Multi-Author Publications: Summer Schools / Intensive Courses)

1. Lecture Notes aim to report new developments in all areas of mathematics and their applications - quickly, informally and at a high level. Mathematical texts analysing new developments in modelling and numerical simulation are welcome. Manuscripts should be reasonably selfcontained and rounded off. Thus they may, and often will, present not only results of the author but also related work by other people. They should provide sufficient motivation, examples and applications. There should also be an introduction making the text comprehensible to a wider audience. This clearly distinguishes Lecture Notes from journal articles or technical reports which normally are very concise. Articles intended for a journal but too long to be accepted by most journals, usually do not have this "lecture notes" character.

2. In general SUMMER SCHOOLS and other similar INTENSIVE COURSES are held to present mathematical topics that are close to the frontiers of recent research to an audience at the beginning or intermediate graduate level, who may want to continue with this area of work, for a thesis or later. This makes demands on the didactic aspects of the presentation. Because the subjects of such schools are advanced, there often exists no textbook, and so ideally, the publication resulting from such a school could be a first approximation to such a textbook. Usually several authors are involved in the writing, so it is not always simple to obtain a unified approach to the presentation.

   For prospective publication in LNM, the resulting manuscript should not be just a collection of course notes, each of which has been developed by an individual author with little or no coordination with the others, and with little or no common concept. The subject matter should dictate the structure of the book, and the authorship of each part or chapter should take secondary importance. Of course the choice of authors is crucial to the quality of the material at the school and in the book, and the intention here is not to belittle their impact, but simply to say that the book should be planned to be written by these authors jointly, and not just assembled as a result of what these authors happen to submit.

   This represents considerable preparatory work (as it is imperative to ensure that the authors know these criteria before they invest work on a manuscript), and also considerable editing work afterwards, to get the book into final shape. Still it is the form that holds the most promise of a successful book that will be used by its intended audience, rather than yet another volume of proceedings for the library shelf.

3. Manuscripts should be submitted either online at www.editorialmanager.com/lnm/ to Springer's mathematics editorial, or to one of the series editors. Volume editors are expected to arrange for the refereeing, to the usual scientific standards, of the individual contributions. If the resulting reports can be forwarded to us (series editors or Springer) this is very helpful. If no reports are forwarded or if other questions remain unclear in respect of homogeneity etc, the series editors may wish to consult external referees for an overall evaluation of the volume. A final decision to publish can be made only on the basis of the complete manuscript; however a preliminary decision can be based on a pre-final or incomplete manuscript. The strict minimum amount of material that will be considered should include a detailed outline describing the planned contents of each chapter.

   Volume editors and authors should be aware that incomplete or insufficiently close to final manuscripts almost always result in longer evaluation times. They should also be aware that parallel submission of their manuscript to another publisher while under consideration for LNM will in general lead to immediate rejection.

4. Manuscripts should in general be submitted in English. Final manuscripts should contain at least 100 pages of mathematical text and should always include

 – a general table of contents;
 – an informative introduction, with adequate motivation and perhaps some historical remarks: it should be accessible to a reader not intimately familiar with the topic treated;
 – a global subject index: as a rule this is genuinely helpful for the reader.

Lecture Notes volumes are, as a rule, printed digitally from the authors' files. We strongly recommend that all contributions in a volume be written in the same LaTeX version, preferably LaTeX2e. To ensure best results, authors are asked to use the LaTeX2e style files available from Springer's web-server at

ftp://ftp.springer.de/pub/tex/latex/svmonot1/ (for monographs) and
ftp://ftp.springer.de/pub/tex/latex/svmultt1/ (for summer schools/tutorials).
Additional technical instructions, if necessary, are available on request from:
lnm@springer.com.

5. Careful preparation of the manuscripts will help keep production time short besides ensuring satisfactory appearance of the finished book in print and online. After acceptance of the manuscript authors will be asked to prepare the final LaTeX source files and also the corresponding dvi-, pdf- or zipped ps-file. The LaTeX source files are essential for producing the full-text online version of the book. For the existing online volumes of LNM see:

http://www.springerlink.com/openurl.asp?genre=journal&issn=0075-8434.
The actual production of a Lecture Notes volume takes approximately 12 weeks.

6. Volume editors receive a total of 50 free copies of their volume to be shared with the authors, but no royalties. They and the authors are entitled to a discount of 33.3 % on the price of Springer books purchased for their personal use, if ordering directly from Springer.

7. Commitment to publish is made by letter of intent rather than by signing a formal contract. Springer-Verlag secures the copyright for each volume. Authors are free to reuse material contained in their LNM volumes in later publications: a brief written (or e-mail) request for formal permission is sufficient.

Addresses:
Professor J.-M. Morel, CMLA,
École Normale Supérieure de Cachan,
61 Avenue du Président Wilson, 94235 Cachan Cedex, France
E-mail: morel@cmla.ens-cachan.fr

Professor B. Teissier, Institut Mathématique de Jussieu,
UMR 7586 du CNRS, Équipe "Géométrie et Dynamique",
175 rue du Chevaleret,
75013 Paris, France
E-mail: teissier@math.jussieu.fr

For the "Mathematical Biosciences Subseries" of LNM:

Professor P. K. Maini, Center for Mathematical Biology,
Mathematical Institute, 24-29 St Giles,
Oxford OX1 3LP, UK
E-mail : maini@maths.ox.ac.uk

Springer, Mathematics Editorial I,
Tiergartenstr. 17,
69121 Heidelberg, Germany,
Tel.: +49 (6221) 4876-8259
Fax: +49 (6221) 4876-8259
E-mail: lnm@springer.com